PRODUCING NATURE AND POVERTY IN AFRICA

Edited by

Vigdis Broch-Due and Richard A. Schroeder

NORDISKA AFRIKAINSTITUTET 2000

Indexing terms

Natural resources
Poverty
Environmental management
Colonial and postcolonial interventions
Africa

Cover: Alicja Grenberger

© the authors and Nordiska Afrikainstitutet 2000

ISBN 91-7106-452-4

Printed in Sweden by Elanders Gotab, Stockholm 2000

Contents

Preface

The essays in this anthology come from a larger collection of papers produced for a conference held in 1997 under the auspices of the Research Programme "Poverty and Prosperity in Africa: Local and Global Perspectives" at the Nordic Africa Institute, Uppsala, Sweden. This conference focused on the politics of poverty and environmental interventions in Africa. To explore this field, the convenors, Richard A. Schroeder, Director of African Studies, Rutgers University and Vigdis Broch-Due, head of the above research programme at NAI, gathered several of the leading specialists in their respective fields, drawn from Scandinavia, Europe, USA and Africa. The list of participants included anthropologists, geographers, historians, sociologists, and ecologists from the academic world, as well as participants who are involved in development work in Africa.

The participants in the conference and editors of this volume share a growing sense of unease about the effects of continent-wide environmental programs on African peoples in the colonial and postcolonial eras. Drawing on case study materials from eight different countries, our essays demonstrate quite clearly that environmental programs themselves often have direct and far-reaching consequences for the distribution of wealth and poverty on the continent.

The following lines of enquiry formed the framework both for the original conference and the individual chapters in this collection:

- How can we theorise the specific forms of intervention that have materialised over the past decade (e.g., strategies of spatial control such as protected areas, buffer zones, and wildlife corridors; commodification schemes centred on ecotourism, extractive reserves, and debt for nature swaps; and the production of diverse scales of intervention via "strategic area management plans", "integrated conservation and development programs", and "National Environmental Action Plans", etc.)?

- What historical dis/continuities exist between contemporary and past environmental policies and practices? How "new" are contemporary interventions? What can we learn from the historical record that will help us understand contemporary environmental politics in Africa?

- What patterns of accumulation have spun out of heavy public and private investment in "the environment" in Africa?

- What kinds of social, cultural and political dislocations have accompanied environmental interventions? What patterns of resistance have grown up in response to them? What are the effects on "the topography of wealth" in the target areas? How has the reinterpretations of the links between poverty and environment affected the poor?

- What effect does the globalisation of the environment discourse have on "the production of locality" in African communities? What kinds of subjects and identities are being produced, what kinds of spaces are created and refigured, and how should these processes be conceptualised and represented?

– Given that the coupling of economic growth with environmental concerns has furthered the process of commodification of nature (e.g., "parks", "resources", "intellectual property rights", etc.), how do capitalistic notions of "scarcity" and "efficiency" that promote privatisation affect and articulate with existing property relations and redistribution systems? What is the effect on land-use patterns? How can we incorporate gender into the theorisation of capital, nature and the global consumption of environmental imagery?

– Given that social life and cultural distinction are often tied to indigenous perceptions of "nature", what effect does the resignification of "nature" as "environment" have on local systems of knowledge and regimes of value?

We hope that this volume will highlight the complex interconnectedness of the conceptual and factual in all these areas and stimulate debate and fresh research into the ethnography of poverty and nature, as well as their historical transformations.

We are grateful to all authors who have been so willing to rewrite and reorganise their papers along the theoretical lines suggested by the editors. We would also like to acknowledge the contributions to the discussions at the conference by those participants who for various reasons could not be represented by a separate chapter in this book. These participants include Signe Arnfred, Hildegarda Kiwasila, Gufu Oba, Tord Olsson , Mohamed Salih, Hans-Otto Sano, and Hanne Svarstad. Thanks are also due to all those at NAI who worked so tirelessly to make both the conference and this publication a success—especially Ulrica Risso Engblom, Karin Andersson Schiebe, Abraham Barmikael and Sonja Johansson. We are also particular grateful to Janet Opdyke for her excellent copy-editing work and to Mike Siegel for preparing the maps.

Last, but not the least, we would like to acknowledge the financial support to the original conference by the Nordic Africa Institute and, by extension, the Foreign Affairs Ministries of Denmark, Finland, Norway and Sweden who jointly fund the activities of the above mentioned research programme.

Vigdis Broch-Due and Richard A. Schroeder

General map of Africa

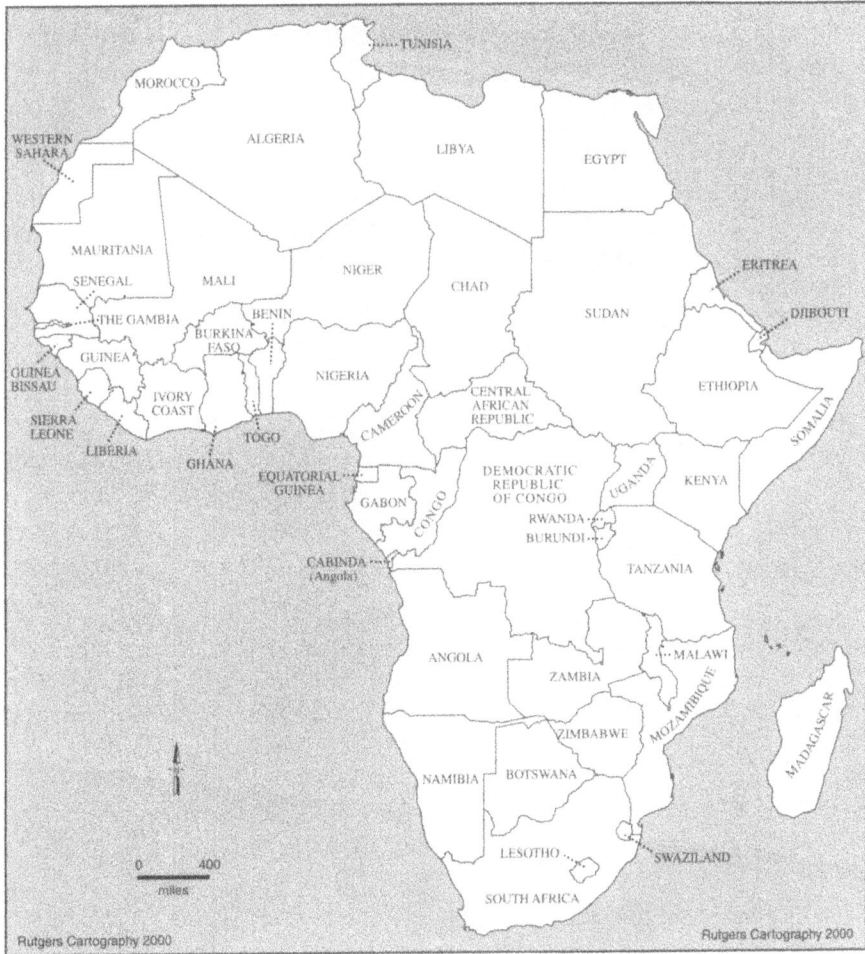

Producing Nature and Poverty in Africa:
An Introduction

Vigdis Broch-Due

... Geographers in Afric-Maps
With Savage-Pictures fill their Gaps,
And o'er inhabitable Downs
Place Elephants for want of Towns.

> Jonathan Swift, 1726, *On Poetry*

The wild peoples of Eastern Africa are divided by their mode of life into three orders. Most primitive and savage are the fierce pastoral nomads, Wamasai and Gallas, Somal and certain "Kafir" sub-tribes: living upon the produce of their herds and by the chase and foray. ... Above them rank the semi-pastoral, as the Wakamba, who, though without building fixed abodes, make their women cultivate the ground. ... The first step towards civilisation, agriculture, has been definitely taken by the Wanyika ... and other coastal tribes.

> Burton, 1872, *Zanzibar: City, Island, and Coast*

To behold the full perfection of African manhood and beauty one must visit the regions of equatorial Africa, where one can view the people under the cool shade of plantains, and amid the luxuriant plenty which those lands produce. ... Their features seem to proclaim: "We live in a land of butter and wine and fullness, milk and honey, fat meads and valleys".

> Stanley, 1899, *Through the Dark Continent*

Man reclaims, disciplines and trains Nature. The surface of Europe, Asia and North America has submitted to this influence and discipline, but it still has to be applied to large parts of South America and Africa. Marches must be drained, forests skillfully thinned, rivers be taught to run in ordered courses and not to afflict the land with drought or flood at their caprice; a way must be made across deserts and jungles, war must be waged against fevers and other diseases whose physical causes are now mostly known.

> Eliot, 1905, *The East African Protectorate*

Economic development is held back by the low standards of productivity of the African peasant, by his unwillingness to adopt improved agricultural methods and his failure to take proper measures for the conservation of the soils.

Creech Jones, 1947, *Development of Colonial Resources*

Africa is dying and will continue to die. Old maps and remnants of settlements and animals show that the Sahara has advanced 250 miles northwards. ... So much of Africa is dead already, must the rest follow? Must *everything* be turned into deserts, farmland, big cities, native settlement, and dry bush? One part of the continent at least should retain its original splendour. ... Serengeti, at least, shall not die.

Grzimek, 1959, *Serengeti Shall Not Die*

Countries do not get on this list by accident. A typical least developed country is composed mostly of peasants or nomads whose consumption mainly comes from the efforts of their own household. One in seven families draws its livelihood from outside agriculture. Agriculture is unproductive and lacking in immediate prospects for improvement, partly because the natural conditions for effective agriculture are poor. The most frequent terrain are desert, savanna, salt pans or mountains. In none of these countries is there a large fertile flood plain. ... It is virtually impossible to find locally the capital, purchaseable inputs, training, transport and so on, essential to increase agricultural productivity. It is not surprising that technology is backward, although it may be well adapted to the harsh environment and the limited input availabilities. In most provinces of a typical least-developed country, there are poor communications with the outside world; ... Few people ever have contact with modern health services and schools are scarce and primitive in quality. ... There is little contact between urban and rural populations. ... Sizeable nomad groups are unreachable by central authorities. Alternatives to the rural subsistence way of life hardly exist.

Development Assistance Committee Review, 1972

It would be a grim irony indeed if just as the new genetic engineering techniques begin to let us peer into life's diversity and use genes more efficiently to better the human conditions, we looked and found this treasure sadly depleted.

Gro Brundtland, 1987, *Our Common Future*

About half the world's poor live in rural areas that are environmentally fragile. ...The poor are both victims and agents of environmental damage.

World Development Report, 1992

The broad morass of lakes and swamps called Lake Kyoga, with its primitive island villages, is utterly roadless and indeterminate in configuration, like some labyrinthine swamp of ancient myth.

Matthiessen, 1991, *African Silences*

Women generally have before them a life of gross injustice, poverty and hardship.

Development Assistance Committee Review, 1995

Africa represented

The chorus of voices quoted above speaks to us about an Africa viewed from a variety of different historical and cultural locations—all outside the continent. Each represents a dominant perspective, a discursive site on the long historical trajectory of Western engagement with Africa from the early era of exploration to the contemporary one of "aid" and development co-operation. While the claims proffered by the more recent voices may sound less jarring than the older ones, they all straddle the gulf between observation and description in a way characteristic of their time. The views they offer are multiple and contradictory, revealing ruptures as well as continuities in the way Africa is perceived. While each voice clearly speaks its time and social milieu, taken together the medley of voices, agreeing on some aspects and dissenting on others, perfectly conjures the ambiguity of Western-derived discourses about nature and poverty in Africa.

This volume takes hold of this ingrained ambiguity, examining the historical and sociological contexts that shaped these diverse perspectives, and charting their shared epistemological ground but also the places where they point in different directions. Our aim is to bring a degree of originality and a critical edge to current scholarly debates about the field in two important ways. Not only do the authors subject different Western discourses and policies with regard to Africa to intense empirical and theoretical scrutiny, but they also bring African voices and perspectives to bear on these. It is a truly interdisciplinary exercise, bringing together within the same horizon of inquiry the perspectives of historians, ecologists, geographers, and anthropologists.

As the simple juxtaposition of the epigraphs signals, the essays collected in this volume group themselves around a single, not extraordinary, but highly fertile proposition: neither nature nor poverty is "natural" given the facts about Africa but has been produced by discourses and activities, many of which have arisen in the Western world and been transplanted to Africa with unintended and sometimes disastrous results. As many of the essays show, these unintended outcomes have usually been the fruit of a clash between imported Euro-American models and local models produced within the African worlds that the West has been so concerned to transform.

The essays range widely over different historical eras, themes, and encounters. Some look at colonial and postcolonial conflicts of resource extraction, displacement, and warfare; others at the effects of contemporary interventions in the name of sustainable development; and still others at new forms of tourism in "partnership" with local communities that combine wildlife sightseeing with so-called cultural shows—all repackaged as conservation. There are discussions on the trajectories of scholarly models, development narratives, and recurrent modes of environmental interventions. Still other essays make forays into the contested fields of nation-state building and beyond, into the transnational terrain, analyzing how interests generated on these levels are projected into diverse localities where they interplay with indigenous theories of place, person, and social identity.

The essays cover an eclectic and extensive ground of scholarly inquiry, but all invoke in one way or another central problematics of scale and perspective to engage these diverse issues. They demonstrate how colonial interests have been hard at work over the past centuries to produce and reproduce a globalizing view of African nature and poverty, "scaling these up" from the specific local circumstances of their occurrence to a vague and generalized condition. This multifaceted project has involved asserting the primacy of the global perspective in the construction and prioritization of specific environmental and economic problems and in the invocation of particular management rationalities. These are often completely out of touch with local realities.

The retreat from the realities of African diversity to the abstract high ground of uninformed supposition so characteristic of European representations, beautifully captured in Swift's sixteenth-century doggerel, has been carried into the contemporary age in various ways. It is reproduced not only in glossy tourist brochures but in the narratives framing scholarly models and the technical paraphernalia and concepts deployed in development interventions. As our essays attest, such positions commonly fail utterly to consider, let alone integrate into their policies, the varied realities of African people's everyday, lived experiences. The production of the global perspective has of necessity involved eclipsing other perspectives, visions, and social struggles. The essays examine how, both discursively and through direct material incursion, the needs, desires, and definitions of people in particular locales have been subsumed into those of external managers in charge of colonial and postcolonial policies. Indeed, most of the essays show how the points at which the "local" meets the "global" are points of fierce contestation between differing worldviews and practices.

Nature, as the volume's title implies, looms large. All localities examined are affected by environmental interventions of some form or other, whether the setting is rural, as in most of the examples, or urban, like tourism in Zanzibar Town. Across the cases, past and present, we can see how perceptions of the physical world and nature are bound up with the creation of meaningful social identities and how these are reconfigured when the politics of nature changes. Not only is nature a source for daily subsistence; it is also the source for indigenous theories of transformation, temporality, creativity, and cosmology—it informs life and living.

Each of the essays shows the diverse ways in which these complex links to landscape are shifting as local people are forced to confront, with increasing frequency, the interests of mining and logging companies, forestry and soil conservation projects, tourism, and the competing discourses of Western environmentalists. All of these different definitions of nature, the global and the local ones, interact and influence one another as they enter into the political struggles that frame resource flows. The essays also sensitize us to a significant dimension embedded in the outcomes of such struggles, notably, the restructuring of gender and class relations, usually to the disadvantage of poor people.

The implications of the volume's central proposition—that nature and poverty are indeed "produced"—are thus not merely scholarly. There is surely no more politically or humanly significant set of questions posed on the African continent today. The ways in which these vexed issues are engaged and negotiated, both by African peoples and in transnational networks, will shape the future of citizens, communities, and countries right across Africa. Knowing how to respond to different types of resource crisis in African arenas demands first and foremost that we pay close attention to local-level realities and conflicts, those junctures at which peoples conceptualize and contest the global and where global ideas about the "nature" of poverty and nature redefine themselves. Our ambition is that this volume will contribute to the growing international scholarship involved in critical inquiries into power, poverty, and environmental equity. Ultimately, we hope it will make a contribution to this highly politicized debate by producing discursive resources for those struggling to better their lives and protect their livelihoods.

Nature and poverty thickly conceived

Historical analysis of the two concepts at the heart of this collection show them to be as changeable and various as they are grand and important. Neither nature nor poverty has ever had a precise conceptual outline in Western thought but has rather had a constellation of meanings in which particular concepts cluster, often temporarily, before moving on to inform other constellations, discourses, and practices.

Before tracing the poverty-nature discourse through its different transformations, it is important to remind ourselves that all interpretations of the world draw on the pool of social knowledge available in any particular time or place. We need to become interested in how new discoveries change the topology of the known and establish new modes of experience and explanation. In this process, social constructs like gender, class, race, and origins—which are all powerfully at work in the conceptualization of poverty and nature—are constantly being produced anew within different and competing discourses. A return to the epigraphs, which are organized chronologically at the beginning of this introduction, gives us a glimpse of the sheer scale, complexity, and contradictions inscribed in foreign agendas for Africa.

In Swift's sixteenth-century prescription for mapmaking, Africa is epitomized by two things; elephants and savages. He makes us laugh at how his contemporaries' image of Africa erases all traces of urban sites and ancient civilizations, reconfiguring the continent as wilderness writ large with few signs of human habitation or history. While Swift's flippant verse seems far removed from the detailed colonial mapping celebrated by the modernist zeal of, for example, an Eliot, and perhaps is even more at odds with the postmodern genetic engineering conjured by Brundtland—he clearly has his finger on something that has been true down the centuries.

Despite the remarkable ethnic, linguistic, and cultural diversity that we know from historians was characteristic of the continent from the first colonial encounters, African peoples have been consistently described in the astonishingly poor vocabulary of savages, primitives, colonized, clients, partners or global citizens—whatever the changing labels may be. Whether subdivided into noble and ignoble tribes, nomads and settlers, traditional and modern, or, in more updated versions, good and bad environmental managers, these distinctions are only the veneer of one essential trait of their "nature". For, as perfectly captured in Stanley's excited celebration of equatorial vistas, where "one can view the people under the cool shade of plantains", natives have always merged with the particularities of their environment in the eyes of European beholders (see also Crosby 1990; Pieterse 1992; and Adams and McShane 1996).

While all these representations differ in some ways, they share the application of a perspective and scale removed from the context of description. While sixteenth-century maps conjured an Africa of pictographic simplicity on a continental scale, the twenty-first-century imagination of Africa has reached planetary proportions. From this height, the world is envisioned as a unitary whole in which all elements are integrated and this global ecosystem is to be conceived as a global "ecocracy", a management system that applies a set of standardized solutions regardless of the geographical or historical context. In all the representations cited, we see how certain "facts" become appropriated as powerful emotional movers for the advancement of particular political agendas.

Discourses about nature and poverty are also discourses on morality and the imaginary. For example, a seductive mixture of fact and fiction is very active in Grzimek's apocalyptic image of an advancing Sahara, its rate specified in miles, and a dying continent. Yet the Sahara is impossibly distant, geographically and conceptually, from the place and project he is trying to promote—namely, the turning of the Serengeti plains in Tanzania into a national park.

Although it is of course true that any piece of writing of whatever genre has to use textual devices like tropes and metaphors to convey its message, what is so telling about the history of representations of Africa conjured by the epigraphs is not only the narrow register of stereotypes recurrently used in the descriptions—one need only slightly edit the oldest voices from their prejudicial excess to produce more modern ones—but also how far removed these narratives are from African realities.

Western audiences have long been accustomed, for example, to a particular version of Africa that casts it as "nature" writ large in all its primordial splendor. Beryl Markham in *West with the Night* ([1942] 1983) captures this typically Western feeling of reverence for African nature. Here we do not have Eliot's wish to reclaim, discipline, and train nature for the good of man but rather a nostalgic desire to merge with nature and preserve it.

> Watch the fence. Watch the flares. I watch both and take off into the night. ... Ahead of me lies the land that is unknown to the rest of the world and only vaguely known to the African—a strange mixture of grasslands. Scrub. Desert sand like the long waves of the southern ocean. Forests, still water, and age-old mountains, stark and grim like mountains of the moon. Salt lakes and rivers that have no water. Swamps. Badlands. Land without life. Land teeming with life—all of the dusty past, all of the future. The air takes me into its realm. Night envelopes me entirely, leaving me out of touch with the earth, leaving me with this small moving world of my own, living in space with the stars. (15)

This celebratory image of African nature often accompanies a less agreeable but equally sedimented Western perspective on African humanity. In contrast to the "abundance" of nature, here "absence" becomes the keyword. African peoples, if represented at all, are seen to be ignorant of the natural treasures they have at hand, the full glory of which is left to a Western mind to appreciate from a privileged distance.

For almost every scenario related to perceptions of Africa and Africans by Europeans, as well as the interventions made on the basis of such perceptions, we can find precedents in European arenas. Thus, in the *Myth of Wild Africa* (Adams and McShane 1996:6), we learn how wilderness was constituted in European minds from early encounters onward as a place beyond human control, where order breaks down, a notion that "stems from sources as diverse as Beowulf and the Bible and ... evokes responses ranging from fear to awe to delight". In *Black on White*, Pieterse (1992) makes the interesting point that the European nation-states were produced though the subjugation, Christanization, and exploitation of regions of Europe in ways replicated in the later colonization of the rest of the world by European empires. Discourses of savagery as opposed to civilization can be traced to the classic worlds of Greece and Rome. While it is beyond the scope of this introduction to encompass all this, it is useful to pause briefly at one site along the trajectory of Western visions of nature—the period of romanticism—since this era was productive of ambivalent attitudes still active in the contemporary visions of African nature.

In the romantic vision, human "nature" was constituted not only as flesh but as spirit and both aspects of human nature were integral parts of the natural world. The romantic spirit of nature in one version is all harmony and rhapsodic beauty. In another, it contains darker, chaotic forces of sexuality and violence. Rhapsody and rapacity have always traveled together in Africa, as is evident in our collection of quotations. While some romantic poets celebrated the wildness in the human character and nature alike, many philosophers were of the opinion that humans should subdue and control this chaotic streak of "nature". In *Nature's Economy*, Worster (1987) shows how this ethic of domination woven into the plot of romanticism was influenced by a particular reading of Darwin's theory of evolution in which "survival of the fittest" became the explanation and raison d'être of savage nature.

One obvious reason why the part of Darwin's model that pictured a ruthless nature became foregrounded in public imagination was that it aligned so perfectly with the Victorian interpretation of "civilization" as a cultural script for the rational and humane management of nature (Kuper 1988). The favored metaphor for this was "the garden: civilization within the walls, Darwinian jungle without" (Adams and McSchane 1996:7). The garden responded to the genteel taste for bringing the broadening knowledge of the world into a small compass—to "the desire, as Bacon put it, to have 'a model of the universal nature made private' at the convenient disposal of civilised man whether in his garden and park or his study" (Hale 1994:530).

From the Renaissance onward, the European garden served as a cabinet version of nature in which the plants and animals enclosed were primarily there as exemplars of what lay outside. Placed in European parklands were not only plants and fishponds, but also deer, wild goats, hares, and rabbits. While the exterior ground of estates was equipped with these more or less "local" species, the interior of houses exhibited exotica from Africa and beyond. Crocodiles, stuffed pelicans, mummified Egyptian cats, and elephant tusks were indiscriminately mixed with fossils, strange roots, and dried nuts and berries. Although often part of the same estate, gardens and "cabinets of curiosities" epitomized particularly well the two ethics projected onto nature, the desire to change and the desire to conserve, but also the ambiguous connections between these apparently conflicting projects.

The garden was the perfect image of that civilizing project with respect to nature that would be projected onto Africa (Anderson and Grove 1987). And, in reality, places like the island of Madeira and Kew Gardens in London served as halfway houses, helping transplanted species to adapt to a new climate and soil. In a very concrete way, the garden served as model for a whole set of symbols of civilization spatialized on different scales and levels of complexity—private gardens, plantations, irrigation projects, and game parks alike. In an extended way, gardening was the technology that tamed nature and "submitted it to (man's) influence", as is ideally sketched out in Eliot's citation. For centuries, gardening has clearly encouraged the enterprises of natural species collection, cross-fertilization, experimentation, and genetic engineering, which now "let us peer into life's diversity" in Brundtland's postmodern vision.

However, as was anticipated by Burton, domesticating African wilderness through cultivation was analogous to the colonial project of domesticating African humanity. Agriculture constituted literally the seedbed that would grow, mold, and civilize the African person as he or she worked the land. In *Colonial Inscriptions*, Shaw (1995) unpacks the myriad meanings that filtered through the garden imagery and brought into one unified equation not only the social tastes and distinctions inscribed in the colonial system of class, race, and gender but the practical knowledge of selective breeding linked to the desire to make something European out of African matter. The imagery of the garden not only came to embody the distinctions between European and African personhood but also informed divisions projected across the entire colonial landscape. Cultivation became the key sign of the distinctions between African farmers and pastoralists, between settlers and nomads, between order and chaos, between reason and ignorance, between modern and traditional, and between good and bad. Most significantly, through the collection of meanings that cluster around cultivation, the concept has become crucial for defining the "poor" as opposed to the "prosperous". The prototype of today's poor is not far from that of yesterday's primitive (Broch-Due, 1995b, this volume). One only needs to replace the string of words *savage-wild-primitive* in Burton's descriptions of peoples on the margin with *poor* to arrive at the now classical description of the "target group" of the development business—a point I shall return to later.

In contrast to the garden, the cabinet of curiosities represented nature conceived in its pristine and pure glory. The fossil, the elephant tusk, and the decorated mask are all artifacts that stand in a part-for-whole relationship with an imaginary original wilderness outside the walls of the garden. Here humans and animals live in their own "natural" habitats outside of time and social history. They provide their own context and can thus easily be collected and moved for exhibition in cabinets on the other side of the world. Sometimes "primitive" specimens of colonial peoples, rather than their images or artifacts, were the curiosities displayed. Since the 1850s, expositions in Copenhagen, London, Paris, and New York re-created "native scenarios" in which exotic peoples were offered for viewing in showcases similar to zoological exhibits (Barkan and Bush 1995). As late as 1931, the French government staged an Exposition Coloniale complete with African villages, mosques, pagodas, and "natives" of all sorts. Most tellingly, the pavilion sponsored by the US government invited spectators "to experience in one day the thrill and excitement of the 'jungle'. African and New Caledonian schoolteachers and civil servants were enrolled to pose as 'authentic' savages" (Shelton 1995:327). Joy Adamson's tribal portraits (Broch-Due, this volume) belong to this tradition of showcasing, as do the "cultural shows" integral to contemporary eco-tourism (Larsen, this volume; and Johnson, this volume).

The pedagogy inscribed in this aesthetic of "othering" (Neumann, this volume) comes from "the past is a foreign country" notion, deploying exhibits, shows, zoos, and parks as a window between self and other and sustaining different inclinations and sensations. The most obvious and conventional reading is the ways in which the image of the spectator is refracted as "civilized" against the "primitive" figure on display, but there is also an alternative, related reading—of a deeper and subtler desire by the spectator to momentarily merge with the primitive other. While the first mode is generated by the evolutionist reduction of *primitive* to *simple*, *crude*, and thus *inferior*, inscribed in the dominant domains of primitivism, the second mode is more ambiguous. It aligns us with the more submerged discursive domains of primitivism that, as succinctly put by Torgovnick in *Primitive Passions* (1997), encompasses "thinking about origins and pure states, ... [and] informs desires for known beginnings and, by extension predictable ends. Primitivism is the utopian desire to go back and recover irreducible features of the psyche, body, land, and community—to reinhabit core experiences" (5). In this particular form, primitivism was heavily implicated in the grand nineteenth-century place-making projects of Europe, particularly the nation-state as it was envisioned by Germanic romantic nationalists, in which specific combinations of body, soul, and soil were bound into one essential equation. This formula is not only redeployed in the construction of settler colonies in Africa, and the creation of tribal territories within each colony (Broch-Due, this volume), but it has found a new discursive site in Western environmentalism, which continues to reproduce these types of autochthonous bodyscapes (see the essays by Benjaminsen, Giles-Vernick, and Schroeder, this volume).

These containers of nature—gardens and cabinets of different scales—have been remarkably persistent devices in the politics of representation that surrounds Africa. What has come out of all this ancient European obsession with the bounding of nature is an ingrained ambiguity that continues to project two distinct sets of values. On the one hand, to subdue wilderness is to establish order and perhaps re-create Eden by cultivating nature, human and nonhuman, in the mold of Christianity and/or capitalism. On the other hand, to feel awed is to celebrate wilderness in its orginary form, a nostalgic mode that still travels with a wish to preserve nature as a way to preserve our common past (or *Our Common Future*) but is no less open to commodification and exploitation. As is amply attested in the essays here collected, most colonial and postcolonial explorers, development agents, and tourists are products of these two kinds of ethics, imaginaries and economies, sometimes convergent, often in conflict.

The crucial point is that these various modalities of Western imagery, past and present, that envision the relationship between African places and peoples are not just "points of view"; they become powerful and coercive for the people they envision. They become realities in their own right, transforming the peoples and places that they were originally intended to describe.

Such imaginaries are powerfully at work on the ground because they are so heavily implicated in the ways problems are framed, which itself informs the outcomes of resource struggles. Power, as we know, consists not only of coercion and force but of "the ability to make others inhabit your story of their reality—even, as is often the case, when that story is written in their blood" (Gourevitch 1998: 48). This point takes center stage in many of our essays, each of which shows in the locality under study that—to paraphrase Neumann (this volume)—historical struggles over geography are not only about military conquest or economic dominance but about ideas, images, and imaginings. The different landscape visions carried by dominant ideas can be analyzed as "virtual realities" (Cline-Cole, this volume), which are shaped and sustained by social forces and specific technologies of representation. While all virtual realities represent a particular perspective that forecloses alternative interpretations, those also equipped with polices and power to intervene may end up reproducing elements of the envisioned scenario in the targeted environments.

We shall see how the conflicting ethics inscribed in European ideas and practices come trailing particular forms of virtual reality and environmental intervention. We shall start with the disaster visions in which African human nature is perceived as having a malignant effect on the nature of wilderness. This negative production usually travels together with the modernist charters to control and manage nature, most of which are (re)created in the Malthusian mold.

Cutting down trees: The terrain of Malthus revisited

In their essay, Fairhead and Leach explore how localities are being reconfigured through a particular forestry vision built upon European experiences and evidence that has been transplanted to West Africa. Colonial and postcolonial forest managers arrived with the firm belief that local peoples have been cutting down the trees, damaging what was thought to be a primordial forest landscape. By a careful reexamination of the archival data used to predict climatic collapse caused by forest cover change, the authors are able to reconstruct parts of the environmental history of the region. They discover that the evidence has massively exaggerated the rate, direction, and extent of recent high forest loss. The contemporary wisdom is that the savanna is spreading southward and the pockets of forest are what remain in the wake of a continual process of degradation. The authors turn that directional logic on its head, arguing that the presence of tree clusters represent the advancement of the forest. These are deliberate plantations of trees created by people to extend the benefits of having forest resources close to hand.

In a particularly instructive way, this essay shows how the erroneous interpretations of forest cover change have been linked to policy formulation

and administrative interventions during the twentieth-century. While there
have been many different experiments in national policies aimed at achieving "sustainability" over this period, there have nevertheless been strong
historical continuities in the science used to frame "the problem". One crucial finding is that this distorted depiction of forest history is not only embedded in natural science but is also heavily implicated within the social scientific canon of the region. Modern works in the social sciences pointing to
the social and economic causes of forest cover loss, and their historical time
scale, generally reproduce and reinforce earlier analyses. Social scientists
have thus supported remedial policies in agriculture, forestry, and conservation policy similar to those of the 1930s and 1950s, depicting "locality" in
ways still framed by colonial conservation policy. Illustrating the argument
with case studies from Ghana and Côte d'Ivoire, Fairhead and Leach
demonstrate that many works within the social, as well as natural, sciences
are now central to a development regime that today produces locality in
such a way as to remove resources from inhabitants' control. In other words,
this essay manages to demonstrate in an exemplary fashion how relationships between local inhabitants, resource use, ecological change, history, and
sociality are all cast in conformity with Western-derived interpretations of
ecological problems.

Cline-Cole's essay probes further into the discourses that African forests
have drawn around themselves. Far from being produced through one
dominant model, contemporary forestry discourses in dryland Nigeria are
produced through many competing or contradictory "virtual realities". The
complexity at play here is partly an effect of scale, a focus on the production
of one product—fuelwood—across different environs and land use systems,
agrarian and pastoral. Given their different production profiles and social
needs, inhabitants come to the problem with different, sometimes divergent
visions of landscape and the place of fuelwood production in it. Far from being the product of a monolithic vision, regional "forestry", or, more inclu-
sively, agro-silvi-pastoral "landscapes" and "fuelscapes", are social products
invested with diverse meanings by different individuals and groups. They
represent sites of contestation for human agents and state agencies engaged
in constructing, maintaining, and modifying wood fuel—and other forestry-
related items. This essay juxtaposes several such contests, their "meanings",
and the discourses of which they are a part. It does so with particular reference to perceived linkages between fuelwood use and production, on the
one hand, and vegetation and environmental "change" and "degradation",
on the other.

The point foregrounded in the essay by Benjaminsen on the history of
forest legislation that surrounds the sparse woodlands of Mali is how perception, power, and coercion were articulated in that legislation and the
forestry management it guided. Colonial and later national forest policies
were not only dominated by a top-down approach, as elsewhere in franco-

phone West Africa, but equipped with a punitive forestry regime created by the French colonial government. A paramilitary Forest Service was made responsible for implementing a stringent policy of permits for use and fines for rule violation. The whole exercise of control came out of the virtual reality as examined in the other essays on the topic, conjuring a vision of severe deforestation caused by local destructive land use. The writer explores the myriad social and ecological effects of various forest laws and regulations from 1900 to the present day. This oppressive policy instated during colonial rule persisted after independence into the postcolonial landscape. Paradoxically, the punitive aspect of forestry management was further encouraged by increased environmental "awareness" during the 1980s, producing even more costs and obstacles to local producers.

As was the case in the coastal forests, empirical studies of the use of forest products in Mali demonstrate that the use of fuelwood and other forest resources locally was on sustainable levels and thus a far cry from the destructive picture that for so long had dominated national forest policy. After a political change in 1991, a decentralization reform was introduced in Mali, which devolved more decision-making power to the local level, also in natural resources management. Although there are promising signs that democratization works to enhance local producers' control over their environment, there are also signs suggesting the countercase, namely, that increased power locally will be co-opted and compromised by the forces of commodification that travel together with decentralization.

The virtual reality highlighted in these essays is a very dominant one in the history of Western representation of African nature. It is premised on the "Africa denuded and choked by sand" scenario evoked by Grzimek. From different angles, these essays question the reality claims carried by this apocalyptic vision. By reconstructing what in retrospect turns out to be a chain of misinterpretations and errors that have given rise to our present-day state of knowledge about forests in West Africa, Fairhead and Leach give us a privileged insight into how this version of nature in decline is being fabricated. By scrutinizing specific instances in which there is an initial misinterpretation of a particular set of data, they are able to show that misinterpretation is repeated and how it gets generalized and amplified. This "scaling up" is achieved by the application of evidence to either a broader geographical area or a longer historical period than that which can soundly be sustained by the information at hand. And, last, they show how the end product—the vision of deforestation and desertification—becomes part of accepted wisdom.

The essay by Benjaminsen retraces much of the same historical trajectory but adds in interesting ways to the equation of misplaced modeling. Highlighted here is a discussion about how hard-core notions of desertification and deforestation as well as high-handed colonial forestry laws and administrative structures interact in unpredictable ways with more recent and softer concerns about poverty and the environment. Cline-Cole's essay brings to our attention the importance of examining how African "virtual" realities are brought to bear on the Western model. By focusing on the operation of power, the essay echoes the finding of Fairhead and Leach: at the end of the day and after much disagreement, those who come out as winners and get their wishes implemented are, not surprisingly, not local peoples but state agents. In the postcolony, the national elites manage and represent the perspectives and policies that have become globally accepted as the "correct" ways throughout a long history of Western dominance that has disqualified, muted, and displaced alternative visions.

Collected around the core image of Africa's "virtual" reality as the "fall from an ecological paradise" (Hoben 1995) are contemporary images produced for global consumption of dark, poor masses of peasants and pastoralists. The mass media invite us to watch as Africans go about destroying forests and mountainsides with axes and machetes, depleting the fish stocks of lakes and rivers with poison and undersized nets, destroying endangered wildlife and plant species, and spreading desertification by means of oversized cattle herds. The effects of these "degrading" practices are featured in a related series of disaster images—the skeletal and swollen-bellied famine victims, the desiccated landscapes, the butchered carcasses of elephants and rhinos. These are emotive images for Western audiences, which call out for action. In other words, the negative images conveyed by such stock in trade notions as overgrazing, deforestation, and desertification not only invite and legitimize interventions into distant communities but crucially convey the image that conservation problems are rooted in the behavior of the African peoples targeted (Anderson and Grove 1987; Leach and Mearns 1996).

The implications of these particular images, mistakes, and misinterpretations are serious. Together the essays demonstrate in sobering clarity how the doomsday imagery of Africa has forced scholars to ask the wrong questions and produce inadequate diagnoses, obscuring the central role that forest communities have had in managing their resources, often under very adverse conditions. Most significantly, because these scholarly distortions empower the emotional charge embedded in the "Africa is dying and will continue to die" idiom, they justify fairly heavy handed external interventions that are equally misplaced. The interventions by governments and development agencies linked to these images have frequently both impoverished Africans and hindered their efforts to enrich their local landscapes.

The imagery of a bountiful nature inhabited by hordes of parasitic primitives and paupers can be seen as a particular version of the Malthusian vision of the land-labor equation in which a large population on the land is not thought to promote prosperity, as Malthus's mercantilist predecessors so positively believed, but rather to produce poverty due to diminishing resources. It is instructive for us to imagine how shocking and bizarre Malthus's vision must have sounded in the ears of his contemporaries long accustomed to the idea that "Fewness of people is real poverty; A Nation wherein are eight millions of people is more than twice as rich as the same scope of Land where are but Four" (Petty [1662] 1963:34). On this premise, William Pitt advised the House of Commons in 1796 to reward large, poor families since they "after having enriched their country with a number of children, have a claim upon its assistance for support" (Ricardo 1951:109). Just two years later, with the publication of Malthus's *Essay on the Principle of Population* in 1798 , the opinions on both poverty and state support were turned on their heads; what captured the public imagination in the nine-teenth-century was the terrifying possibility of an ever-increasing popula-tion of paupers eating its way into the nation's wealth and turning nature into a wasteland (Broch-Due 1995b).

In establishing population growth as a "problem" in the minds of his audiences, Malthus drew on the negative imagery that had collected around the notion of population as a net "consumer" rather than "producer". A con-sumer in the definition of the time was simply somebody using up every-thing produced (see Williams 1976). The common person in this sense was constructed in terms of a belly rather than two hands, a machine for eating rather than manufacturing. And while children and reproduction had been seen as a multiplication of busy hands at work in the mercantilist model, in the post-Malthusian world they came to mean the multiplication of hungry bellies and a natural world devoured. This shift from a producer- to a con-sumer-dominated model of the human relationship to natural resources was also symptomatic of the larger transformation that was occurring from mer-cantilism to capitalism. With the advent of the market and modernity, a more neutral pairing and abstract use of the terms *producer* and *consumer* be-came commonplace, but the negativity inscribed in the idea of consumption was to linger at least until the nineteenth-century. Interestingly, in connec-tion with the ecological sensibilities of recent times, this disapproving vision of consumption has cropped up again in the term *consumer society*—a criti-cism pointing in the direction not of the poor but of the prosperous, to the wasteful and throwaway features of modern lifestyles in the West.

Whatever the case, these two completely opposite conceptualizations of land-labor dynamics in the topography of wealth and scarcity are instruc-tive. For they remind us just how contradictory the imagination and model-ing of the relationship between people and nature have been. Much of this ambiguity is sedimented in the contemporary models projected and imple-mented through the process of globalization—as the essays in this volume make abundantly clear. Yet what is probably not so clear in the minds of most modern audiences is how far the terms of this debate stretch back through layers of European history and how deeply entrenched they have been in orchestrating internal European affairs.

"Too few Africans" was once effective rhetoric in a mercantilist world that *supported* colonization simply because, as it saw things, the more people

added to the equation the more wealth could be extracted from nature. This was radically turned around with Malthus's theories linking a large population no longer with prosperity but with poverty and decline. In the aftermath of this modern vision, there were too many Africans making inroads on nature's resources, a state of affairs that provided the rationale for the foreign interventions now labeled development. And while African nature was *excessive* for the mercantilists, in the nostalgic moods of their post-colonial counterparts it is *deficient*, always on the brink of disappearing in the haze of modernization and impoverishment.

Recently, the imposition of environmental imperatives has been especially dramatic in Africa, where sharply increased development and private donor investment over the past decade have sought to reinvigorate environmental conservation and protection programs through a major round of new investments. Linked to this is a shift in the discourse of the causes of poverty. While the ecologists' explanation for dwindling natural resources in the 1970s centered on economic growth coupled with uncontrolled industrialization, in the 1980s many of them came to perceive poverty as a problem of great ecological significance (Escobar 1995).

This reinterpretation prompted a new strategy, which promised to eradicate poverty and protect the environment as parts of a single package fleshed out in the report *Our Common Future*, commissioned by the United Nations in 1987. Labeled as "sustainable development", it responded to the heightened international interest in biodiversity maintenance, habitat protection, and environmental rehabilitation. Integral to its assumptions, however, is the reinvention of the Malthusian idea that poverty and the problems of population are the direct cause as well as the direct effect of environmental problems. Given this diagnosis, economic growth is needed for the purpose of eliminating poverty and the elimination of poverty is needed for the purpose of protecting the environment. Under the World Bank banner "sound ecology is good economics", sponsored by public and private capital alike, the strategy has given rise to a broad pattern of interventions related to the environment and ecology.

Conservation trouble

Despite being revamped and scaled up to planetary dimensions, the recent wave of projects to land on African communities is loaded with the same old crisis imagery that has been a characteristic of conservation policies since their inception. In this context, it is not particularly surprising that many conservation measures have left not sustainable communities but endless conflicts, dislocations, and poverty in their wake. This is not only the case with those operating with virtual realities that portray African humanity as inherently "against nature", and thus in need of radical reform, but also with projects that are constructed around the alternative ethics—a wish to preserve and perceive the native as a "natural" resource manager.

The politics of parks illuminates this problem particularly well, as many essays in this volume show. Entitled *No Room for Animals* (1956), the significant but highly sentimentalized advocacy of Grzimek for protecting wild animals against humans contains the rationale for erecting privileged environs for wildlife. It also carries the seeds of their contestation and the disputes created by the implementation of their explicitly antihuman agenda.

> Perhaps one day in the future the new park could be fenced in. Then the animals would have to remain inside it. They would be protected from settlers near the park and prevented from dying from hunger and thirst when all the timber around their water holes had been felled and all their pastures are over-grazed by native cattle.

"Eating dust" is a Maasai metaphor for hunger, expressing in one of our essays their experience of decades, indeed centuries, of efforts by countless external agents to mold and remold their environment. Finding themselves on the legendary Serengeti Plains, Maasai have been powerless in the face of such awesome symbolic capital and its appropriation by the Western imagination. For Serengeti stands as a part-for-whole for the East African savanna, an icon of a pristine landscape shaped by elephants and other precious species—a template that takes the human mind back to time immemorial before human greed entered the equation. In other words, Serengeti represents a particular image of nature filled with spirituality and power: "I speak of Africa and Golden joys", marveled Roosevelt in *African Game Trails*, "the joy of wandering through lonely lands, the joy of hunting the mighty and terrible lords of wilderness, the cunning, the wary, and the grim" (quoted in Adams and McShane 1996:25).

Roosevelt's celebration of nature did not in any way prevent him from destroying it. Hunting of wildlife—for leisure, commerce, and army rations—has been integral to European exploits in Africa since colonization. The sheer volume and ferocity of the slaughter finally prompted critical voices in the West and thus motivated the conservation movement. However, the most striking element evoked in Roosevelt's citation is the widely held perception of the landscape as empty of human presence and activities other than those of the privileged narrator. This is a persistent misconception of African landscapes dating back to the era of exploration and fueling a dominant image of Africa as nature writ large without humans (Adams and McShane, 1996).

In her essay about the creation of the Ngorongoro crater on the Serengeti Plains to suit the virtual reality of a dehumanized "natural" wildlife reservoir, Johnsen shows how this "place making" for animals has meant dislocation for Maasai peoples, who are watching their cattle herds being gradually replaced with herds of elephants and zebras. Conservation and the process of "parking" Maasai pastoralists has transformed the district as a whole into a Mecca for tourists and a "cash cow" for investors in the local tourist industry. The pastoral community, in contrast, has been reduced to "eating dust" and destitution—as vividly expressed by the Maasai metaphor. Eating dust has taken on an eerie tangibility as along the roads leading to the Ngorongoro Conservation Area Maasai parading their tribal finery are regularly coated in dust by the passing safari vehicles. These impoverished peoples are desperately trying to entice the tourists to throw a few shillings their way in exchange for a snapshot of the noble savage in his or her habitat.

The revenues from tourism have not been channeled back into the community, leaving the impoverished Maasai not only without a fair share of the capital but on the brink of destitution. Despite the fact that the park was set up as a multiple land use zone in which the wildlife sanctuary and cattle herding would coexist, few resources have been invested in pastoral development. On the contrary, despite their initial compliance, Maasai have experienced a history of broken promises, continued land alienation, and further restrictions on herd movements. This has resulted in a drastic decline in the pastoral economy. Pauperization produced through conservation management shows up in uncontrolled cattle diseases, resulting in smaller cattle holdings and less milk to feed the family. It is also about the restructuring of herds to include more goats and sheep to sell—itself a sure sign of worsening poverty since their value is not reflected in the low prices they fetch on local markets (Talle 1988, 1999). These processes combined have led to a gradual collapse of clan-based systems of mutual assistance organized around cattle exchanges and an increase in Maasai cultivation within the crater.

Based on a long history of erroneous assumptions that pastoralists are pure (male) herders who never cultivate (see Hodgson 1999), the spread of these sorghum fields has particularly angered the park management. This is not only because wildlife conservation and cultivation are not regarded as a sound "product mix" but because these external agents have come to interpret the fields as signals of Maasai obstructiveness and their failure to become "partners" in joint development ventures. The preferred solution to what is labeled "the human problem" by management and Western environmentalists alike is an eviction of people and livestock from this landscape in which they have lived for generations.

What has happened, the essay concludes, is that the power of the extraordinary topography of the Ngorongoro crater has gone hand in hand with the power of translocal agents, which rests on a bedrock of belief that Maasai subsistence pursuits are antagonistic to conservation objectives. This idea is deeply entrenched in Western discourses of poverty, as a glance at our citations from Burton to the World Bank evidences. Malthus, for example, penned a warning against dispensing aid to the poor on the grounds that they would spend the money on meat at the expense of cereals, thereby fueling a demand for cattle and increasing the amount of good arable land turned over to grazing. In his words: "A fattened beast may in some respects be considered as an unproductive labourer: he has added nothing to the value of the raw produce that he has consumed. The present system of grazing tends ... to diminish the quantity of human subsistence in the country" ([1798] 1976:107). Colonialists and postcolonialists seem to have shared the contention that the "livestock-labor-land" equation is intrinsically an "unproductive" one. In fact, this constitutes the recurrent "problem" that their policies have sought to redress. However, the perception of whether pastoralists were essentially prosperous or poor has changed dramatically over time. Colonial policymakers were of the opinion that pastoralists had too much animal wealth, performed too little work, and were thus uninterested in the "progressive" and "civilizing" effects of selling their labor and livestock. Colonials chose taxation as their instrument for transforming idleness into industriousness (Waller 1999; Broch-Due 1995b).

During the postcolonial period, pastoralism has become integrally linked to poverty as policies have focused more on the troubled livestock-land relationship. In modernist reworkings of the Malthusian model in which cattle "eat" men, nomads and their beasts typify both backwardness and a threat to the land itself. In "The Tragedy of the Commons" (1968), Hardin constructs cattle as voracious destroyers of common land, which could be more productively used for agriculture. Moreover, the lack of private property means that there are no checks on the tendency of the cattle population—and thus the human one—to grow beyond the carrying capacity of the pasture. Applied to Africa, this has led to the discouragement of herding and attempts to shift pastoralist labor into alternative, and allegedly more secure, forms of production and work (Anderson and Broch-Due 1999). Central to this larger project has been an ingrained agrarian bias against nomadism, with its porous and shifting boundaries, geographically and socially (see Giles Vernick, this volume; and Broch-Due, this volume). The development agencies' charter for a change in livelihood and labor thus has come to embody an extension of the nineteenth-century colonial project of specifying new social forms of living for the poor and marginalized (Broch-Due 1995b).

East African pastoral communities have fiercely opposed these perceptions and policies. From their perspective, the claim that pastoralism is an unproductive enterprise and pastoralists are idle is outrageous. Indeed, domesticated animals compare remarkably well with almost any form of capital, monitory or otherwise. They multiply by themselves and thus generate wealth without the medium of markets or other exchange mechanisms. This, in turn, reproduces family and community, physically, socially, and symbolically. Yet the pastoral enterprise is always faced with the possibility of rapid growth and decline. Thus, within the pastoral world of the Maasai and Turkana, wealth in children and calves is a sure sign of industriousness and skillful management. Their shortage, however, is an equally sure sign of self-inflicted impoverishment in the minds of the successful.

For, although it is acknowledged that misfortune may strike anyone, the dominant assumption among Turkana, for example, is that the prudent person will be able to recover while the imprudent will not. This moral twist is part of the conceptual and moral universe of pastoralism, which profoundly ties together human life and the life of herds. Poverty is interpreted as a negation of this universe, and it seems to be the result of not managing these two vital assets—herds and humans—in a proper way (Broch-Due 1999).

While pastoralists are conscious of the manifestations of poverty in their midst, the terms around which they choose to describe and articulate their attitudes toward impoverishment remain incomplete and refractory, tending to obscure the real material and social processes that lead to exclusion from the pastoral economy itself. Instead, pastoralists prefer to highlight the linkages between poverty and more sedentary pursuits and lifestyles. In other words, they recognize readily the end result of exclusion but not the paths that lead to it. The subtle, ideological misrecognition generated by pastoralists themselves has been uncritically reproduced in the scholarly discourse, reinforcing the normative stereotypes of pastoralist egalitarianism (Anderson and Broch-Due 1999).

This muting of the realities of poverty in pastoral societies forces us to acknowledge that power is not only an effect of "global/local" encounters but is of course equally operative in promoting certain interests against others *within* the communities investigated in this volume, be they agrarian, pastoral, or urban. Likewise, the revisionist bent of many of our essays that argue against dominant models of resource destruction by local populations on the grounds of recent research findings does not necessarily imply that "local" peoples are "naturally" equipped with sound environmental strategies. Sometimes they are, and sometimes they are not.

Whatever the case, as argued in the essay by Neumann, so-called indigenous peoples bear an extremely heavy burden, which is to continue practices that are always, and in every instance, conservative or even curative of environmental problems. In other words, planners expect them to accomplish what immigrants and Westerners, including the colonial powers, were themselves not able to do: "To continue to produce and reproduce themselves in an environmentally benign fashion", as the author so succinctly formulates it. The essay critically evaluates integrated conservation and development programs in Africa, focusing on protected area buffer zones. Despite the emphasis on participation and benefit sharing currently in fashion in conservation circles, these revamped projects often replicate in different ways the coercive conservation practices they are meant to replace. Entitled "Primitive Ideas: Protected Area Buffer Zones and the Politics of Land in Africa", the essay traces the reason why good conservationist intentions so easily turn sour to the tenacity of Western imagery of the "other". Buffer zones are liminal spaces literally and metaphorically. They are put in place as a protected zone between a park exclusively reserved for wildlife and the ' wider surroundings, their inhabitants being allowed to stay as long as they guard the boundaries between these spaces physically and symbolically, for buffer zones are produced in the tension between the two conflicting ethics surrounding images of primitive Africans either as "bad news for nature" or "natural-born nature managers".

Good natives are those having a "traditional" livelihood sustained by "indigenous knowledge". They are perceived to be closer to nature and thus consistent with the environmental managers' designs for parks or protected areas. Bad natives are those who are in some sense "modern", and thus removed from nature, their modified lifestyles and greed for consumer goods representing a particular threat to the natural treasures enclosed. Good natives are invited to participate and comanage the resource in buffer zone projects, being rewarded with benefits that include rights to access, social services, and political empowerment. Bad natives, in contrast, are summarily displaced. They are routinely forced out of parks and protected areas.

Both the construction of spaces at play in Neumann's essay and the cast of natives are prefigured in the conventional European iconography projected in literature and painting. The production of particular landscapes in contemporary conservation in Africa are adjusted reproductions of romantic ideals of nature conjured in landscape painting—the placid pasture, the wild forest, the threatening mountain range. Even those painters inspired by naturalism had to adapt to such sedimented expectations when composing their canvases. Painters during the era of romanticism produced visionary landscapes, often in the forms of pastoral idylls in which the linkage between cultivated and uncultivated lands was typically the gentle shepherd with his domesticated flocks of sheep and goats (Hale 1994). One notes how this placid European icon is not only recast in the same mediating role as the "good native" of the African buffer zone but how far this homegrown "pastoralist" is removed from the fierce Africans who animated nineteenth-century accounts of the savage nomad, now earmarked for expulsion by the contemporary conservation scenario of Serengeti. Again the larger point is that both figures, whether malevolent or benign, came into Arcadia not from life but from literature, in the past as well as the present. African nature, though explored and exploited as never before, remains close to Western visual economies and iconography. From this foreign perspective, African nature has, through a long tradition, been thrown up on a metaphorical screen, recontextualized, and displaced.

Neumann's essay not only exemplifies particularly well how this narrated dimension of nature is put to play in the buffer zones with dramatic effects but also how unstable its casting of good versus bad natives is when re-created in African realities. He examines the process through which conservationists alternatively invoke images of the good and bad native and in doing so define "legitimate" claims to land in protected areas. His, like other essays, shows that primitivism is not only a set of ideas but a set of forces of power played out in very concrete struggles over scarce resource flows. The stereotypes projected result in misguided assumptions in conservation programs, which have important implications for the politics of land in buffer zone communities.

The ingrained irony here is that those cast as good natives in conservation are cast as bad natives in the discourse on civilization and modernization. The paradoxes and moral ambiguities produced in the interstices between the two conflicting ethics of transforming and preserving are given a particular twist in Larsen's essay on tourism in Zanzibar Town—typically the next stop after Serengeti on the itinerancy of safari travelers. Wildlife with a dash of exotic spice is sold as part and parcel of the same package of eco-tours or what is labeled "alternative tourism". However only the older quarters, called Stone Town, has become a tourist attraction precisely due to the "thousand and one nights" images that collect around its ancient walls. Thus, the "expandable landscapes" of tourists, like those of conservationists, are structured by iconography that has been popular in the West for centuries. Indeed Stone Town was recently declared a conservation site rated as comparable in significance to Victoria Falls and the Ngorongoro Crater.

Yet this appreciation is not innocent, as the author points out, but comes trailing a particular image of locality that highlights the "Arabicness" while downgrading the African influence on Zanzibar—which has throughout history had a multicultural society. This is not only a highly politicized representation, but it is also a highly paradoxical one. The other side of town, the modern quarters, is shunned by the tourists as a squalid signpost of triviality and Third World poverty. These blocks of apartments were built after independence in order to "modernize", "westernize" and rid Zanzibar of a colonial and primitive past. Ironically, the new town perceived so proudly by the inhabitants as a sign of betterment is by today's tourists and expatriates found aesthetically undesirable—a sign of poverty even. Stone Town, in contrast, conjures, in the virtual reality of the tourists, exotic and erotic images of former grandeur, affluence, and sensuality. Local people, most of whom are Muslims, find that privacy can be upheld much more satisfactorily in the new section of town. To them, these spaces of "ubiquitous modernity" constitute a site where they can maintain a sense of identity—one that relates to the possibility of having social interactions in morally correct ways. To most peoples of Zanzibar Town, this is a more attractive site for their identity management than entertaining the fixed identity ascribed to them by Westerners, circumscribed by the stones and the carved doors of the old town.

The author explores how women and men in Zanzibar Town conceptualize and adjust to present changes within their urban environment—changes caused by a rapidly expanding tourism that are represented by the government as a prime generator of economic activities and benefits. In doing so, Larsen focuses on local discourses on culture, self-perceptions, and hence questions of morality and self-presentation in a situation in which women and men see their landscapes invaded not only by tourists but by "other ways of life". In the articulation between local and global perspectives, the so-called cultural show constitutes a site particularly contested. Such performances are staged in response to the insatiable demand by tourists for the erotic inscribed in the exotic—of "primitive" nudity, dancing, and drumming. While tourists perceive such displays as being true to the "nature" of the place, the local peoples are of the opinion that they reflect an invasion of "modern" lifestyles of an immoral kind, corrupting and embarrassing to Muslim ways. Within the local community, there is thus a tension between the desire to maintain privacy and the desire to benefit economically from the tourist trade. Employment opportunities entice people into activities, moments, and encounters that are problematic at best and downright degrading at worst.

Insofar as tourism is about people moving to distant corners of the earth to expand their experiential, imaginary, and ideological landscapes, it is also about the effects this form of travel has on the landscapes of those who receive them. In this sense, tourism involves an "interaction of landscapes", a productive encounter in which contrastive identities play themselves out against one another in unpredictable ways.

The essays in this section all help to bring home one of the central points of this volume, notably, that we are confronted with a situation in which global and local models each come to the problem of poverty and nature with its specific ideological bias. This is not to say that poverty is merely a matter of divergent ideas, but that even where its social manifestations are blatantly evident and concrete its meanings and interpretations are inevitably culturally and conceptually mediated. Far from being a straightforward condition of deprivation and destitution that is easily defined empirically, poverty is in fact a contentious and complex construct that encapsulates a vast range of social and historical struggles and constantly evolving cultural values. As our essays make clear, poverty is deeply embedded in conflicting perceptions of nature (Broch-Due 1995a).

Nature, poverty, and modernity

Taken together the case studies in this volume encompass the colonial era and the rise of contemporary networks of unequal global relations. In other words, the historical coordinates of the volume coincide with those of modernity, which in Africa is nowhere complete and hardly anywhere fully absent. Modernization is a process of accelerating social change at work on a world scale, the complexity of which is reflected differently through the range of topics dealt with in the essays. Notions of nature and poverty are central to the production of modernity as a cultural project formed in the "West" in the wake of industrialization and transported to the "rest".

Modernity is a shorthand concept that distills the complex and highly problematic processes of transformation that cover the whole trajectory of international relations. Its temporal and spatial frames begin with early explorations that gave rise to the trans-Atlantic trades in slaves, ivory, and minerals. As slaves were removed from the commodity ledger and other valuables added, the colonial enterprise perfected its techniques of resource exploitation and control before, in the twentieth-century, transforming itself into the "development world" and erecting the postcolonial aid structure (Hobsbawn 1994) through which huge amounts of capital have been injected into African arenas ostensibly to create viable economies, eradicate poverty, and protect the environment.

A common misconception is that development as a social phenomenon signifies the break in Africa between predatory colonial regimes and a more benign postcolony promoting economic growth and social equity through global partnerships. As Cooper (1997) makes clear, however, development, which was socially constructed in the 1940s, is well within the frames of colonialism. Yet there were many sites for its production. Development arose as a radical charter for social reconstruction in response to social and political unrest in an impoverished Europe in the wake of the Second World War. It signaled the construction of the welfare state. Key elements were imported from America, where cooperate capitalism had already embarked on the process of replacing class struggles with consumerism as an instrument for social homogenization.

Consumerism has gradually emptied the poverty field of political content by harping on its central idea that everybody can share in prosperity through the machinery of modernity—work, education, private property, shopping, and the family. Launched into its global orbit, this recipe for prosperity (alas, though, few of its products) has been promised to the world's poor. Thus, the idea of consumerism is reproduced throughout the web of the expanding donor network, turning the development industry into one gigantic "antipolitics machine" in the terms of Ferguson (1990).

Given this rough sketch of development, it may seem a contradiction of terms that in the context of Africa development was inaugurated to strengthen colonialism, although it turned out to be central to the process by which colonialism was terminated. This reminds us that development is a double-edged discourse. Its potential for promoting economic growth and discipline had a strong appeal to the colonizers. In colonial Africa, the modernization project embedded in development was only a more effective extension of the much older project of civilizing the "primitives". Seen from the perspective of those in power, development appeared as a discourse of control along lines similar to civilization. Yet development also contained new discursive resources formed around civil rights and equality, which were drawn upon by the educated elites in their political struggles for national independence. From the perspective of the colonized, development has sometimes been a discourse for liberation, equality and empowerment. By reconfiguring some hegemonic claims, development opened numerous points for contestation within the colonial powers' own discourses, which are highly significant for the topic of this volume. Cooper succinctly sums this up as follows:

> What colonial governments—in hitching their aspirations for a politically legitimate and economically productive empire to the idea of development—had given up was their old claim to be presiding over immutably distinct peoples, of providing order to savage peoples and slowly bringing them into civilisation. The acultural concepts of development and industrial relations presumed that Africans could function as producers, merchants or workers much like anybody else. (1997:75)

Thus what development helped to make manifest at this particular historical moment and particularly in the settler colonies was the fragility of the colonial nation-state structure. Confronted with the enormity of the development enterprise—its long list of changes to be implemented and the difficulty of getting them done—colonial officers were persistently frustrated. They had recourse to the discourse of primitivism that, as our earlier discussion has established, is always bound into modernity, almost as its dark twin. The architects of development in the colonial offices in 1947 referred to Africans as "a great mass of human beings who at present are in a very primitive moral, cultural and social state" (Cooper 1997:74).

We see these tense and fraught shifts of discourse played out in my own essay about the dislocation of the Isiolo Turkana in 1958, involving the forced deportation of settlers from Isiolo Town back into the drylands of Turkana District. The almost three decades between the planning and execution of what the colonial regime in Kenya labeled a "repatriation" perfectly fit the time span in question here. In both its conceptualization and implementation, this tragic deportation case was rife with ambiguities. Its uneasy positioning at the interstices of many agendas implied that everything and everybody involved in the move were from one perspective or another "matter out of place". As far as the colonial officers were concerned, the most obvious matter out of place was the Turkana victims themselves. The whole rationale behind the operation was precisely to move them back into place—to their tribal territory—although when they arrived there they were still matter out of place from yet another perspective, notably that of the pastoral host community.

Their already problematic identity in terms of indigenity was confounded further in the colonial discourse by the fact that the context of this particular dislocation was the Northern Province, the territory of the traditional "primitives"—the nomadic pastoralists (see also Giles-Vernick, this volume). Thus, the repatriation argument came up against another equally compelling aim, that of sedentarization and modernization. Decades earlier, their nomadic movements had made the Turkana into matter out of place from a sedentary perspective. To return settled, urbanized people, most of whom were drawing a living from wage labor, to a rural subsistence economy as nomadic herders seems to have been a step back in the linear process of "progress" deeply ingrained in the evolutionary ethos of the time. This tension between the colonial desire to create tribal constituencies and fix tribal identities, on the one hand, and the desire to civilize by transforming nomads into settlers, on the other, permeated the case from the planning stage to its implementation.

To these, we can add layers of tension within the realm of the colonizers, brought to the case by officers and other characters directly involved in orchestrating the repatriation. Their views and voices represent widely different interpretations of the vexed matters of identity and place that come out of the various traditional-modern constructions in circulation at the time. Central in the efforts to construct the colony was a concern with the shaping of bodies and their spatial belonging. This constituted another conflicting arena, as the corporeal politics incorporated the reproduction of particular styles as embodied signs for fixed tribal constituencies—like Turkana versus Somali—but also the capacity to signify social transformation along the axis of primitive-civilized and nomad-settled.

At one level, this case was an object lesson in the way meanings diverge and in the fact that the semiotics of landscapes and bodies are never unambiguous and can never fully be controlled, even by a dominating colonial power. This case aligns itself with the insights elaborated in the essays of Larsen and Neumann—notably, that bodily signs and signposts in the landscape are always potentially subversive and troublesome, lending themselves easily to manipulation and contestation. As my essay exemplifies, these contested meanings are not only produced at the interface between colonial and local models but also within the groups of both colonizers and colonized. Those of different gender, generation, and social positioning, with differing motivations, wealth, and access to power, disagree within the pastoral communities, as they do in the colonial offices and beyond in the settler community. In all these arenas, there was a chorus of voices at play, but they were not all equally loud or authoritative in determining the action taken; weaker voices were muted and disregarded.

The global reconfigured: Connection and disconnection

The same canvas of modernity covers the construction of the nation-state and its gradual demise on a world scale. One decisive turning point located by many scholars is the 1960s, a decade during which private capital managed to restructure and move operations across geographical borders to escape the regulatory frameworks of the nation-states. This new form of globalization in which finance became liberated and occupied a transnational space where nation states remain functional but have a subordinate position, is a very significant factor for the topics discussed in this book. For, although there are many continuities with the perceptions and policies formed around poverty and nature during colonial and postcolonial times, as evidenced by the essays in this collection, the ways in which private capital has established a network globally and contemporary states have abandoned any pretense of welfare and equality have precipitated a defining rupture that provides the context for completely new reconfigurations of nature, poverty, and power in Africa (Berry 1993; Geschiere and Meyer 1998; Appadurai 1998).

Structural adjustment polices imposed by the World Bank to speed up privatization channeled private capital into the domain of nature through a range of new commercialized ventures formed around conservation, parks, and bio-prospecting. This has not only propelled the commodification of nature to extreme degrees, drawing even the genes of flora and fauna into the property domain, but it has also meant that what used to be common natural resources—firewood, fodder, building materials, berries, game, and sometimes even fields and pastures—have been turned into commodities too. In a situation in which many African states' welfare provision is shrinking, we can see the contours of a vast field of poverty in the making, and a process in which critical resources will be concentrated in fewer hands, creating new divides between wealth and want (see the essays by Östberg, Neumann, and Benjaminsen, this volume).

Globalization has not just been about integration and the flexible deployment of capital through privatization and democratization. It has also been about the withdrawal of the "global" from the "local", about disinvestment, abandonment, and despair. Thus, integral to globalization is the experience of countries or communities, previously connected through investments by private capital and donors, of being subsequently "disconnected" (Ferguson 1999). As private and public capital moves on to more promising sites, corporations search for more profit and aid agencies target new groups in search of development success stories. Those left in the wasteland created by this reconfiguring of commodification and concern find themselves with less than when they connected with these transnational forces and flows. Moreover, their hopes and aspirations for a better future, formed through earlier waves of modernization, are now being crushed. Many African peoples currently embroiled in conflicts and warfare share the social experience of economic contraction and impoverishment.

The essay by Katz takes up one such exceptionally tragic case, contemporary Sudan, which has been beset by crisis for decades. Embroiled in a civil war, wracked by "structural adjustments" imposed by the International Monetary Fund, and almost uniformly ostracized in international arenas, the state and Sudanese population struggle to survive saddled with debt, inflation, and the effects of a corrosive war. The author shows how these circumstances have fostered a peculiar form of self-reliance that revolves around clearing old growth and other forests in southern Sudan. Woodcutting for charcoal production has become a key means by which the state secures tax revenues, one of the only reliable sources of dry season income for much of the disenfranchised and otherwise marginalized rural population of central Sudan, and a hedge against the importation of costly cooking fuel for those in the towns. The fundamentalist Sudanese state authorizes and even fosters this deforestation with patriotic appeals, assuring would-be woodcutters from the Islamic North that their work helps to root out southern guerrilla fighters hiding in the forests. Woodcutting has thus become an accessory to war. This essay draws on extensive field research in Sudan to examine the political ecology of charcoal production. The author demonstrates how the everyday social, economic, and environmental practices of differently placed social actors work to destroy particular forests. She convincingly argues that the confluence of the interests of the state, the rural poor, and the urban elite has conspired to destroy the regional environment in potentially disastrous ways.

This case, which is not uncommon, brings out starkly the ways in which environmental conflicts are increasingly a part of African "warscapes" (Nordstrom 1997). Donors and nongovernmental organizations (NGOs) are everywhere involved in poverty assessment and alleviation programs, yet their activities frequently assume peaceful situations. While it is obvious that full-scale armed conflict causes poverty through the destruction of natural and human resources, perhaps less clear are the ways in which interethnic and intercommunal strife threatens the livelihoods of communities and undermines strategies designed to alleviate poverty and promote social well-being. These latter processes have been poorly understood.

Templates: The problem with totalizing models

There is another emergent feature of modernity at the core of this volume. It has to do with processes of bureaucratization and the particular modes of knowledge production linked to it. The success of any state, whether colonial or not, hinges on its ability to comprehend local communities so that they can be organized to satisfy the classic state requirements of taxation, conscription, and control. As argued by James Scott in *Seeing Like a State* (1998), this "legibility" of local communities is produced through a set of "state simplifications" designed to reduce not only the "opacity" of local peoples but also the complexities of their livelihoods and landscapes. A typical simplification device is the template.

Templates are built by drawing on the techniques of storytelling, being equipped with a narrative structure that is comprehensible within the horizons of experience and expectations of the target readerships. Certain templates and blueprints have long been at work in European representations of Africa, not only as technical devices for storytelling but as filters selecting certain kinds of stories and not others. Templates are born out of the particular constraints placed on the communicative relationship between authors and audiences when something unfamiliar is to be portrayed. As with any story, a template weaves itself from the real in different ways and with different levels of opacity. Which narrative gets picked up, which truth claims are conveyed, and which genre of evidence is evoked depend largely on its location in the wider political economy. Colonial and postcolonial templates differ according to a noticeable pattern due to the different sets of economic interests at play, the different locations of key audiences, and the scale of the social landscapes portrayed through the template.

During colonialism, the social and natural worlds of Africa were gradually inscribed in a new template organized around money and markets. Toward the end of colonialism, this terminated with fullfledged ledgers in which everything could be reduced to units and numbers. Given the fact that real extraction was at the heart of the colonial enterprise, the scale of the map and its template generalizations could not move too far from the ground where the resources were located; otherwise, the effect would be not legibility but chaos. Most colonial states employed microtechnologies of power through which we can see the contours of something that could have developed into the type of "power-knowledge" regime Foucault (1980) envisioned for Europe.

Colonial governance rested on the principle of indirect rule—to mobilize local political structures in order to reduce the cost of governing. Anthropologists worked within the context of a large colonial bureaucracy, which occupied itself with the self-conscious project of collecting and organizing knowledge. In the target regions, fieldwork combined with surveys and censuses aimed at counting, describing, and dividing up the population into manageable units so that administrative routines could be established and procedures standardized. The reportage retained the descriptive dimension of differences. At the same time, it also produced a taxonomy that translated diverse case materials into broader templates in terms of bands, tribes, chieftainships, and kingdoms. And this grand project of typologizing societies aligned with the broader aim of colonial policy. For colonial regimes were trying to define the constituents of a certain kind of civil society, even as they camouflaged the fact that they were inventing them beneath the idea that society was a "natural" given and the state a neutral observer and regulator (Cooper and Stoler 1997).

The study of specific cultures—of local manners, meanings, and points of contestation—was at first a low-key affair. The participant observer focused on the few customs he needed to know in order to function in a competent way. In this scheme of things, cultural variation was something to overstep in order to make general models that could explain more "economically" the empirical diversity of life and living. The colonial ethnographers searched for the universal in the local, the general in the particular, the whole in the part. Thus, to a large extent, they created their own object of inquiry: stable societies and stable categories of difference. Whatever currency these constructs had at the time of colonial contact, they certainly took on a material reality through the actions of the colonial government. The larger point is that while the colonial regimes usually got the data directly linked to their revenue system reasonably right, they got other kinds of more culturally specific, fluid, and contested data totally wrong. In these cases, ideology and observation merged. This ideology-laden and refractory nature of colonial knowledge production is echoed in many of this volume's essays (Fairhead and Leach, Benjaminsen, and Broch-Due).

However, in the postcolonial era, with the global proliferation of private capital and the streamlining of financial institutions with the advent of the development industry, templates and modes of knowing African worlds have moved not closer but further from the empirical ground. The development industry has been a major generator of preconstructed frameworks like templates and blueprints that can be deployed as tools to simplify and control across continents a range of diverse and complex environments (Hirschman 1968; Roe 1991; Hoben 1995; Leach and Mearns 1996; Cooper and Packard 1997). The centralization of power globally and the sidetracking of particular states mean that policies and plans destined for the reformation of African realities are made in Washington, London, or Stockholm. The simplification, schematization, and standardization of knowledge about African realities has reached a scale at which—to paraphrase Marx—"all that was solid has melted into thin air". The development industry churns out plans and applications for global consumption offered for implementation in places very distant from the site of production. And in order to grow, proliferate, and fulfill the demands of accountability the global development industry needs to appeal to the public imagination, not of its aid recipients in rural communities in Africa, but of its audiences in the West.

What has happened in this reorganization of the social is that one no longer speaks of placebound entities as societies or tribes but of cross sections of certain activities, sectors, and individuals, target groups that are assumed to be similarly constituted in terms of need across communities and continents, as if we were all citizens of a single global society. Social scientists are central in this postmodern form of knowledge production as project consultants on short-term and instant assignments. Consultants are forced to take shortcuts, sometimes to the extent that previous reports are used as blueprints; bits and pieces of new information are added and old ones removed. Those less scrupulous who sincerely seek to address the subject population on its own terms are forced to tailor their observations to the limited timetable given, which often means that simple surveys are chosen at the expense of more in-depth, qualitative case studies and systems analysis. The outcome is often fragmented and random information, on the basis of which the consultants feel obliged by the "terms of reference" to make sweeping generalizations, which in most cases are poorly justified by the evidence.

Planning for places distant both geographically and conceptually means that the level of uncertainty rises and unintended outcomes abound. All this gives rise to an important question. Should these interventions, planned in the centers of globalized capital and implemented in a myriad of peripheral settings, best be understood as the expansion of a Western power-knowledge regime into African arenas (Sachs 1992) or, alternatively, as "the growth of ignorance" pure and simple (Hobart 1993)?

Interestingly, postmodern development templates have much in common with medieval templates and those of the age of discovery that grew out of them. Perhaps part of this resemblance can be understood by considering the reward structure. During both the era of early encounters of Africa and the contemporary "development" world, funding for expeditions and projects has depended on sponsors. Whether the funding comes out of multilateral and bilateral donor budgets, medieval patrons, or audiences of travelogues, the search for these funds requires mobilizing the imagination of a home constituency. The "fabulous" nature of medieval travelogues is clear to us today. Medieval writers did not travel widely and relied on their imaginations and long-established narratives to describe other peoples and places. What is more surprising, however, is that when Europeans did resume long-distance traveling to Africa and elsewhere during the so-called Age of Discovery the writing did not change much at all.

Despite regaining firsthand observations from distant places, the style of reporting continued revisiting the same tropes and fictional tracks established during medieval times (Pieterse 1992). While travelers had been on the spot, changes in the technologies of representation and the expansion of audiences this entailed, meant that they had to employ artists who had *not* been in Africa but whose illustrations were vital to the popularity of the text and thus its profitability. Indeed, whatever was written on the page was given a newfound tangibility and truth value through such visual aids (Bucher 1981). Thus, like the authority inscribed in the field photography, the news reportage, and the World Bank's annual report, the drawing served as a signpost to authenticity of the author's knowledge based on his having actually been there.

In a fascinating study of travel engravings, Steiner (1995) shows how the very lack of firsthand experience forced artists portraying foreign peoples and places in Africa and elsewhere to draw on familiar iconography already in circulation. Most significantly for the topic of this book, engravings of exotic peoples were modeled on particular representations of a European peasantry that was seen as equally "primitive" and unknown. The core template embracing the European and African savage was the dance around the golden calf, a reproduction of the ancient biblical image of heathenism.

Images were either copied from earlier sources or picked out of a conventional ledger of notions about what savages "look like and are known to do", at home or abroad. This pastiche of ancient European symbols of the primitive was replenished with additions like elephants, palm trees, and carved idols, deployed as conventionalized signs to the readership that the drawing was not representing European peasants but people far more exotic, uprooted, and savage. The supposed realism of these surreal representations was assured by the fact that neither the producing artists nor the readers were likely to "have been there". Indeed, the popular appeal that comes with recognition of something familiar combined with advances in the printing media ensured widespread circulation. Steiner (1995) demonstrates how this engendered a shared set of criteria that effectively created the very standards by which the subject of the picture could be identified and verified as primitive.

The quintessentially negative primitive in the European convention of colonial times was the nomad, who inhabited Burton's 1872 classification of East Africa and continues into contemporary times to inhabit accounts of what characterizes a "least developed country" according to the international donor community in 1972. Yet this African-derived type of the unruly and impoverished nomad had already a decade before Burton's evocation been put back into the European landscape by Mayhew to describe the London poor. Mayhew believed that everywhere on the "entire globe", from Scandinavia to Arabia, from South Africa to Britain, almost every settled "race" was surrounded by hordes of predatory nomads—the "paupers" of their environment.

> Such are the bushmen and the Songuas of the Hottentot race—the term "songua" meaning literally pauper. ... That we, like the Kafirs, Fellahs, and Finns, are surrounded by wandering hordes—the "Songuas" and the "Fingoes" of this country—paupers, beggars and outcasts, possessing nothing but what they acquire by depredation from the industrious, provident, and civilised portion of the community. (1861:2)

The character profile of the English version of the nomadic type was based on "extensive observations in South Africa" by the anthropologist Dr. Smith, who inspired this detailed description rendered by Mayhew, the first sociologist to study the poor in Europe "from the lips of the people themselves". Introducing himself in the preface as being the first "traveler" in the undiscovered country of the poor, he returned with a large number of "facts".

> The nomad then is distinguished from the civilised man by his repugnance to regular and continuous labour—by his want of providence in laying up a store for the future—by his inability to perceive consequences ever so slightly removed from immediate apprehension—by his passion for stupefying herbs and roots, and, when possible, for intoxicating fermented liquors—by his extraordinary powers of enduring pain—by an immoderate love of gaming, frequently risking his own personal liberty upon a single cast—by his love of libidinous dances—by the pleasure he experiences in witnessing the suffering of sentient creatures—by his delight in warfare and all perilous sports—by his desire for vengeance—by the looseness of his notions as to property—by the absence of chastity among his women, and his disregard of female honour—and lastly, by his vague sense of

religion—his rude idea of a Creator, and utter absence of all appreciation of the mercy of the Divine Spirit. (2)

In terms of the politics of representation, this is striking only for the fact that men like Smith, Mayhew, and Burton, who had exerted themselves to see the world afresh, returned with such stock observations. For what these nineteenth-century scholars did, unlike their contemporaries in the colonial administration, who actually had to grapple—however refractorily—with an African reality, was to revisit an ancient European terrain populated with savages and paupers, which, having been out of public circulation for awhile, was now being brought back to be "rediscovered" by new audiences. The templates of primitivism helped to weave African peoples together with domestic groups also situated on the margins of the sociopolitical geography and thus equally "unknown", notably the poor living in the slums of the Western cities emerging in the wake of the Industrial Revolution (Comaroff and Comaroff 1992). The two iconic figures, the pauper and the nomadic primitive, were juxtaposed in reportage throughout the nineteenth century and during the twentieth became subject to the same reforming measures of their lifestyle and livelihoods (Broch-Due, this volume, 1995).

Through series of such complex reworkings of geography and chronology (Fabian 1983), this poverty imagery has continued to the present day. Take the following extract from the World Bank's 1980 development report, which, despite being couched in more sober language than Mahew's, still reproduces many of the same assumptions and story elements.

> The poor are a mixed group ... the marriages and ceremonies after the harvest are in stark contrast to the hunger and illness that often precede it. The poor have other things in common, apart from their extremely low incomes. More than three-quarters of them live in (often very remote) rural areas, the rest in urban slums ... poor people live mainly by working long hours—men, women and children alike—as farmers, vendors and artisans, or hired workers. As much as four-fifths of their income is consumed as food. The result is a monotonous, limited diet of cereals, yams or cassava—with a few vegetables and in some places a little fish or meat. Many of them are malnourished ... the physical and mental development of their children is impaired. ... They are often sick. ... The great majority of poor adults are illiterate. ... Unable to read a road sign, let alone a newspaper, their knowledge and understanding remain severely circumscribed. (WDR 1980:33)

Like mediveal templates and those of Mayhew, this abstract portrait of the poor invokes simple images of rural village life as isolated from the dynamism of world history, its poverty a "natural" feature of geography and culture, demarcating an area distinct from a "real" economy and the city, where rationality, measurability and adaptability prevail. Left out of the equation is any and all location of the description in a specified reality, thus precluding any historically specific and socially situated analysis of the ways in which poverty is produced and maintained through a whole set of social, political, and economic factors. In these modern versions of "the stories we tell ourselves about the African other", nature and poverty figure in a predictable blend, conjuring up conventional spatial signposts, binarily paired as country versus city, which is simultaneously a temporal contrast between traditional and modern—all translated into a topography of wealth that conjures a spatial divide between poverty and prosperity. Such representations not only reinforce old stereotypes but, more importantly, they demarcate a space for a host of interventions to reform the target group.

While certain Western templates are further from reality than others, all convey a general message of otherness while at the same time letting the reader believe that each locale has genuinely been captured. The price the template constructor must pay to create such standardized conventions is indifference to local detail. What Steiner so succinctly says in relation to travel engravings also holds true for contemporary templates, for

> the authority inherent in the very conventions so established yields, by a somewhat curious and ironic twist, a substantial return; tipped off by a caption, the reader is more likely to believe that the unique, local identity of the scene has been portrayed with fidelity not to convention but to reality itself. (1995:211)

The problem of ever-expanding agendas

Template production is the symptom of another problematic effect of globalization that comes out of our essays. This is the way colonial and postcolonial agents of change have come to subscribe, either implicitly or explicitly, to totalizing ideologies of development, modernity, and progress that assume a uniform approach to what are in reality rather disparate needs of "the poor" in Africa. Linked to this is the shifting nature of policies through the decades, where agendas either shift rapidly from one set of generic "issues" or "problems" to another or continue to develop an ever-expanding agenda that is so all-inclusive that it suffers ultimately from complete analytic and practical paralysis. International development institutions often justify their actions with broad but vaguely defined agendas that include everything from the state to capacity building, democratization, political liberalization, women, and environment. These wide-ranging agendas are often so broad in scope that they are totally ill suited to real and adequate policy responses. Paradoxically, then, when viewed from a more historical perspective, the recurrent shifting and expanding nature of development policies since colonial times may actually reinforce or even create many of the poverty scenarios, conflicts, and difficulties those policies are meant to eliminate by inadvertently fostering conflicts of various sorts at the local level. Two essays focus on this problem and assess its environmental effects.

Östberg's essay on the history of failed soil conservation in a Tanzanian region adjacent to Serengeti forces us to ask, to paraphrase Oscar Wilde,

whether the only thing worse than environmental interventions is not having environmental interventions. By means of drastic measures, namely, the
removal of all livestock from the so-called Kondoa eroded area, the colonial
government successfully rehabilitated a landscape that was seriously
degraded. The first round of Swedish developers continued, in collaboration
with state agencies, a fairly coercive top-down project of soil conservation
by extending its scope. The result of decades of firm control on resource use
was a dramatic regeneration of vegetation, with the area in the 1970s and
1980s able to support more people than ever before. A shift in development
thinking during the 1980s toward softer models of "community participation" and "partnership" made the Swedish agency withdraw its support.
The quarantine system lapsed, and livestock filtered back into the area.
Come the 1990s with the promotion of neoliberal ideology and charters for
individual rights, privatization, and increased commodification, the land
was up for grabs not by local farmers but by private investors. The tenure
system earlier imposed by the state collapsed, turning it into a case of open
access and a "tragedy of the commons" in the making, where the wealthy
used their power to prohibit any interference with their accumulation from
this now very fecund reclaimed area.

The essay thus sensitizes us to the intersection of identity, tenure, and
power. By applying a historical scale, we get a sense of how environmental
interventions, and the ideology that underpins them, shift along with the
effects that they have on politics, locally and nationally, over the decades.
The geographic scale applied by the author is principally focused on the
Kondoa eroded area and the events that transpired there. This raises some
interesting questions. What would happen if we looked beyond the boundaries of the eroded area? Where did the livestock expelled from the area by
the colonialists end up? We see here the contours of a widespread problem
of the displacement not only of peoples but of degradation itself, a prime
force in the complex connectivity between poverty and environmental interventions.

At the heart of the Kondoa drama was the fact that landscape rehabilitation renders marginally productive land resources more valuable to a
broader set of users. The question of who gets access to rejuvenated lands is
often highly political, as Schroeder's essay on land reclamation in The
Gambia demonstrates eloquently. Environmental managers "reclaim" land
resources by rehabilitating them, but they simultaneously reanimate struggles over property rights in the process, allowing specific groups of resource
users to literally and figuratively "re-claim" the land. Relying on data gathered during fieldwork conducted between 1989 and 1995, the author analyzes the openings created by environmental policy reforms introduced over
the past two decades along the Gambia River Basin, and the tactics and
strategies rural Gambians have developed to manipulate these policies for
personal gain. Specifically, the essay explores how women market gardeners

pressed "secondary" usufructuary rights to great advantage to ease the economic impact of persistent drought conditions for the better part of a
decade, only to have male lineage heads and community leaders re-claim
the resources in question through donor-generated agroforestry and soil and
water management projects. This is thus a study of the responses different
community groups have made to a shifting international development
agenda centered on environmental goals. It is simultaneously an analysis of
those environmental policies and practices and their impact on gendered
patterns of resource access and control within a set of critical rural livelihood systems.

Modernity and the problem of identity management

While templates and ever-expanding agendas produce conflicts and confusion at their sites of implementation, something else is also produced, something at the heart of identity management under modernity: generic categories and identities.

Conventional "target group" thinking is part of this generic scenario. In
contemporary discourse on the environment, the most popular target
groups are women and the poor, often forcefully combined in the slogan
"women are the poorest of the poor". Whatever the case, targeting assumes
that the category ("natural" or "statistical") of people identified make up a
bounded social group, which one can reach directly with project inputs. Real
life, however, is far more complex and chaotic. Not only are persons from
different communities who have been assigned to the same standardized
target group in fact very different, but they are also tied by a nexus of relationships to individuals not included in the statistically defined target group
within their own communities (and beyond). As our essays show, it is the
claims, rights, and obligations crisscrossing the artificial boundaries of the
target group that in fact determine the capacity of the targeted to successfully compete for resources. Clearly, the better such processes are understood the better the chance is of finding key entry points for aid investment
that can reverse the poverty-producing process or increase the options so
that women can advance their own interests. This may help us identify what
it is possible to achieve by means of "development", and what can only be
achieved by other means, for instance, politics. Despite such commonplace
insights, the generic continues to grow and proliferate.

Generic entities are useful at the level of planning and policy formulation
precisely because they effectively strip those entities of any messy traces of
contextual specificity, yet through being brought into real encounters they
help to bring the equivalencies presumed by the generic entities into existence. This reproduction of the generic either changes the ways in which resources are distributed or results in a situation in which recipients begin to
define themselves in generic terms in order to gain access to resources.

Either way, we are confronted with a defining feature of modernity that Berman, building on Marx, describes as repeated transgressions of limits, cultural pluralism, and "the fissuring of little worlds" (1983:51). Implicit in this reconfiguration is a new form of complexity, which consists of a new relationship between the individual and his or her environment: The individual is brought to mirror a world of a larger scale, an international world, rather than the kin group and the small-scale local world. The "splitting" of small local worlds means that persons are brought to mirror the international in addition to themselves. In other words, the individual as well as the local are reorganized. Thus, along with the streamlining of private capital, financial institutions, and policies globally, there has been the somewhat paradoxical proliferation of novel forms of self-production (Broch-Due and Rudie 1993).

These new forms of self-production not only respond to external labeling and targeting but embody identity projects conjured by marginal groups in claims against the state, church, and capital. Indeed, these counterproductions of selves often deploy the same labels previously assigned by external agencies, but they reinscribe these within different registers of meaning. However, whether ascribed and/or asserted, these identities are all produced on the grounds of the generic; they all appeal to essentialistic notions. Whether or not they are freshly assertive ideologically and newly rephrased, even the most radical identity projects forged in opposition to external agencies pick up on the historical experience of unequal treatment and oppression on the basis of gender, race, language, ethnicity, and tribal identity. Interestingly, while historizing their identity claims, many marginal groups appeal to some sort of primordialism. In doing so, they inadvertently show the way out of the generic trap precisely by pointing to the ways in which the translocal forces actively participate in the social production of cultural differences. Despite being largely the product of social constructions through naming, generic groups, identities, and categories are not ephemeral and thus not easy to replace and reconfigure. Names never cease to stand in a conventionalized relationship of sorts with the things named.

Several essays explore the construction and contestation linked to generic identities. For example, as many of our essays show, the idiom of global social differentiation is replicated at all levels so that "traditional" and "modern" become a way in which not only nations and communities are classified but also persons within those communities. While clearly reflective of economic disparities and distinctions, these idioms express the widespread sense that external interventions, colonial and postcolonial alike, to some extent require a new kind of social person (Pigg 1997). These encounters produce a need to have someone locally who can understand and manoeuvre in the world beyond the community. One such generic identity is typically the "educated person" who can arbitrate and communicate with external agents on behalf of the community. The educated elite became particularly empowered as a mediating force at the end of colonialism with it attendant new national identity. In the postcolonial era, by contrast, it has often been the generically "indigenous" who have been favored and presented in certain situations as those linked not to "backward" and delegitimized nationstates but to the planet itself. It is they, not the educated elite, who are "global citizens".

While the recent surge of globalization of private capital and the environmentalist discourse has sidetracked nation states, it has enabled marginal peoples in the South American rain forest, for instance, to cash in on ecotourism by maintaining the autonomy of their own community. The success with which these Indian groups have managed to mobilize the benign strands of primitivism for the enrichment of their landscapes and cultures has not been replicated in African arenas. The factors contributing to the variable success of indigenous groups in Africa and the Americas in appealing to an international stewardship are probably complex.

In our context, it seems clear that the specific merging of the discourses of environment and poverty in Africa seems to have reduced the discursive space for identity projects appealing to cultural difference and tradition even for such cultural icons as the Maasai, who have loomed large in the European imagination since the early encounters. The poverty discourse applied to Africa has incorporated the modernizing charters of the colonial enterprise and with that a particularly negative version of primitivism. Its dominant ambition is certainly not the preservation of specific peoples and places but a charter for grand social reform. Most communities on the margins are keenly aware of this.

Thus, contemporary forms of conservation interventions in Africa that seek to reach beyond the "educated person" and get hold of the "indigenous person" are often frustrated in their benign efforts to conserve "traditional" livelihoods and "traditional identities"—since so few of the people targeted want to take on that role, let alone perform it in the ways expected. Conservation programs, however, are embedded in wider cultural politics. While a person who stands for "tradition" is in a weaker position in a widening power differential, a person who can dress up as "modern" has a ticket into other arenas of power and rewards, particularly those that cluster around the public rhetoric of citizenship and nationalism. "The poor" is precisely another such generic identity, straddling the ambiguous space between traditional and modern. In some situations, to inhabit the category of the poor holds up the prospect of provision and empowerment; in other situations, being labeled poor may lead to punishment and impoverishment.

The larger point here is that the local context in which projects unfold already contains ideas about the experiences of earlier encounters and interventions. This is why we need to understand interventions in the name of poverty and nature as parts of *recursive* processes continuously at work across the colonial and postcolonial landscapes. Thus, the representations carried by new models recombine with signs, values, and social identities already in circulation, creating new reconfigurations out of existing and imported elements. Any intervention in the name of nature or poverty is inserted into a contested terrain where it intermingles not only with one "local" model but several, often immediately producing new models as some people start to define themselves in and through the terms that traveled with the project.

A development project, to paraphrase Stacy Leigh Pigg, "starts out as a plan but turns into a context " (1997:270) in which people are brought together to interact around some activity, bringing with them diverse forms of knowledge and practices. These different perspectives act to insert the development idea into different localized meanings, which in turn weave themselves into fields of the social fabric not strictly defined as development. Interventions in the name of poverty and nature are social activities; they bring certain modes of engagement and certain social identities into being. Development engagements tend to produce and reproduce the generic identities and categories in global circulation. As the essay by Giles-Vernick so elegantly discusses, the generic identities evoked are sometimes formed around "the indigenous", which sounds so specific to a particular place but in fact is an integral cast of the generic register—and which, as our next example shows, was created only after new spatial boundaries were inserted by developers.

Giles-Vernick's essay examines the historical assumptions underpinning categories of migration and indigeneity in the Sangha river basin of the Central African Republic. It argues that these categories of movement and stasis have been part of long-term debates over movement and settlement, state building, and productive resources. Early-twentieth-century explorers, French colonial administrators, and more recently a World Wildlife Fund conservation project have sought to divide Sangha basin populations, particularly Mpiemu speakers, into bounded categories and to attribute particular histories of movement or stasis to them. These bounded categories helped to justify various environmental interventions. Giles-Vernick shows how Mpiemu speakers have actively appropriated these categories and historical visions from other Africans and from explorers, administrators, and contemporary conservationists. As a result, they now define themselves as "a dead people".

At the crux of this process is the concept of "origin" and its meanings in practical terms. This essay exemplifies a struggle between two different spatial models, a nomadic-pastoral and a settled-agrarian, and the different ways identities and relationship to place are mediated within these opposing modes (see also the Johnsen and Broch-Due essays). The idea of place-based identity has been invoked in the aftermath of the creation of the national park, turning different local groups and livelihoods against one another. The managers of the World Wildlife Fund and the various state agencies working with them have sided with those who easily conform to ideas and images in circulation about "indigenous peoples" whose "traditional way of life" per definition is vulnerable and "in need of protection". Clearly, these generic indigenous are particularly attractive clients because they seem more bounded and rooted than other groups that also claim a primordial link to the land—for example, the migrants who by definition are difficult to control and are thus perceived as destructive and disruptive.

The problem of generic identities and categories generates a whole string of policy questions. Who is entitled, for example, to the benefits of park revenues that are supposed to go to local people? Do migrants fit that category in any way? Whose residual claims are seen as legitimate and on what grounds? Whose traditional patterns of resource use are acknowledged and sanctioned? Is it the migrant peoples or is it some other group that's now designated as indigenous?

The key point Giles-Vernick makes is that, alongside the creation of the park, its managers, in this case the World Wildlife Fund and the state, become the arbiters of those decisions and definitions. Thus, those external to the community determine who is a migrant and who is not, who has legitimacy and who does not, whose traditions will be followed and whose will not. In this sense, there is a direct continuity in many respects with the colonial period, when these various ethnic distinctions were first produced, created, and in many ways cemented in place. The practical effect of this line of erroneous historical logic and reasoning is that the World Wildlife Fund is in fact promoting migration but in one direction—out of the park.

Nature and poverty thinly conceived

In the commodified form that dominates contemporary reading, nature reduced to "resources" and "inputs" has been propelling economic growth either through extracting the fruits of foreign lands or through reforming indigenous land use systems and livelihoods. Yet nature transformed to resource units also makes it possible to monitor cycles of growth and decline. In the matching commodified interpretation, poverty is reduced to a measurable state of calories and cash, the lack of which is by definition intrinsic to those located on the margins of the market.

At first glance resources, coins and calories seem to be perfectly neutral components—indeed, even highly productive ones—for the scientific project of counting and classifying, enabling extrapolation and comparison on a grand scale. Through the compilation of an avalanche of statistics based on scientific techniques of counting and remote sensing, the facts of widespread poverty and environmental problems have been made visible and accepted as objective parts of African realities. For the many Africans who find themselves in a situation of suffering and scarcity, this global awareness may open up new avenues for empowerment and entitlements, but paradoxically it also may put the brakes on political struggles and processes of redistribution.

Here we touch on a very significant tenet of this volume. For, although popular representations of the "facts of poverty and nature" draw on everybody's daily experiences and serve a didactic purpose, the knowledge they produce is inadequate and misleading. I am not suggesting here that standard economic and ecological models are necessarily "wrong", although some of the dominant ones investigated in our essays are devoid of much factual relation to reality and can be refuted on empirical grounds. Indeed, mainstream poverty research is still very thin and economists continue to dominate larger discussions on their own terms.

My larger point, is that the economistic discourse on poverty and environmental degradation that dominates the popular imagination has become a seriously reductive one. It appears to give a simple and revealing categorization of areas, populations, and their needs. However, by making such entities as income, nutrition, and soil erosion the only factors standing for complex social realities, it can just as easily come to conceal and misrepresent important social and political processes defining and creating poverty and environmental pressures.

This is because such "thin descriptions", in Glifford Geertz's' terminology, model social phenomena in minimal and measurable terms. They give a comforting appearance of objectivity and seem to travel with ease across cultural and historical boundaries. Through their endless reproduction in diverse discourses, thin descriptions often assume a taken for grantedness that escapes critical scrutiny. They seem context free and commonsensical and, for their audiences, apparently free from the contamination of authorship and agency. Such thin descriptions and images have a tendency to turn into very thick, politicized, and controversial ones the moment they move out of global speech-space and become localized and situated in social reality.

In African arenas, the target groups for poverty alleviation programs have not so much been poor consumers generally but whole communities, often rural, defined as "traditional" and thus antithetical to "modern" lifestyles and patterns of aspiration. This definition of *poverty*, loaded with an ideological baggage of primitivism, prompts external interventions to reform local forms of sociality to produce a more profitable match between labor, land, and capital. As their models have become globalized and successfully linked to the aid of bureaucracies worldwide, these reductionist definitions of *poverty* and *nature* have become important parts of the makeup of "development"—one of the dominant faces of modernity.

This is what this collection of essays aims to call into question by taking hold of some of the thick descriptions and practices, global and local, that poverty and nature have drawn around themselves. They show how these thick descriptions become discourses that, as Foucault reminded us, are "practices that systematically form the object of which they speak" (1979:49). Focusing these discourses are what Williams (1976) defined as "keywords"—the sites at which the significance of social experience are encountered, evaluated, and established in culturally specific ways. Poverty and nature are exactly such keywords at work in the world, engendering social effects, juxtaposing the politics of emotions and the politics of economics, reshaping social identities, and, in ways that we must come to understand, remolding the future of the African peoples and places with which we are all concerned.

References

Adams, J.S., and T.O. McShane. 1996. *The myth of wild Africa: Conservation without illusion.* Berkeley: University of California Press.

Anderson , D.M., and V. Broch-Due, eds. 1999. *The poor are not us: Poverty and pastoralism in Eastern Africa.* Oxford: James Currey.

Anderson, D.M., and R. Grove, eds. 1987. *Conservation in Africa: People, policies, and practices.* Cambridge: Cambridge University Press.

Appadurai, A. 1990. Disjuncture and difference in the global cultural economy. *Public Culture* 2:1–24.

——. 1998. Dead certainty: Ethnic violence in the era of globalisation. *Public Culture* 10(2):225–47.

Barkan, E., and R. Bush, eds. 1995. *Prehistories of the future: The primitivist project and the culture of modernism.* Stanford: Stanford University Press.

Berman, M. 1983. *All that is solid melts into air: The experience of modernity.* London: Verso.

Berry, S. 1993. *No condition is permanent: Social dynamics of agrarian change in sub-Saharan Africa.* Madison: University of Wisconsin Press.

Broch-Due, V. 1995a. Poverty and prosperity in Africa: Local and global perspectives. In *Poverty & Prosperity.* Occasional Papers, no. 1. Uppsala: Nordiska Afrikainstitutet.

——. 1995b. The "Poor" and the "Primitive": Discursive and social transformations. In *Poverty & Prosperity.* Occasional Papers, no. 5. Uppsala: Nordiska Afrikainstitutet.

——. 1999. Remembered cattle, forgotten people: The morality of exchange and the exclusion of the Turkana poor. In *The poor are not us: Poverty and pastoralism in Eastern Africa,* ed. D.M. Anderson, and V. Broch-Due. Oxford: James Currey.

Broch-Due, V., and I. Rudie. 1993. An introduction. In *Carved flesh/cast selves: Gendered symbols and social practices,* ed. V. Broch-Due, I. Rudie, and T. Bleie. Oxford and Providence: Berg Publishers.

Bucher, B. 1981. *Icon and conquest: A structural analysis of the illustrations of the Bry's Great Voyages.* Chicago: University of Chicago Press.

Burton, R.F. 1872. *Zanzibar: City, island and coast.* Vol. 2. London: Tinsley Brothers.

Cooper, F. 1997. Modernizing bureaucrats, backward Africans, and the development concept. In *International development and the social sciences: Essays on the history and politics of knowledge,* ed. F. Cooper and R.M. Packard. Berkeley: University of California Press.

Cooper, F., and A.L. Stoler. 1997. *Tensions of empire: Colonial cultures in a bourgeois world.* Berkeley: University of California Press.

Comaroff, J., and J.F. Comaroff. 1992. *Ethnography and the historical imagination.* Boulder: Westview.

Crosby, A.W. 1990. *Ecological imperialism: The biological expansion of Europe, 900–1900.* Cambridge: Cambridge University Press.

DAC. Annually, 1961–1995. *Development Assistance Committee Review.* Paris: Organisation for Economic Cooperation and Development.

Eliot , C. 1905. *The East Africa Protectorate*. London: Edward Arnold.

Escobar, E. 1995. *Encountering development: The making of the Third World*. Princeton: Princeton University Press.

Fabian, J. 1983. *Time and the other*. New York: Colombia University Press.

Ferguson, J. 1990. *The anti-politics machine: Development, depoliticisation, and bureaucratic power in the Third World*. Cambridge: Cambridge University Press.

———. 1999. *Expectations of modernity: Myths and meanings of urban life on the Zambian copperbelt*. Berkeley: University of California Press.

Foucault, M. 1979. *Discipline and punish: The birth of the prison*. Harmondsworth: Penguin.

———. 1980. *Power/knowledge: Selected interviews and other writings, 1972–1977*. Brighton: Harvester.

Geertz, C. 1974. *The interpretation of cultures*. New York: Basic Books.

Geschiere, P., and B. Meyer. 1998. Globalization and identity: Dialectics of flow and closure. *Development and Change* 29:601–15.

Gourevitch, P. 1998. *We wish to inform you that tomorrow we will be killed with our families: Stories from Rwanda*. New York: Picador.

Grzimek, B. 1956. *No room for wild animals*. London: Thames and Hudson.

———. 1960. *Serengeti shall not die*. New York: Dutton.

Hale, J. 1994. *The civilization of Europe in the Renaissance*. London: Fontana.

Hardin, G. 1968. The tragedy of the commons. *Science* 162:1243–48.

Hirshman, A.O. 1968. *Development projects observed*. Washington, DC: Brookings Institution.

Hobart, M., ed. 1993. *An anthropological critique of development: The growth of ignorance*. London and New York: Routledge.

Hoben, A. 1995. Paradigms and politics: The cultural construction of environmental policy in Ethiopia. *World Development* 23(6):1007–22.

Hobsbawn, E. 1994. *The age of extremes: The short twentieth century, 1914–1991*. London: Abacus.

Hodgson, D. 1999. Images and interventions: The problems of pastoralist development. In *The poor are not us: Poverty and pastoralism in Eastern Africa*, ed. D.M. Anderson, and V. Broch-Due. Oxford: James Currey.

Kuper, A. 1988. *The invention of primitive society: Transformations of an illusion*. London and New York: Routledge.

Leach, M., and R. Mearns, eds. 1996. *The lie of the land: Challenging received wisdom on the African environment*. London, Oxford, and Portsmouth: IAI, James Currey, Heineman.

Lowenthal, D. 1985. *The past is a foreign country*. Cambridge: Cambridge University Press.

Malthus, T. [1798] 1976. *An essay on the principle of population: Text, sources and background criticism*, ed. P. Appleman. New York: Norton.

Markham, B. [1942] 1983. *West with the night*. San Francisco: North Point.

Matthiessen, P. 1991. *African silences*. London: Harvill.

Mayhew, H. 1861. *London labour and the London poor: A cyclopedia of the condition and earnings of those that will work, those that cannot work, and those that will not work*. London: Griffin, Bohn.

Mitchell, W.J.T., ed. 1994. *Landscape and power*. Chicago and London: University of Chicago Press.

Nordstrom, C. 1997. *A different kind of war story* . Philadelphia: University of Pennsylvania Press.

Petty, Sir W. [1662] 1963. *The economic writings of Sir William Petty*. New York: Kelly.

Pieterse, J.N. 1992. *White on black: Images of Africa and blacks in Western popular culture*. New Haven and London: Yale University Press.

Pigg, S.L. 1997. "Found in most traditional societies": Traditional medical practitioners between culture and development. In *International development and the social sciences: Essays on the history and politics of knowledge*, ed. F. Cooper and R.M. Packard. Berkeley: University of California Press.

Ricardo, D. 1951. *The work and correspondence of David Ricardo*. Cambridge: Cambridge University Press.

Roe, E. 1991. "Development narratives", or making the best of blueprint development. *World Development* 19(4):287–300.

Roosevelt, T. [1910] 1988. *African game trails*. New York: St. Martin's.

Sachs, Wolfgang, ed. 1992. *The development dictionary: A guide to knowledge as power*. London: Zed.

Scott, J.C. 1998. *Seeing like a state. How certain schemes to improve the human condition have failed*. New Haven and London: Yale University Press.

Shaw, C.M. 1995. *Colonial inscriptions: Race, sex, and class in Kenya*. Minneapolis: University of Minnesota Press.

Shelton, M.D. 1995. Primitive self: Colonial impulses in Michel Leiris's *L'Afrique fantome*. In *Prehistories of the future: The primitivist project and the culture of modernism*, ed. E. Barkan and R. Bush. Stanford: Stanford University Press.

Stanley, H.M. [1899] 1988. *Through the dark continent*. Mineola, NY: Dover.

Steiner, C.B. 1995. Travel engravings and the construction of the primitive. In *Prehistories of the future: The primitivist project and the culture of modernism*, ed. E. Barkan and R. Bush. Stanford: Stanford University Press.

Talle, A. 1988. *Women at a loss: Changes in Maasai pastoralism and their effects on gender relations*. Studies in Social Anthropology, no. 19. Stockholm: University of Stockholm.

——. 1999. Pastoralist at the border: Maasai poverty and the development discourse. In *The poor are not us: Poverty and pastoralism in Eastern Africa*, ed. D.M. Anderson and V. Broch-Due. Oxford: James Currey.

Torgovnick, M. 1997. *Primitive passion: Men, women, and the quest for ecstasy*. Chicago and London: University of Chicago Press.

Waller, R. 1999. Pastoral poverty in historical perspective. In *The poor are not us: Poverty and pastoralism in Eastern Africa*, ed. D.M. Anderson and V. Broch-Due. Oxford: James Currey.

WDR. Annually, 1978–96. *World Development Report*. Washington, DC: World Bank.

Williams, R. 1976. *Keywords: A vocabulary of culture and society*. London: Croom Helm.

World Bank. 1975. Assault on world poverty: Problems of rural development, education, and health. World Bank.

World Commission on Environment and Development. 1987. *Our common future*. New York: Oxford University Press.

Worster, D. 1987. *Nature's economy*. Cambridge: Cambridge University Press.

A Proper Cultivation of Peoples: The Colonial Reconfiguration of Pastoral Tribes and Places in Kenya

Vigdis Broch-Due

In July 1958, colonial officers expelled thousands of protesting Turkana from Isiolo town in central Kenya and forced them to return to Turkana District in the northwest corner of the country. The reasonable tones of the administrative rationale for this relocation masked a long history of brutal misunderstandings. The roots, labyrinthine complexities, and tragic effects of this history are the subject of this essay.

The colonial authorities portrayed it as "repatriation", a return of exiles to their "roots" and "habitat "—a wholly positive restoration of a "natural" order of things—"tribal" peoples returned to their "tribal" territories. It was a disaster for those targeted. Their hard-earned goats died en masse, and when they finally arrived on foot weeks later they found themselves in a territory to which they were meant to belong but in which they were strangers. They were soon impoverished and marginalized. And this was not the first round of dramatic dislocation and pauperization experienced by this particular people. For the bizarre irony behind the "repatriation" was that the community of Isiolo Turkana had been created only decades before by pastoral refugees desperately impoverished in the wake of the colonial military campaigns of the 1920s and the failed agricultural schemes of the 1930s, both launched to "civilize "and control Turkana pastoralists and promote their sedentism.

So, within the span of two generations, the group's social life had been completely convulsed—twice.

After an initial period of hardship and hunger caused by the first round of upheaval, most had managed to build relatively stable lives and incomes from the labor markets in Isiolo, reinvesting any surplus in goats. No sooner had they done this than the colonial government engineered another dislocation, that of 1958, which once again left them poverty stricken and marginalized. The first dislocation was imposed on them in the name of civilization, pushing them out of Turkanaland and into Isiolo town. The

To Todd Sanders and the staff at the National Archives in Nairobi, I am very grateful for the help and support in excavating, coping and binding all the dusty files concerning the dislocation of the Isiolo Turkana. To Dismas Karenga who has been my co-worker, interpreter, and friend on many field trips in Turkanaland since the early 1980s, I am particularly indebted. To Graham Townsley, I am very appreciative for his editing work.

second round, in the name of repatriation, forced them in the opposite direction, out of town and back to the plains into a specialized pastoral system to which they no longer had access, their economic and social pathways into that world having been severed generations earlier.

In this essay, I want to explore this extraordinary case in depth because it epitomizes in a special way the hallucinatory logic of ideas, plans, and effects—many of them unintended—that characterized much of the colonial administration and continues to characterize many aid programs today. There is a complex and important web of causality here, which has repeatedly powered the collision of apparently bland and well-meaning administrative concepts with the intractable otherness of Turkana existence, generating tragic outcomes such as the Isiolo debacle: Nomads forced into sedentarism then back into nomadism and at every turn left more disorientated, impoverished, and traumatized.

Of equal importance, but less well known, is the way that this chaotic administrative momentum has been instrumental in reconfiguring ethnic identities and topographies in northern Kenya according to its own, very European, discourse of "tribal identities". An analysis of this multifaceted project of administrative control and redefinition of social identities seems long overdue, and I hope to contribute to that in the following.

The location of Isiolo town in this contradictory geography of "civilisation" and "repatriation" is very significant for the unfolding of events related to the rise and fall of the Isiolo Turkana community. This is because Isiolo is situated at the crossroads of several significant divisions: ecological, political, and conceptual. The township stands at the southern margin of the vast low-lying and dry savanna of the Rift Valley, sprawling northward from just beneath the fertile highlands surrounding Mount Kenya. At the time of the eviction in 1958, Isiolo was the center of governance of the unruly Northern Province, stretching into the contested ground where the Ethiopian and British Empires intersected. Isiolo was a frontier in the larger process of nation building in which boundary making was of paramount concern in a very tangible, geographical sense. Yet subtler boundaries were even more at stake, being constructed along several other "us-them" axes, physically, socially, and symbolically.

During colonial times, Isiolo was the site of a large-scale reconfiguration of rights in land. Some pastoral groups were removed altogether—the Laikipiak Maasai and the Isiolo Turkana. Others, like the Samburu, were pushed to the outskirts, while the Borana escaped the reshuffling. The space vacated was settled by the more urbanized "Alien Somali". Coming from Aden and Kismayu, the Somali community, having served as soldiers in the British army during the First World War, had the choice of being compensated with "blood money" or foreign land (Hjort 1979). Isiolo also served as a boundary that spatially separated the white settler community, with its wheatfields and ranches, from indigenous, nomadic, subsistence herders, and, even further back, formed the climatic marker between contrasting agrarian and pastoral livelihoods, with their very distinct lifestyles.

Thus, Isiolo, has always brought into spatial proximity different kinds of peoples, cultural styles, and livelihoods and this turned it into a particularly productive site for the crafting of a whole series of contrasting identities central in constructing the colony of Kenya. No social biography embodies this whole series of distinctions—primitive and civilized, nomad and settler, colonizer and colonized—better than that of the Isiolo Turkana. The trajectories of this group through the colonial landscape provide an excellent historical case for exploring the role played by a particular political regime, the colonial state of Kenya, in the *production* of ethnic boundaries between social groups (Barth 1969, 1996) and the reconfiguration of not only groups but also bodies in the process.

Throughout this chapter, I shall focus first on the clash between colonial and indigenous geographies of groups and territories. Second, and perhaps more interestingly, I shall explore the way in which the colonial model was formed and then how it re-formed reality in its own image. In the course of this chapter, I want to explore how the colonial tribal model was produced and the particular logic of each of its stages—both its internal structure and its interaction with local Turkana models. I want to show how the implementation of each model comes with a particular inscription in physical space—a sort of figure-ground construction—which is added to, and articulates with, earlier inscriptions.

Inspired by Appadurai's disaggregation of space into its different dimensions (1991), I toy with his term *scape* because it sensitizes us to see the plurality of perspectives not only in terms of physical space but in terms of what I have elsewhere called a bodyscape (Broch-Due 1990, 1993). The concept of "scape" helps to situate the significance of the social positioning of spectators within it. In the context of this essay, the particular points I want to tease out from the complexity of the colonial "scape making" process are these: each stage and layering came with its particular practice and style, which articulated in determined ways with local models and practices—in each case modifying the topography of wealth, producing different forms of impoverishment. It also brought into crisp focus the problem of definitions of identities and rights; colonial scape making evoked the whole question of who qualifies as indigenous, immigrant, emigrant, or simply a temporary refugee to be "repatriated". Who came out as winners and losers in this "game of origin" had much to do with who was constructed as good and bad natives in the colonial moral economy. And, finally, the dominant site of deliberation around these social and moral distinctions was the poverty discourse itself. Imported from Europe, it was equipped with a flexible list of definitions and interventions handy for the larger colonial project of reshaping the social lives of the colonized.

In all its complexities, its eccentric characters, motives, confusions, and accidents, the repeated dislocation of the Isiolo Turkana is a perfect case for situating the larger canvas of ideas, discourses, and practices that went into the production of place, people, and poverty in the Northern Province of the colony of Kenya.

Proper tribes

The colonial regime in Kenya set itself the goal of dividing up the drylands into "tribal" territories and concentrating all efforts on enforcing tribal boundaries. In the Northern Province, the production of confined entities commenced just before the turn of the century when efforts to pacify the local populations by the empire intensified. The territorial boundaries were officially inscribed and acknowledged by the Carter Land Commission in the 1930s. In a comparative case from the neighboring Baringo basin, Little succinctly sums up the aim and effects of the work of the commission as follows:

> To investigate "native" land rights and to resolve land disputes. Leaders of numerous ethnic groups were asked to provide evidence. For many groups it was an opportunity to plead for additional land and to mimic notions of territory and identity that were consistent with a European worldview. For pastoral peoples the colonial government was persistent in rigidly associating ethnicity with particular pieces of space, even for nomadic populations. Thousands of pages of documentation and evidence were produced, a perverse testimony to how Africans had openly to negotiate their cultures and identities with British administrators. (1998:449)

The eviction of the Isiolo Turkana was a direct result of the scenario described by Little (1998). They were grazing their own animals alongside those they were hired to graze by the Somali owners in the so-called Leasehold area set aside to accommodate the so-called Alien Somalis. Under the heading "Reasons for Removal", a District Commissioner (D.C.) memorandum from 1948 explained:

> The Carter Land Commission set aside the Native Leasehold area for the Somalis and it is accepted that they have prior claim there. The Turkana will never intermarry with and become absorbed by their neighbours ... but will always remain a separate and prolific community. If they and their stock continue to multiply at the present rate they will in a very few years have crowded out the Somalis from the Leasehold area and will begin to press on their Meru, Dorobo and Samburu neighbours. There is already friction between the Somalis and the Turkana over grazing grounds and water and the danger of serious incidents and bloodshed will increase as the Turkana grow in numbers, wealth and self-confidence.[1]

The fact of the deportation demonstrates in grim terms that the Isiolo Turkana did not manage to successfully negotiate their links to the Isiolo

---

[1] 13–5–1948, Adm 15/7, DC/ISL/1/77.

area itself. In a last-minute effort, however, quite a number of them managed to come up with a convincing enough case en route, even according to
the strict colonialist definition of pedigrees and genes, to be able to stay put
in Baragoi in Samburu District. In order to appease the unhappy deportees
and quell the potential uproar, the officers in charge of the "repatriation"
agreed to arrange a hearing of the claims.

> On August 16th District Officer Chambers and the Baragoi Turkana elders com-
> menced hearing claims from Isiolo Turkana to reside in Baragoi, there [they] were
> later joined by D.C. Samburu. A total of 250 Turkana heads of families applied for
> residence to Baragoi. ... A claim was generally acceptable if the claimant had not
> left the Baragoi area prior to 1948. If during the claimant's absence from Baragoi,
> his mother or father, wife or children continued to reside in Baragoi and were still
> doing so, then the claimant stood a fair chance of his claim being granted. The ex-
> istence of sisters and brothers in Baragoi was not considered to be a strong
> enough case.[2]

While the sister-brother relationship is at the core of Turkana kinship, the
bridewealth for sisters allowing brothers to marry, as many as 491 of the deportees passed the even narrower European kinship test and, together with
587 cattle and 4,427 sheep and goats, left the procession of deportees on their
way to Turkana District. Despite the fact that this group constituted only
half of the claimants, even this limited number of successful petitioners
came as a great surprise to those in charge, for in the "summing up" section
of the report of repatriation we learn that:

> The early reactions of the Turkana made it plain that many had claims and strong
> claims at that to stay In the El Barta region and it looked at one stage as though
> they might stage a "sit down" strike on reaching that area. ... But above all the
> willingness of the Samburu Admin. to listen to Baragoi claims in the knowledge
> of the unfortunate consequences which might have occurred had not this policy
> been adopted was the biggest single contributing factor to the success of the
> move. ... Any action which falls short of this could only have produced a large
> body of extremely disgruntled tribesmen having to be pushed every inch of the
> way from Baragoi to Turkana.[3]

This colonial lapse of memory about the places from which the immigration
to Isiolo had originated is instructive, for it illustrates the ongoing changes
in the political geography of colonial place making during the span of two
decades. This not only questioned the very foundation of their fixed tribal
models but it posed further problems for those caught up in them, who were
subject to fresh rounds of negotiation about origin and identity in the wake
of the colonial reformulation. Most telling, the older records clearly show
that few of the "repatriated" came from the part of Turkanaland that the

---

[2] G.W.L. Pryer and M. Wasilewski, "Turkana Repatriation, Isiolo-Kangetet, 6 July–31 August
1958".
[3] Ibid.

colonialist called Turkana District. The first mention of measures to control the influx of Turkana people into Isiolo is a letter from 1939, written by the local D.C. and addressed to the D.C. in Samburu District. Under the heading "Ref. Turkana from Samburu", we learn that:

> There are as you know a large number of Turkana from Wamba and other parts of Samburu who are employed an herd boys by the Alien Somali of Isiolo District. ... They are the poorest class ... and they, and their relations who tend to join them and live clandestinely in the District, have long proved an embarrassment on account of their incurable poaching proclivities. ... More recently there have been numerous complaints by Alien Somali stock owners of organised thefts by bands of Turkana; such complaints are certainly exaggerated but ... I am therefore considering the feasibility of refusing O.D. permits to Turkana from the end of this year; ... [with] the return of these men and their unlawfully resident dependants to Samburu.[4]

By 1943, the policy of pushing the Turkana back to Samburu had clearly failed:

> Periodical round-ups of vagrant Turkana and their return to El Bata, nearly aggravate the problem and cause a great deal of hardship. ... The time has now come when something of a constructive nature should be done to help these people. ... The Turkana, although mainly pastoralists, are an adaptable people and appear readily to take to agriculture. A scheme for agricultural settlement on the Tana river might be worthy of consideration.[5]

Thus, fifteen years before the "repatriation", nobody in the colonial offices seemed to regard Turkana District as Turkanas' proper homeland. Indeed, the suggestion put forward was to relocate them in the opposite direction— to Tana River, close to the coast, where we know for sure that no Turkana had ever set foot, not surprisingly given the presence of flies that would have killed their cattle! Most interestingly, when the idea of moving them all the way back to Turkana District was conceived, the considerations were clearly in terms of control rather than repatriation.

> The trouble with the Turkana is that they are so destitute that two months in jail is no deterrent, and if escorted over Uaso whence they came, they are back again almost as soon as their escorts, so trying to keep down their numbers is like trying to stem the tide. ... The scheme for returning the Turkana to Turkana district would help temporarily if put into operation, but I think we are always going to be faced with Turkana immigration from El Barta so long as the government, traders, and Somalis are employing members of the tribe.[6]

When the plan to solve the problems faced by the colonial administration in Isiolo was put to its counterpart in Turkana District in 1944, it was flatly re-

---

[4] 2-10-1939, Adm.1 5/16/2285, PC/EST2/11/19.

[5] Letter of 21–8–1943, George Adamson to DC/ISL/1/75.

[6] 16–6–44, ADM/15/52/96, DC/ISL/1/75.

fused by the acting D.C. there, who backed up his arguments for not receiving the Isiolo Turkana by questioning their origin in his constituency. Asking first for more precise information to evidence their previous areas of residence in Turkana proper, which he seems to have doubted, he continued:

> It would be unfair to saddle the Turkana with another 2000 odd mouths to feed and very considerable help in the way of food will have to be given by Government. To my mind the main point is—do they themselves want to come back here?[7]

The D.C. in Isiolo was clearly baffled by this democratic attitude, to which he replied:

> Even if they do not want to go back, I do not see how they can remain here forever. ... I think it is pointless to ask people like this whether they prefer Isiolo or El Barta or Turkana. Their primary aim is to endeavour to avoid starvation.[8]

The administration in Turkana was not inclined to incur further cost by having to feed destitutes from other districts and continued to refuse cooperation. In desperation, the D.C. of Isiolo wrote to complain to the provincial commissioner and to the central office in Nairobi, but his counterpart in Lodwar did not change his mind. In resignation, he wrote to provincial headquarters:

> In view of the uncompromising attitude taken up by the District Commissioner, Turkana ... there seems little use in pursuing the matter further at least until there has been a change of District Commissioner in Turkana. ... Should the Tana River irrigation scheme ever be started, the Turkana might be moved there as a tribal unit ... but there seems to be no likelihood of any work being undertaken on the scheme for many years.[9]

However, the plan came alive again later the same year when there was an unexpected turnover of staff. The new D.C. of Turkana District had earlier served in Isiolo and had taken part in the formulation of the plan to move the Isiolo Turkana. In a letter addressed to the Honourable Chief Secretary, The Secretariat, Nairobi, dated 15 September 1944, he formally agreed to receive those expelled from Isiolo town in Turkana proper.

Most significant from the day the plan was formalized and agreed upon by the districts involved, and until its final execution in 1958 after several abortive attempts, the origins of the victims of dislocation were gradually enveloped in a sort of administrative haze and then emerged on the other side in a quite different form. In 1939, everyone was sure that the Isiolo Turkana were from Samburu. By 1958, this had changed to the firm conviction that they had come from Turkana District. This move from one linguis-

[7] 22–7–1944 Ref. No. L&O.17/6/31. DC7ISL/1/75.

[8] 11–8–44, ADM/4/371237—DC/ISL/1/75.

[9] 15–12–1944, ADM. 15/52/1890. DC7ISL/1/75.

tic location to another in the imaginary political geography of "tribes" and "territories", was effected by a long process of ever-vaguer ethnic mapping. For example, in 1949 the question was whether "to move the Isiolo Turkana back to their habitat",[10] while in 1957, close to the final date of the dislocation, this had been redescribed in the minds of the colonial officers as moving the Isiolo Turkana back to "the home of their fathers".[11] This institutional forgetting had of course not wiped out the memories of the paths on which they had previously journeyed in the minds of the Isiolo Turkana. Having been subjected to increasing attention, policing, and surveillance by the colonial authorities in Isiolo town, they had been forced to learn to negotiate their identities in a European idiom, as they showed by their successful claims in Samburu country.

Path versus place: mapping pastoralists

The confusions about tribes, origins, and territories so evident in colonial minds had a long pedigree in East Africa. Faced with their country's "mosaic" of Europeans, Africans, and Asians, Kenyans, like those in other settler colonies, have never been able to entertain simple ideas about the nation (Ashcroft, Griffiths, and Tiffin 1995:151). The European "root" metaphor so central in nineteenth-century nation making (Schama 1995) and the ways in which it weaves together a particular people's habitats and histories, all with their particular poetics, could hardly be put to work in colonial Kenya. Inasmuch as any nation is an "imagined community" in Anderson's sense (1983), there was of necessity a more foregrounded self-consciousness about the *constructedness* of nationhood in colonial discourses than elsewhere. In the ambivalent process of imagining the Kenyan colony, *place*, but also *placelessness*, played an important part.

Place making, but also putting everybody and everything "in place", became the defining features of the colonial regime in both words and deeds. European ideas of nationalism were creatively drawn upon in the colonial place-making process. The juxtaposition between identity, landscape, and jurisdiction was reproduced not in the singular and large-scale form of a nation but in a multitude of small-scale versions—the tribal territories—each imagined as a nation in miniature and each solidified in reality through a string of bureaucratic decrees, acts, and interventions. Indeed, the vocabulary deployed in the colonial archival documents abounds with slippage between the terms *tribe* and *nation*. These nation-tribes in the plural were ranked hierarchically and subsumed under the governance of a singular white mother nation, the Kenyan colonial state.

From this perspective, those most "out of place" were the nomads of the Northern Province. The colonial discourse of tribes, territories, and fixity

---

[10] ADM.15/52/95—DC/ISL/1/75.

[11] ADM.15/52/117167—DC/ISO/3/1/7.

was the result of a sedentary vision bound to a very European vision of nature. In this vision, nature is essentially predictable and controllable, generating notions of stable boundaries and identities. Relatedness in the European sense is constructed, first, around the passing on of a given set of genes from parents to children, which generates kinship, and, second, around the investing of personal rights to negotiate relatedness to a place. The intertwined notions of place of birth and rights of birth, or mother country and mother tongue, bring out very clearly this widely shared Western juxtaposition, at least on the ideological plane, between an essentialistic definition of *personhood* and *place*.

This European vision was totally opposed to the Turkana vision and those of other pastoralist peoples of the Northern Province. These pastoralists' production of nature and identities is a much more fluid one, presuming a constant flux of elements and identities in the production of both individuals and groups (see Broch-Due 1990, 1993). Pastoralist identities are "identities on the move" to borrow a phrase from Schlee (1989).

Indeed, unlike its European version, the very idea of Turkana personhood and relatedness is founded on anti-essentialistic ground, almost perfectly resembling the ecological idea of elements being constantly reconfigured as they pass though a food chain. Humans, like cattle, in Turkana thought are temporary stages in a continuous process of the reconfiguration of food elements—grass, water, milk, meat, and blood. These food "paths", as Turkana call them, grow bodies and relationships. Path and process, not bounded place, are everything. In the idiom of "path" (*erot*) the cross-cutting pathways linking camp and countryside are physical signposts conjuring up a social landscape—a web of relations and potential transactions that might come along these paths. They are the routes to building real and symbolic capital in terms of human relationships, herds, and procreation (see Broch-Due 1990, 1993, 1999).

In this nomadic vision, path, motion, and exchange are the keys, not the fixity and boundedness of an agrarian world. The Turkana relation to land is a fluid one, constantly re-created at every step along the way. At every new campsite on the migration route, for example, each Turkana person will take a handful of soil, eat part of it, and smear some on the forehead—an act intended to make "this place agree with us". The constant re-creation of human relatedness to the places through which their paths wind is also very literal and visceral.

Perhaps surprisingly, this pastoral model of place making also comes with it own cartography. It is instructive to compare Turkana mapmaking with that of the colonialists. As I have explained in more detail elsewhere, the Turkana body, mapped on the inside and the outside alike, is a microcosmic landscape—a *bodyscape*—of persons, artifacts, cattle, plants; the seasonal landscape is a socially differentiated, macrophysiological body turned inside out. The dominant physiological idea that the world is composed of a

series of microcosmic and macrocosmic bodies, and that these bodies inter-
penetrate each other in some substantive way, is enacted and made real for
the Turkana by a host of practices and everyday acts. Indeed, the main
method of discovering what the divinity Akuij is up to at any given moment
works precisely on this idea of places, groups, and bodies contained like
concentric circles within each other. Divination finds the perfect image of
these multiple worlds in the goat intestines curled upon themselves within
its belly that the Turkana read to gain information about events in camp,
clan, and cosmos. For goats bear in their bodies a complete configuration of
the totality that comprises everything Turkana. All paths, places, and
"bodies", both physical, social, and cosmic, are nested in a map of landscape
and living graphically represented in the image of an animal's belly as a
microcosm of the world in which they move (Broch-Due 1990).

In contrast to the colonial map, however, the Turkana belly cartography
is not seen as a representation or metaphor for the world but as a microcosmic body substantively linked to it, which really registers its fluctuating
states in the transient texture of blood bubbles, grains, and the quality and
texture of the digested grass in the animal's stomach and guts. These are
visual signs of events and transformations in the external world. The data
elicited help in the evaluation of pasture and the planning of herd movement. They are also employed as a basis on which to formulate actions in the
human world: the timing of weddings and birth and burial rituals and the
state of affairs in relationships. After the animal belly map has been inspected and made known to everybody; it is cooked and eaten to ensure that
the knowledge is stored in everybody. Turkana thus comprehend the world
by literally taking it in.

What I want to emphasize in this context is that both models, the colonial
and the local, operate with the assumption of identity mediated through
links to a landscape. Both thus share an immense investment of culture in
the production of nature and place but in radically different ways. Aligned
with its sedentary bias, colonial mapping emphasized cartographic enclosure. The mapping of the territories in the Northern Province was part and
parcel of the establishment of the high degree of coercion and containment
that are historically implicated in the colonial enterprise.

Turkana mapping in soft animal matter, in contrast, perfectly represents
their vision of flexible cross-ethnic patterns. This is a shifting map whose
contours change as pastoral journeys proceed along seasonal migration
orbits. This paradoxical Turkana notion of a map as shifting ground comes
close to ideas discussed by Deleuze and Guattari (1987) when they draw
attention to the fact that the map is potentially "rhizomatic". In other words,
maps are open and connectable in all their dimensions; detachable, reversible, and adaptable. The idea I want to communicate is that maps are
essentially plastic devices, which can be deployed for inscription of power in
any political economy but also as a means of countermapping and resisting

Turkana goat-belly map

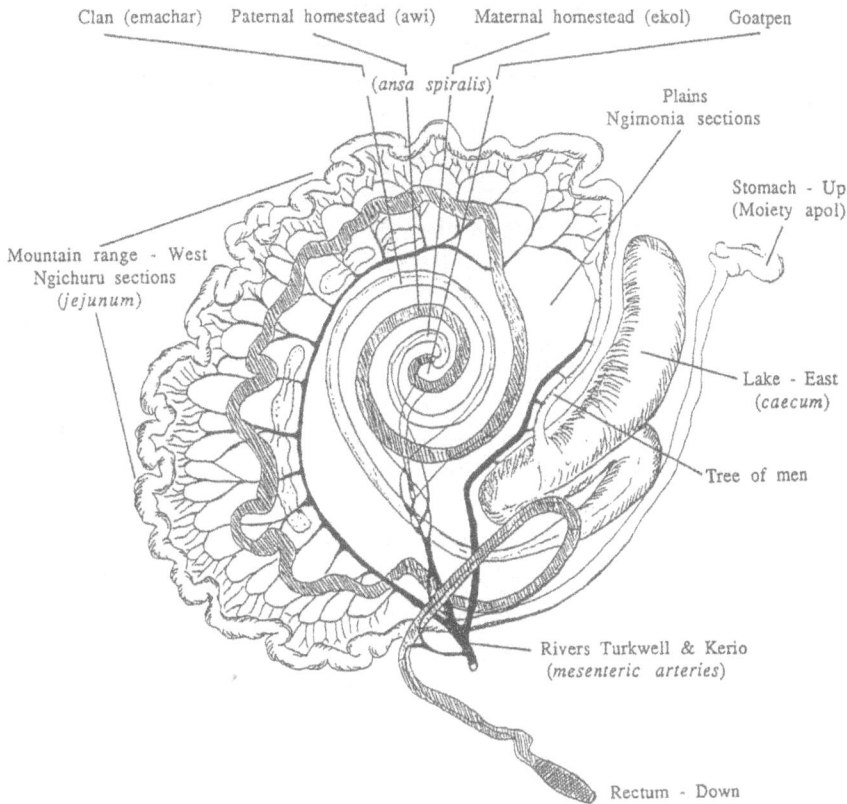

the powerful. Perhaps the moving pastoralist territories and the mosaic, pre-colonial pattern of crisscrossing identities can be grasped in terms of such "open" mapmaking as shifting grounds that are themselves subject to transformational patterns of de- and reterritorialization as the power and wealth of particular groups rise at the expense of others.

For the Turkana, the hills, rivers, and ridges showing where one territory ends and another begins are signposts that are not forever fixed within one territory. Rather, landforms serve as fluid markers of spaces that are constantly renegotiated and may, for instance, be seized by a group that for the time being can muster the strength necessary to either win public approval or seize it by force. Most significantly, the physical territory is overlaid with an "alliance territory" (Broch-Due 1990), a social topography that includes the areas of associated groups. Here host groups will provide visitor groups access to pastures, provided the visitors have the courtesy to ask for permission in advance or to announce the reasons for their presence.

In precolonial times, the full exploitation of the flexibility inscribed in the idea of alliance territory gave Turkana a competitive advantage over many of their pastoral neighbors, like the Samburu, who were absorbed in great numbers into this evolving regime (Lamphere 1992). Turkana herds participated in an expansionist project whose aim was precisely to extend the Turkana social reach over wider areas rather than to anchor and fix its scope to one stronghold narrowly defined in physical space. Indeed, the content of the bovine belly map, which we have seen as an icon configuration of Turkana country and the social and natural relations within it, is appropriated in the animal through its journeys far and wide.

To recapitulate the comparison: contrary to the colonial sedentary model of the essential and exclusive linkage between person and place, the nomadic model shared by the pastoralists in the Northern Province was one between person and path, a fluid, expandable and flexible construction. Scaled up, the regional picture at the time of the first encounter with the colonialists reflected these person/path mediations, resulting in a mosaic of clans, ethnicities, and identities equipped with ill-defined territorial boundaries. In precolonial Turkanaland, there were constant regroupings of larger polities, like Turkana and Maasai, along with the ups and downs of the local livestock economies, climate, and shifting power relations. Smaller enclaves of cultivators, foragers, and fisherfolk served as "halfway houses", helping the needy to adjust their livelihood and identity. As elsewhere in East Africa, accounts bear evidence that droughts, famine, and epidemics clearly contributed to the reconstruction of societies, binding different economies together through the movement of people and produce between them (Waller 1985; Broch-Due 1999). For example, during the rinderpest outbreak in the 1890s Turkana cattle were the only herds spared, contributing to the southward expansion of some clans into the Isiolo area. Many Samburu and other local people were absorbed into Turkana clans through adoption, patronage, marriage, and herding arrangements (Lamphere 1992). The British, whose system of indirect rule required locating tribes and fixing them to a territory, clearly faced trouble in these corners of the colony. In the absence of the kinds of "tribes" they had in mind, however, the British went about creating them (Gutkind 1970). During the early twentieth century, the colonial state demarcated the region and mapped it into tribal reserves and districts. These efforts to tape ethnicity to territory and to keep these separate entities in place were the dominant measures designed to ease the control and governing of native people (Sobania 1988; Anderson 1988).

The sad destiny of the Isiolo Turkana was clearly produced in the clash between these conflicting perspectives, the colonial and the local, of both the natural and the social worlds. They were, of course, not the only victims of this clash. The relocation of the Laikipiak Maasai in 1904 and 1911 from Isiolo surroundings to the southern region of Kenya was on an even grander scale. Since the Maasai were landed in a foreign territory to which they had no prior ties of any conceivable kind, the eviction of these nomads from the environs of the Rift Valley was not so easily repackaged as repatriation.

To evoke the tribal homeland discourse in this instance would clearly have been politically counterproductive, for the clearly stated aim was to make space for the creation of a white settler colony on the fertile soils of the Naivasha-Nakuru basin, which also happened to be favorably located for accessing the Uganda-Kenya railroad and commercial markets. The subsequent settling by the British of their allies, the Alien Somalis, on former Maasai ground just added more layers of foreign politicking to the place. However, the discourse of territoriality entered into the justification of the Maasai move as well, albeit in a slightly different fashion. Remarkably, the Maasai were evicted on the grounds that they had stolen the land from other native "tribes" in the first place (Knowles and Collett 1989). This image of nomadic peoples like the Maasai and Turkana as aggressive marauders always expanding their "nations" at the expense of their peaceful agrarian neighbors dominated the mind-sets, rhetoric, policies, and actions of nineteenth-century colonialists.

Control and predation

As the colonial regime figuratively ate its way into the pastoral communities in northern Kenya, it also did so quite literally, expropriating huge herds for its own provisions. No less than 250,000 Turkana livestock were seized by the King African Rifles in the 1920s, causing widespread starvation and suffering. From its first days, the colonial regime was a voracious one, eating its way into the backbone of the Turkana political economy. Since every peripheral jurisdiction on the drylands in colonial Kenya had to pay its own way, the creation of a civil administration reinforced this predatory behavior. Through taxation schemes, animals continued to be appropriated not only to feed the growing administration but as the only local resource available that through export sales could be converted back into cash to cover salaries, stationery, housing, and other expenses. Indeed, this bureaucratic logic of building smaller, self-generating constituencies of taxpayers out of the larger canvas of fluid nomadic communities became the overriding colonial concern in the region.

The colonial production of landscapes and livelihoods in Turkanaland, as elsewhere in the Northern Province, had at least three discernible layers, which I will (following Appadurai 1991) refer to as scapes. First, there was a militaryscape created around mapping, routing, and punitive raiding. This was the network of roads and forts constructed by local labor and it constituted the dominant spatial inscription. Colonial road building clearly got in the way of local path making. The rerouting it entailed, physically, socially, and symbolically, was fiercely resisted by the Turkana, on whom it had a catastrophic effect.

The Turkana path model, as outlined earlier, contains a spatial history, a calendar of memorable events linked to past generations of people and cattle inscribed in the paths through the landscape (see Broch-Due 1990). One is

ekaru na Nayece, the "time of Nayece", the "tribal mother" who followed the
footprints of a stray ox down the escarpment from Karamoja and thus found
the first path into Turkanaland. The making of the contested path of this
chapter, the one across the Uaso Nyiro River and into the plains around
Isiolo, is dated to *ekaru lo Loparala*. It is named after a leader who in precolonial times led Turkana famine refugees into this new "land of survival",
as Isiolo Turkana named it (Hjort 1979:28).

Less agreeable memories are recalled by *ekaru ka Aperitit*, "the time of
scattering", when British forces launched a series of punitive raids, throwing
people off their regular paths and forcing the refugees to flee north. After
this campaign, the Turkana entered a critical phase in their history, with
most of their cattle gone. Epidemics, drought, and famine followed, making
things even worse. Left behind were empty camps and corrals, silent signs
of the suffering brought about by extreme hunger and hardship. The time
events linked with the ways each group of Turkana reopened and reclaimed
older paths, after having been scattered, thereby reconstructing itself, are
vividly imprinted in the collective memory of contemporary Turkana people. It is interesting that the prophecy announcing the arrival of the British
made by one of the great Turkana war prophets was likewise cast in classic
"oxen on the path" imagery (Broch-Due 1990:448).

> Koletiang dreamt that a white bull was coming to the land to dominate the cows.
> He told people about the dream and asked them to choose between two reme-
> dies. He spoke: "Either we knock the bull on the nose, so that it returns, or we
> throw a spear in the leg, so it stops." The crowd discussed the puzzle put in front
> of them, and finally chose to immobilise the bull and performed the purification
> ceremony. This was wrong, and *Nangolenyang*, "The yellow with a white metal
> crown" settled in Lodwar. Then it was too late to hit the black bull on the nose
> which later came along the same path, so *Engolekinom*, "The black with a white
> metal crown", took over in Lodwar.

The resistance against the European intruders focused on obstructing the
construction of new networks of routes and roads. References to the logistics
of traveling, whether by foot or by vehicle, is the major topic running
through early colonial reporting; the counting and rerouting of raided cattle
is another. This central concern with the movement of troops and things
refers to the policing role as the prominent aspect of the British presence in
the region. They took upon themselves the task of confiscating what they
conceived of as stock stolen by the Turkana (sometimes rightly, sometimes
wrongly), and bringing it back to the rightful owners of neighboring tribes,
not so hostile to the civilizing missions as the Turkana. The Turkana commented, with good reason, that the British, while professing the desire to
stop the raiding between groups, were the biggest raiders of all.

For the cattle-keeping Turkana, as elsewhere in the region, this tampering with cattle movement interfered with the making and managing of the
multiple social paths between people needed to reproduce wealth and well-

being for these communities. The movement of herds from one homestead and pasture was then, as it is today, necessary for ensuring animal feed and the ensuing milk flow and thus human livelihood. While Turkana prophecies of the coming of the colonialist were cast in imaginary terms of paths and oxen, real animals were commonly deployed in the region to prevent the misfortune that followed on the heels of colonial moves with the practice of "burying alive [in the road] a wretched bull", as was reported in the 1930s in neighboring Pokot by a shocked officer (cited in Bianco 1991).

The second layer, superimposed on this grid of contested routes and roads, was made by the addition of structures around barracks in the form of separate living quarters for colonial officers, office blocks, and shops. This constituted the space of Nangolenyang, "The yellow with a white metal crown", which, in Turkana terms, referred to the colonial headquarters in Lodwar, the same icons being placed outside the camp of each of the government-appointed chiefs as well as on the helmet of the tribal police. This "bureaucracyscape" gradually expanded around the tasks of counting the subjects and defining their social identities in preparation for the continued extraction of the only local resource of value—livestock—appropriated in the form of taxes, licenses, and army rations. Given that every peripheral jurisdiction in colonial Kenya had to pay its own way, enumerating, essentializing, and extracting were *the* core activities that created the contours of a local constituency from which revenues could be generated to cover the costs of civil administration.

It is often claimed that the administration of the Northern Province was conducted at minimum cost. This is certainly true in terms of investment by the central office of the colony. However, the cost imposed on the local population was heavy and, given that the revenues extracted were largely used to control and police their movements to the detriment of the pastoral economy, the burdens of colonialism were substantial. The balance between revenues from taxation and expenditures on famine relief was jealously guarded, and this explains why the colonial administration in Lodwar originally was unwilling to accept the immigration of the Isiolo Turkana, who they were told consisted mainly of paupers. Most telling, when the party finally arrived in 1958 it was met by the district officer, Gardener, whose main mission was to inscribe the newcomers into the tax register. He reported to his superior, the D.C. on the meticulous registration performed at the reception. Since they had been counted three times en route, he concentrated on the tax ledgers:

> Each man's name was entered in his chief's tax register under the appropriate section and the number of his last tax receipt and the place where he paid it under the appropriate year; I initialled every entry. His new number was then entered on the back of his latest receipt, so we can find him more easily next year, and also in the "remarks" column of the Isiolo register, so that the Isiolo administration will know he is off their hands (and if necessary know to which chief a particular individual has gone). Finally all Isiolo O.D.O. passes were withdrawn

and burnt. ... I told the immigrants that they would be allowed to go outside the district to look for work but they must first obtain a pass. ... Less than a dozen did so ... I have accordingly told the D.C. West Suk, that he need not expect any significant increase in the number of Turkana passing through Kapenguria to look for work in the settled areas.[12]

Each district was thus organized as a semi-independent serfdom and run very much according to the whims of the colonial officers in charge. The records are full of cases that point to smoldering, sometimes fierce competition between districts based on the economics of oppression. In interdistrict affairs, it required a fair amount of diplomacy to get things done on this level. For example, in his letter of 1939 informing the D.C. in Samburu that the Isiolo Turkana would be returned there, the D.C. in Isiolo hastened to assure him that this would not adversely affect the economics of the competing constituency:

> Before taking any action I shall naturally await your views. I should like to add that such an exclusion of Turkana will not necessarily result in a loss of wages which would otherwise circulate in Samburu since money is rarely paid by the Somalis who prefer either to cheat their herds openly or pay them in stock.[13]

His D.C. colleague a decade later received an altogether different response when he tried to organise the movement through Samburu District for three Turkana pastoral families who willingly wanted to return to Turkana District. The exchange of telegrams is self-explanatory:

> *12 November 1955, from Districter Isiolo to Districter Maralal:*
> "Three Turkana families with stock wishes return Turkana via Yours STOP. Approx. 60 head of cattle 550 sheep and goats and 15 donkeys STOP. We are delighted STOP. Will you signal permission for them to pass."

> *20 November 1955, From Districter Maralal to Districter Isiolo:*
> "Permission positively not granted and will confiscate every head of Turkana Stock crossing from Isiolo through Samburu country STOP."

> *24 November 1955, from Districter Isiolo to Districter Maralal:*
> "Reference your T 795 of 24/11 STOP. You may recollect Policy has been laid down to remove Turkana from Leasehold Area back to their District STOP. Emergency has held this up STOP. If we provide escort to some point chosen by you are you then agreeable to this voluntary move query."

> *28 November 1955, From Districter Maralal to Districter Isiolo, reply 28–11–55:*
> "ref your E 1019 STOP. Suggest this matter discussed at next Boran Border meeting". (To this is appended a note scribbled by pen by D.C. Isiolo: "By which time no trek will be possible!")[14]

---

[12] 15–9–1958, ADM.15/28/55 , DC/ISL/3/1/77.

[13] 2–10–1939, Adm.1 5/16/2285, PC/EST2/11/19.

[14] DC/ISO/3/1/76.

These angry exchanges clearly bring out the ways in which the consolidation of pastoral constituency was linked with the growing concern with control of land and boundaries.

The third layering of colonial control was packaged with lots of plans and lofty ideas for the creation of a "businessscape", interventions to expand the economic resource base of the constituency by redirecting local activities from herding into fishing, farming, and trading. When the newly appointed D.C. agreed to accommodate the immigrants from Isiolo in 1944, he came up with the following options for employment:

> On a cursory examination of the problem the following suggestions present themselves: a) Distribution amongst the Tribesmen. b) Employment In the Fish Industry. c) Employment on water supplies. d) Employment on the Omo Development Scheme. e) Engagement In the Frontier Tribal Police. f) Engagement as Locust Scouts. ... With regard to (a); the men will probably become serfs to the wealthier Turkana and the women will join the crowded ranks of the manyatta prostitutes, neither occupation need be regarded as carrying any stigma. With regard to (b); little can be done in this direction until we acquire boats, nets and European supervision. With regard to (c); a grazing policy should be formulated and enforced before additional waters are provided. It does not, therefore, present an immediate solution. With regard to (d); If the Merille problem can be satisfactorily settled it will be possible for a large agricultural settlement to be established on the Omo [river]. European supervision will be necessary. With regard to (e); up to fifty single men could be enlisted. With regard to (f); work could be found for fifty to one hundred men provided they could speak a little Swahili. ... It will be appreciated that until such time as occupations have been found for them, these repatriates would require to be fed by Government.[15]

On their arrival in 1958, none of these projects, apart from public work, had materialized in any sustained way. An abandoned jetty, now sitting high and dry, miles away from an ever-retreating lake, a slaughterhouse, a school, and a small hospital are the only signposts left of the very modest colonial (re)investment in the local economy. While very little had happened in terms of welfare for the populations enclosed, these "boundaries in the making" had become remarkably solid during the five decades between the Maasai and Isiolo Turkana relocations. They had changed from "imagined communities" (Anderson 1983) to being also "enumerated and essentialised communities" (Appadurai 1991), spatially locked into delimited territories.

By the time the Isiolo Turkana were "repatriated", the whole population along with its livestock had been classified, enumerated, taxed, and issued special passports to prevent "trespassing" across district boundaries. The huge volume of documents, reports, and letters that surrounds the repatriation demonstrate that the colonial reformation of tribal reality was undertaken with clinical efficiency. This excessive reporting turns the case into an

---

[15] 24–9–1944, DC/ISL/1/75.

ideal microcosm of a much larger colonial canvas of classifying, counting, and containing pastoralists within specific tribal identities and territories.

Poverty politics

Paradoxically, the party of Isiolo Turkana that was "repatriated" represented people whose biographies had been dramatically shaped and impoverished by colonial policies in Turkanaland and beyond. In addition to those whose clan genealogies included the migrants from precolonial times, most of the elderly had been pushed out by the famine that followed two decades of military campaigns, lasting into the 1920s. Many of these destitutes, joined by Turkana from Baringo and Baragoi, had been caught up in a colonial settlement scheme in Samburu before moving to Isiolo in search for wage labor in the 1930s.[16] In the early 1940s, this was still acknowledged history in the colonial offices, as the following citations evidence:

> The Turkana are one of the most virile and prolific of the tribes of East Africa. They were unfortunate that the Pax Britannica arrived just when they were in the process of moving South and caught them in very unfavourable country. Because they were a fighting race, they soon came into conflict with authority, which in order to subdue the tribe as quickly and cheaply as possible, adopted the policy of imposing large scale communal fines. The present poverty of the tribe is due to that fact.[17]

> The Turkana settlement was originally indirectly caused by an Administrative Officer. Mr. Sharpe, when he started his agricultural experiments with Turkana and Wandorobo at Wamba. This project was not successful, it seems, and so a number of the Turkana filtered southwards.[18]

When the plan of the move was conceived in the first place, most of the Isiolo Turkana lived in abject poverty, and the ways in which misguided colonial policies had shaped their trajectories was very much foregrounded in the social consciousness of the corps. Then an assistant game warden, George Adamson painted a portrait of the disruptive consequences of poverty on the fabric of Turkana lives in a letter to the district commissioner in Isiolo in July 1943. Apologizing for writing the letter he justified his intervention given that:

> The Turkana in the Isiolo District are a growing menace to Game. ... Fines and imprisonment have no effect, simply because the majority of the men are destitute and in order to live and support their large families are compelled, either to steal stock or poach game. ... The present state of affairs is having a very demoralising effect on the tribe. They feel themselves to be more or less out-laws, their marriage customs are breaking down as few of the young men have sufficient

---

[16] 21–5–1947, DC7ISL/1/75.

[17] Letter of 21–8–1943, George Adamson to DC/ISL/1/75.

[18] 20–11–57, DC/ISI/3/1/76.

wealth to pay for a bride, many of them, today do not know the names of their fathers.[19]

His boss forwarded the matter to the colonial headquarters, elaborating on their plight:

> I do not think that it is sufficiently realised that the Turkana employed by the Somalis are if not slaves then serfs in all but name. They are often not paid wages and then only a pittance, often held up for months on account of some alleged stock theft; often their wages consist of a sheep part of the natural increase of the Somalis herd. They are seldom fed and depend on the milk of their employers stock and the meat of the sheep which die for food. … I therefore feel that now that schemes of post war reconstruction are being considered the Turkana should find a place in them. … I have no doubt that they would make most useful and hardworking citizens if given employment which would implement one of the conditions of the Atlantic Charter and help to give them freedom from want.[20]

It is reasonable to believe that the repatriation plan was, at least in the early stages, motivated by earnest concern combined with the wish to alleviate poverty among needy Turkana immigrants. However, it remarkably soon developed into an instrument of control and punishment of those targeted, who, after two failed efforts in 1951 and 1957, were finally rounded up and marched for five weeks on foot to their "homeland"—a place few of them had ever seen before. To develop this point, let us first explore the social mobility of the immigrant Turkana from the dismal state described in the D.C.'s report in 1943 to the time of their deportation.

Most interesting, the very meticulous preparation of the move, which involved detailed surveying and counting, provides privileged access to the ways their settled careers evolved. From fragmentary pieces of information from the colonial archives, an intriguing image of the social profile of the group emerges. It is clear that they were gradually moving from misfortune to modest prosperity, accomplished without any provisions from the government.

Starting at the bottom as herders for Somali on subsistence terms, it was reported in 1946 that many Turkana had left their meager Somali employment in the hands of another batch of paupers from Borana, turning instead in increasing numbers to public labor in "locust, road and other government works". From this report, we understand that the men were investing their improved earnings in goats, which multiplied fast; more Turkana women had immigrated from Baragoi in Samburu District; and families and corrals were being built.[21] In other words, we get the sure signs of a community in the making. By 1947, we learn that their overall situation had considerably improved:

---

[19] Letter of 21–8–1943, George Adamson to DC/ISL/1/75.

[20] 25–7–1943, DC/ISL/1/75.

[21] 15–7–1946, DC7ISL/1/75.

> The Turkana possess over 5000 sheep and goats and a few donkeys. The numbers increase rapidly because Turkana purchase stock with every spare shilling. They have also got away with a fair number from Livestock Control and Somali herds. Thirdly they hand feed their kids allowing the female stock to breed at a high rate.[22]

Although often distorted, and biased, we can read out of such colonial descriptions the ways in which the Isiolo Turkana grew extremely skillful in well-established pastoral strategies for the swift rebuilding of herds. From the colonial perspective, the terms of employment they initially entered into with the Somali herd owners seemed like "slave contracts" since wages were paid in stock rather than coin. However, it was a very wise move by the Turkana to insist on being rewarded for their services with a source of wealth that was both edible and renewable. Due to their pastoral past, these people were intensely aware that domesticated animals compare remarkably well with almost any form of capital, monetary or otherwise. They multiply by themselves and thus generate capital without the medium of markets or other exchange mechanisms. Moreover, the reproductive rates in livestock populations exceed those among humans, thus creating high returns on herd investment through simple demographic processes. The growth rates are particularly rapid among goats and sheep, which are thus favored starting capital for poor pastoralists. Small stock can be converted to cattle and thus ultimately can provide their owners an entry into the most prestigious pathways in the pastoralist exchange system. The cattle circuit is where social power is encoded, and it also provides the bridewealth with which wives and fresh cohorts of children and calves can be obtained. These are the most valuable assets in an economy in which procreation among humans and herds is simultaneously production.

Clearly, some Turkana community members had managed to embark on the cattle stage in their pastoral careers. Seemingly, they had taken advantage of the growth potential built into the notion of alliance territory, outlined earlier, by establishing a partnership with Samburu cattle owners. In return for herding their partners' surplus cattle together with their own, they would have got not only the milk from their udders but typically also a share of the calves born in their care. Perhaps due to a lack of understanding of local modes of animal husbandry, or due to increasing irritation with the presence of the industrious Turkana, the livestock officer in charge did not appreciate these arrangements. Whatever the case, from behind the screen of his recordings in 1954 clues to the herd-building strategies employed by them emerges:

> In 1951, there were known to less than 300 Turkana owned cattle in the Isiolo Leasehold. At the end of last year I was informed by the Turkana headman, Adome, that there were about 800 such cattle. I have little doubt that this colossal

---

[22] 21–5–1947, DC7ISL/1/75.

increase has been caused by the Turkana agreeing to graze Samburu Stock here in return for the milk they provide. I propose to stop the rot by asking the Veterinary Department to brand all Turkana cattle forthwith. Any such branded cattle found thereafter in the Samburu district can be confiscated under the Diseases of Animals rules and unbranded cattle in Turkana herds will be similarly dealt with. ... (Better still would be to press the Provincial Commissioner, Rift Valley Province, to have all Samburu cattle branded. This could stop both Turkana and Isiolo Somalis from receiving their cattle on loan or by sale.)[23]

At the time of the move, the group was well positioned on the social topography of wealth in the Isiolo surroundings. According to the colonial administration's own confidential assessment, they compared well with other ethnic communities like the Samburu, Meru, Borana and Somali. The full-fledged census of 1957, undertaken in preparation for the "repatriation", revealed the following breakdown of the count of the Isiolo community of 3,407 members and their animals (excluding donkeys): pass holders, 642; wives, 611; children, 1,848; dependants, 308; cattle, 1,939; sheep and goats, 13,562.

Their achievement was even more impressive given their troubled background as descendants of Turkana paupers who had filtered into the township during the preceding decades. After struggling as low-paid herders, domestic servants, and government laborers, most of them had succeeded in building viable livelihoods, evinced in their livestock herds, which, although of modest size, were sure signs of prosperity and good management in the eyes of any pastoralist beholder. As is to be expected, not all had fared as well as others. Individuals would find themselves in different stages in the recovery cycle, but unequally developing resource portfolios probably also reflected the fact that success in dryland stock husbandry requires sophisticated skills and personal capacities. Along with the general betterment of the community as a whole, there was thus increasing social differentiation among them. In fact, the contours of this mixed profile are already evident in the colonial categorization of 1947:

> The Turkana can be divided into three categories: (a) Those in employment of Government or of Somalis. They live at their work and unless checked gather all their relations around them. (b) The elders ... who live with their stock between Lingishu and the Uaso Nyiro. (c) Those who make a bare living by carting water and wood and sweeping for the Indians, Arabs and Somalis of Isiolo town. This category includes the orphans, bastards and beggars.[24]

Paradoxically, parallel to their new-found welfare and confidence as a settler community, the colonial reporting was gradually replaced with negative imagery and labeling. Most interesting, despite the fact that many internal memos demonstrate (albeit incidentally) that the Isiolo Turkana were doing

---

[23] 25–1–1954, DC/ISI/3/1/76.
[24] 21–5–1947, DC7ISL/1/75.

remarkably well on the socioeconomic front, this fact was disregarded in representations for wider circulation.

The "poverty card" continued to be played by the Isiolo District office to the limit, not only in the correspondence with Lodwar but right up to the top of the central office in Nairobi.

It was as if pastoral wealth—the flocks of well-tended sheep and goats together with the gleaming flanks and swollen udders of tattooed cattle—were invisible to the eyes of the colonial beholder. Or maybe it was that wealth on the hoof was considered "bad wealth"—the very antithesis of agrarian wealth and a visible sign that the Turkana remained "bad natives" who even after decades in town refused to respond positively to the civilizing medicine of discipline, cultivation, and settlement. From the colonial counting cited, combined with the tone of the reporting, we can deduce a severe conflict between the Isiolo Turkanas continued pastoral aspirations (although not nomadic ones) and the hostile attitude toward pastoralism widely shared by the colonial corps.

To anticipate the argument somewhat: the fact that the poverty discourse came so easily into play in this environment and era was no coincidence. The European discourse of poverty created immediate and unavoidable links in colonial minds between images of the primitive, nomad, pauper, and criminal (Broch-Due 1995). It was an emotive discourse. The correspondence and the practical measures taken perfectly demonstrate the ways the Turkana fell victim to this discourse and were progressively redefined in its image—with very little reference to any reality.

Combing through the correspondence linked to the move, it has become clear to me that poverty was the only political instrument that could legitimate a relocation operation that legally was on shaky ground. Sentimental imagery of the sorry state of the Turkana stranded far from home and hearth started to circulate in colonial channels. "At this moment", the D.C. for Isiolo conjured, "one may see at a road camp near Nanyuki the unedifying spectacle of Turkana families with little children squatting in shanties beside the road and trying to eke out an existence in this way". In case that was not charged enough, he continued, "last year a number did actually die of starvation—for Somali employers seldom pay or feed them".[25]

The D.C. in Turkana seemingly grew tired of this stereotyping, and although he felt obliged to receive the deportees he replied poignantly to his colleague in the Isiolo office, with copy sent to the chief secretary in Nairobi:

> We should not however delude ourselves that they will be exchanging a life of serfdom and penury for one of ease and plenty or that when they arrive in Turkana their future will be assured; they will merely be swelling the ranks of the half-starved Turkana masikinis.[26]

---

[25] 11–8–44, ADM/4/371237—DC/ISL/1/75.

[26] 24–9–44, Ref. No. L&O. 17/6/23 Vol. 1DC7ISL/1/75.

Perhaps taking this point to heart while adjusting his rhetoric of persuasion, the poverty card soon thereafter became entwined with the social exclusion card, as the D.C: fro Isiolo emphasized how different and deviant the Turkana community was from everybody else.

> As I see it, there is no future for these Turkana in the N.F.D. [Northern Frontier District] as they are so radically different from N.F.D. Tribesmen that there is no question of their being absorbed into any local tribe. If they were allowed to remain ... they would be doomed to a life of helotage and semi-starvation and would become completely detribalised and a menace to law and order.

And once the social exclusion card had been played, it took only paragraphs before the Isiolo Turkana person became inscribed in the vocabulary of social pathology as an unemployable, untrustworthy criminal, and even murderous native:

> Formerly some farmers at Timau employed them (one farmer paid 4/.- a month for their labour) but since a band of Turkana hunted and murdered a Meru herdsman, and owing to constant stock thefts etc. The Nanyuki and Timau people are getting rid of them.[27]

Editing slightly the rougher style of this low-ranking officer, the Provincial Commissioner (P.C.) echoed the gist of his argument in his accompanying letter to the central office. Starting out on a more gentle note, assuring the addressee that he was "concerned not so much with the present generation but with the children for whose future there appears to be no provision", he penned, almost as an afterthought, this query to the chief secretary: "Would it be practicable or wise, I wonder, to start a school for Turkana children?" Whatever the case, the softer, more polished version contained the same stark message: first, of the cultural "difference"; second, of their inability to be "happy, self-supporting and productive"; and third, of "criminal records", making it difficult to develop their careers as migrant laborers because "neither the farmers nor the Police desire them to remain as squatters in the settled areas". He concluded his petition:

> I realise that if they were forcibly returned to their own country many of them would have to be maintained in some way by Government but I feel it would be better for it to be rendered in their own country rather than here.[28]

Once stereotyped, the negative iconography of poverty and criminality began to grow. Turkana people were turned into a community of assertive infiltrators, women and animals alike were said to be "breeding like rattlesnakes", trespassing on the grazing of others, and so on.

Having once been a most sought-after labor supply for private and public employment, from the point of the plan to relocate them onward requests

[27]11–8–44, ADM/4/371237—DC/ISL/1/75.
[28] 30-9-1944, DC7ISL/1/75.

for the recruitment of Turkana labor to other parts of the colony was turned down by the administration on absurd grounds:

> I would not recommend their employment in other Districts because (1) They are raw, their language difficulties would be that, they are unaccustomed to prolonged and/or regular hours of work and they would be incapable of performing ordinary tasks. (2) They are accustomed to quantities of meat and milk which could not be provided. (3) They would probably contract pneumonia or other chest trouble if employed in the Nyeri District.[29]

Again, the more polished P.C. endorsed this view, being "rather doubtful if they would stand the very damp climate in that area" and repeating the overall frame in which all policies toward Isiolo Turkana should be considered:

> [I]f these tribesmen are not to become mere wanderers, then the scheme put forward by the Officer in Charge, N.F.D. for their repatriation should be seriously considered. It would not be for their ultimate welfare, I feel, that they become detribalised.[30]

Another officer, perhaps not as well versed, came up with the following spectacular reasoning against a request for Turkana labor from a ranch in the so-called White Highlands. Remember, Turkana, if nothing else, are astonishingly skilled herders and animal handlers. Yet:

> I regret that I cannot conscientiously recommend Turkana as Syces [sic] to look after horses. They own no horses themselves; the majority of them have never seen a horse; and they have no word for horse in their own language. Although the Turkana are a purely pastoral and stock-owning tribe, they are not particularly adept at caring for their animals. They own large numbers of donkeys, but they make little attempt to look after them. Except when they are required as baggage animals, the donkeys are permitted to roam wild unherded.[31]

But it was the charge of poaching that weighed most heavily against the Turkana and constituted the ultimate pretext for their "repatriation". This is a fascinating fact that deserves closer examination.

The savanna of the Northern Frontier Province was then, as it is today, the site of a remarkably varied wildlife, which since the arrival of the colonialists had been vulnerable to commercial exploitation. Horseback riding and sport hunting were among the favored recreational activities and a source of adventure for the district officers and their superiors in the settler colony. Indeed, the identities of officer and hunter tended to merge in the memoirs of those who served in these environs. Adamson's book entitled *Bwana Game*, is one example. *My Life as Colonial Officer and Big Game Hunter,*

[29] 29–9–1945 LAB.1/52/6(11), DC7ISL/1/75.
[30] 20–9–1945, LAB.27/1/A/182, DC7ISL/1/75.
[31] 13–10–1948, DC7ISL/1/75.

by Von Otter, who served as D.C. in Turkana in the 1930s, is another. What is important to note is the small detail of labeling; the colonizer was a hunter while the colonized was a poacher. Moreover, while hunting by pastoralists and foraging communities, which in precolonial times had been only a sustainable bush-meat trade for subsistence, was outlawed, hunting by Europeans was permitted and even encouraged. The blood of resilient lions, elephants, and gazelles figuratively drips from the pages of these accounts, and even allowing for a fair amount of exaggeration it would be difficult for any native to match the degree of slaughter managed by the colonialist themselves. In fact, game meat, hides, trophies, and ivory constituted major currencies in the military economy of colonization (MacKenzie 1988), and, along with the large-scale appropriation of livestock mentioned earlier, they provided funds and food rations for the operation of the colonial offices on the northern drylands of Kenya.

Yet the poaching card was more and more frequently played against the Isiolo Turkana. It was reproduced from one document to the next, finally emerging as a taken for granted fact about "what Turkana do". There is only circumstantial evidence in the archives to support the claim. For example, when Adamson, on one of his reconnaissance trips to determine the best route for the relocation, failed to spot the amount of game he "remembered" seeing in Suguta Valley (an extremely hot, arid desert that would never support large game populations) when traveling through twenty years earlier, his inference was that the local Turkana had killed the animals off and that "a similar fate awaits the game of Isiolo unless the Turkana are moved out".[32] The casting of Turkana as the ultimate poachers was echoed in a draft survey of fall 1947, marked "for the Turkana File". In it, the excited officer reported that he had discovered a new brand of Turkana primitives, the "wandering men of the bush", who, he claimed incredulously, subsisted without "home, money and work". Indeed:

> They lived by hunting and stealing goats. (Adome) disowned this species of his genus, and said he would co-operate in catching and evicting them. The 3 prisoners who escaped from Jail on 6/5/47 were apparently of this type, and all the Turcs [sic] said they would not harbour this type of man, though that may well have been wool over my eyes. It would be interesting to know how many of this type of Turc are now wandering about Isiolo district. Adome confirmed that it was their intention and practice to kill and eat all and every type of game they could get hold of, regardless of age, season, sex or species.[33]

From the privileged perspective of the present, such excesses of colonial reportage would be almost entertaining were it not for the chilling fact that these fabrications became charters for procedure, legitimating a series of

[32] 28–5–1949, DC7ISL/1/75.
[33] J.H.C. 8–5–1947, DC7ISL/1/75.

punitive interventions into the lives of those so disastrously and unfairly labeled.

We have already seen the contours of such actions in the citations above, the confiscation of Turkana cattle and the brakes put on their earning capabilities by curtailing Turkana employment elsewhere in the colony. These poverty-producing policies were steadily scaled up and broadened in scope. Notices were circulated to all public offices in the district and beyond, first requesting that officers take stock of Turkana workers and, second, ordering them to dismiss all Turkana. The police withdrew their work permits and passports, and it became illegal to hire any Turkana person, either in private housework or herding contracts for Somalis and other locals. The image created of these people as notorious thieves, coupled with the complete misconception by the colonial livestock officers about local pastoral management models and arrangements, spurred large-scale confiscation of Turkana livestock. As the officers saw it, the cattle had been stolen from the Samburu and the grass from the Somali. In one such government raid in October 1955, dramatically scripted by the officer, a staggering 655 cattle and 824 small stock were seized, 10 percent of the bounty was kept as a fine, and the rest were auctioned by the government at dismal prices. In the "summing up" section, we learn:

> The Turkana do not admit that any of the stock belonged to the Samburu. They claim that they bought them all from the local Somalis. I sent a telegram on 15.10.55 to the District Commissioner Maralal to enquire whether he had received any complaints from the Samburu about their stock being seized. I have not yet had a reply. ... I am of the opinion, from the strength of feeling among the Turkana about this seizure, that the stock belongs to them, as they claim (and possibly some to their "Jamaa" working on European farms). The removal and sale of this stock appears to have been a severe blow to the Turkana and a punishment in itself.[34]

These impoverishing measures were engineered by the same officers who had claimed compelling concern for the plight of the poor Isiolo Turkana as grounds for their "repatriation". Indeed, internal memorandums linked to the case demonstrate clearly that impoverishment and economic sanctions were parts of a *deliberate* policy of intimidation designed by the Isiolo administration in the hope that the harsh medicine would make the subjects leave by themselves.[35] Forced removal of Turkana livestock was suggested as a means of removing people, allowing only the chiefs the privilege of keeping a few goats (to be specified in their passports) in return for their recruitment of Turkana labor. Perhaps for good reasons, these plans and policies were concealed from their colleagues in Turkana District, the projected destination of the deportees. From their perspective, as the D.C. office in Lodwar

---

[34] 8–10–55, DC/ISI/3/1/76.

[35] 21–5–1947, DC7ISL/1/75.

repeatedly emphasized, it was vital that the deportees be able to live on their herds on arrival and not become a liability to the government there.[36]

As this confused administrative project grew, Turkana became subject to increasing harassment from the police and were frequently fined, arrested, and imprisoned for minor offenses. For example, when the repatriation plan was put in print and the Isiolo Turkana thereby officially became a "problem", although in the pen of the D.C. still a benign one, the police department almost immediately launched series of punitive interventions into the most intimate parts of Turkana lives. In 1945, for example, the assistant superintendent sent a notice to all public departments containing a detailed scheme for the layout of Turkana settlements spatially. In preparation for this, there would be a "clean out" among the "riff-raff of El Barta" on the basis "no work, no claim to residence". After this purge, workers and their families would have their houses reorganized in five separate lines: "1. Administration 2. Police 3. P.W.D. (Public Work Department) 4. Livestock Control 5. Headmen & Accepted Elders". If the lines are "sensible situated" he explained, "the labour is at hand, is not so often late, and is controllable to a degree to which it was not when their houses were scattered over the countryside".[37] Most significantly, this spatial policy of policing was simultaneously a way of gaining information about other social aspects. For the officer planned "to make a new census of all Isiolo Turkana and collect their tax but shall not do it until some order is created out of chaos!"

This tidying up of Turkana living soon got more refined, penetrating to the core of sexual and family relationships. Signed "Your obedient servant", one Forman reported to the D.C. that housing for Turkana workers had been built as directed. But, since he had found that the "close proximity of Turkana women to the men is not conducive to obtaining the best work from them", he had decided to "discharge the married men". He has thus kept the single men, who were equipped with "tin discs" with specific series of numbers stamped on them.[38] The reaction of the Isiolo Turkana to this Foucaultian "power/knowledge regime" was to move their compounds out of town and live in clusters scattered over the countryside.

Yet the treatment of Turkana citizens in Isiolo at any one time was clearly influenced by the personality of the D.C. in office and not always appalling. Although the plan of relocation had been formally agreed to all the way to the top of the colony, there were many dissenting voices. A recurrent argument against the move was couched in terms of identities and rights. "It is debatable", one D.C. pointed out in 1943, "whether the Somalis have any more right to live in the Isiolo Native Leasehold Area than the Turkana, both pay their taxes and neither pay any fees or dues to government for the privi-

---

[36] 13–5–1948 and 30–8–1948, Adm15/7, DC/ISL/1/77.

[37] 27–2–1945, Ref. No. ADM. 15/52/312, DC7ISL/1/75.

[38] 4–9–1945, Ref. No. ADM. 15/52/312, DC7ISL/1/75.

lege".[39] By 1957, the modernist politics of rights interpretation had become more prolific as evidenced in this confidential memorandum from the acting D.C. to the headquarters of the province.

> It is not, in my submission, enough for us to declare that the Isiolo Turkana have no right to be here and that they must therefore all go back whence they came; lock, stock and barrel. It is estimated that about one third of those who are in Isiolo with no other home, having been born here or brought here whilst still children. As a group I have found them easy to handle, pleasant and, above all in days such as these, loyal to the Crown. During the Emergency we called the Turkana volunteers to work on and defend European farms. Their response was immediate and from all accounts they acquitted themselves very well. They trust their Government to be fair and it is for us to ensure that advantage is not taken of this trust. Were these people unpleasant, politically conscious and Bantu-like we could not embark on the proposed move without getting ourselves involved with petitions, lawyers and busybody MPs' questions in the House.[40]

From this telling statement, we can deduce that there were serious legal questions attached to the "repatriation". On account of the fact that the colonial regime never enrolled any Turkana in higher education despite the heavy taxation levied on them, the "tribe" lacked political representation in the colonial power arena. However, the politics of rights here was not motivated by any notion of general rights as citizens, but on account of the Turkana being good or bad natives and of their loyalty to the Crown. The emergency mentioned was the so-called Mau-Mau uproar against colonial rule staged by the Kikuiye and other "Bantu-like" peoples, which eventually resulted in independence. Ironically, on the receiving end of the deportation the opinion among officers in Turkana District was that the El-Barta and Isiolo Turkana had secretly supported the Mau-Mau.[41]

Whatever the case, good-native, bad-native moralizing was integral to the ethnic politics of playing one "tribe" against another. The Somali-Turkana contrast was a case in point. While Turkana were cast as bad natives in several registers, at least they were industrious. The Somalis, in contrast, were constructed as more settled and civilized but inherently lazy, a weakness related to their reliance on drugs. Indeed, in one document the D.C. argued that allowing some of the Isiolo Turkana to stay would put the brakes on the slothful excesses of the Somali lifestyle. Opposing the common view that removing "badly treated" Turkana laborers, would teach the Somali to do their "own dirty work", for, as the D.C. continued, this tribe is

> so indolent, so full or miraa, that they would not rise to the occasion and would make the loss of the Turkana an excuse to bring in [other] Somalis. ... [This] can

---

[39] 25–7–1943, DC/ ISL/1/75.

[40] 20–11–1957, DC/ISI/3/1/76.

[41] 7–2–55, PC/NFD2/8/1.

best be stopped by allowing a number of Turkana to stay in Isiolo to meet the needs of the immigrant Somalis.[42]

The explanation for the final eviction of the Turkana, put up in all the public places in Isiolo town, blamed it on the Somali community! The notice read: "Because of the many applications and requests made by the Isiolo Somali stock owners to Government asking that the Turkana be removed from this District, Government has decided that the Turkana will be sent back to Turkana"![43]

While colonial voices sometimes disagreed on whether the Isiolo Turkana were good or bad natives, the rationale put forward for either position was clearly drawn from a poverty discourse imported from Europe, which had been shaped in the long European project of "reforming" the social life of the domestic poor. Paradoxically, the reason why the European poverty discourse slipped so easily in place in the Isiolo scenario was because it seemed so familiar. It contained a casting and rhetoric that on the discursive surface ran smoothly with the reality confronting the colonialist eager to create "some order out of chaos". The metaphoric bridge between the savanna of East Africa and the slums of England had been constructed a century earlier, a time when the slum and the savanna were habitats for equally unsettled and unsettling peoples at the center of colonial empire.

In the preface to his detailed field study *London Labour and the London Poor*, published in 1861, Mayhew promised to "supply information concerning a large body of persons, of whom the public [has] less knowledge than of the most distant tribes of the earth"(1861:III). Most significantly, in his efforts to constitute poverty as an object of social knowledge Mayhew invoked the information already at hand about these "distant tribes" as comparative tools in order to comprehend the exotic habits of the home-bred poor. He quoted ethnographic studies from South Africa concerning the relationship between nomadic and agrarian groups as a model for his own account. In every society, he wrote, wanderers are distinct from settlers, vagabonds from citizens, and nomads from civilized people; in every society, elements of each "race" were to be found. Like their counterparts in Africa, London laborers had a "savage and wandering mode of life".

Paradoxically, the discursive figure that brought these unconnected places and peoples into proximity, the poor in the slum and the pastoralists on the savanna, was that of the nomad. It was curious, Mayhew continued in his preface, "that no one has yet applied the above facts to the explanation of certain anomalies in the present state of society among ourselves" (1861:2). Once "discovered", however, the figure of the nomad became increasingly cast as the negative "other"—the poor, the primitive, the savage,

---

[42] 20–11–1957, DC/ISI/3/1/76.

[43] 27–6–51, 15/52/11/25; DC/ISI/3/1/76.

the childlike—turning it into a well-versed trope in much of the writing of the time, scholarly and popular.

Mayhew himself constantly evoked the equation between the street traders in the slum and the nomads on the savanna. Other authors followed suit. For example, a popular travel book entitled *The Wilds of London* (Greenwood 1874) cited many "eyewitness" reports endlessly reproducing the equation of the primitive-poor-nomad, claiming that the lack of a settled homelife among these destitute youths made them seem like the "wandering tribes" of "unknown continents". Their plight justified "the growing moral impetus towards the education, reform and civilisation of the working-class masses". Another example, which could have been the blueprint for the survey of Isiolo Turkana mentioned above, was Thomas Archer's *The Pauper, the Thief, and the Convict* (1865). Its subtitle, *Sketches of Some of their Homes, Haunts, and Habits*, brought together hints of the naturalist's notebook, the traveler's tale, and the erotic gaze. Echoes of surveys and statements made by the colonial administration in Isiolo and beyond could not be more audible.

The so-called civilizing mission informing colonialism was a multiple process, at once seeking to cultivate the primitive and engage in the restructuring of society at home, most explicitly in domesticating the home-bred poor and the foreign pastoralist by changing their trades. The reconstruction of social relationships presupposed a reconstruction of livelihoods—a change from any form of roaming to a settled life. This logic was not only inscribed as a subtext in Mayhew's rally against the "nomadic races", but was also integral to Andrew Smith's ethnographic work in South Africa (see also, Comaroff and Comaroff 1992). In Mayhew's elaborated equation, the relationship between the poor and the prosperous in London is equal to the relationship between the bushman and the Hottentot, a contrast that in livelihood and lifestyle terms participated in a series of binary oppositions between nomads and the settled, primitive and civilized, child and adult. Woven into this binary web was a whole series of ratings in terms of morality, intelligence, language, and body shape. In evolutionary terms, those identities and livelihoods placed on the negative and backward side of these lopsided equations have to be grown, molded, and moved to the positive and progressive side.

In other words, the poor and the primitive could only be shaped into proper citizens by putting down roots, taking possession of the land by investing themselves in it, and anchoring themselves to it through their labor and the weight of their domestic possessions. While a change from streetlife to wage labor was the ideal sought for the domestic nomad, the ideal sought for the African nomad was a turn from herding to crop production. In this huge civilization project, cultivation was seen as the seedbed that would cultivate those who worked it just as they cultivated the land. Thus, in the same way as the colonial government encouraged African "primitives" like

Turkana on the savanna to replace nomadic pastoralism with farming, the poor in the slums of London and Liverpool were provided with small parcels of land for the joint production of food and mind.

Thus, the settler colony saw itself as being rightfully in charge of administering to all these other kinds of unruly subjects a standard recipe of the disciplining, civilizing and educating measures necessary for creating a more conventional nation for the future. Just as ideas central to romantic nationalism had moved into the "imagined community" of colonial Kenya, the practical measures required to mold it were also imported. Elsewhere I have elaborated in more detail on the broader history on the uncanny part-for-whole relationship established between the figure of the English "pauper" and the figure of the African "primitive" through the medium of the "nomad". Characteristics attributed to persons of a particular social class within Victorian society were scaled up to continental proportions and came to characterize all Africans in the eyes of the colonial beholder (Broch-Due 1995).

Similarly, the veil of social pathology drawn around the domestic poor, conceptually grouping them with the criminal, the crazy, and the sexually careless, legitimated a broad range of intrusive measures directed at reforming the "poor" in behavior and body. The same techniques and interventions were deployed in the management of entire populations in the colonies. The treatment of the Isiolo Turkana is an instructive case. In other words, a social margin at the center of the empire became transposed to its spatial margins, casting these peripheral populations as entirely different from the colonizer and therefore entirely problematic. This politics of difference is of course the defining features of what Edward Said in his famous book termed *Orientalism* (1928), which applies equally to the colonization of Asia, Australia, the Americas, and Africa, as a growing scholarship on the subject confirms (Appadurai 1991; Rabinow 1989; Bhabha 1990; Comaroff and Comaroff 1992).

Thus, the poverty discourse, evoked by European colonialists and contemporary developers alike, travels together with a sedentary vision of the world. Together they merge into a modernizing discourse that concerns itself with the practical domestication of nature—the molding of soil, body, and mind. From colonial to contemporary times, the settling of pastoral nomads has been the dominant governmental strategy on the African drylands designed to produce in one "move", so to speak, the favorable outcomes of civilization, commodification, and control. It has been promoted particularly as a means of replacing what outsiders to pastoral communities persistently label an outdated subsistence livelihood based on mobile milk herds. Since the first encounters and up to the contemporary situation, sedentism has been an instrument to radically change disruptive pastoral modes of behavior such as raiding, the evasion of the governmental "gaze", and tattooed bodies whose leather and beads have conjured images of a "primitive" past first in the eyes of colonial reformers and presently in the minds of modern African city dwellers.

In short, "sedentism" has been the only idiom of change and civilization in arid areas since the arrival of colonialists. The practical conclusions have also stayed the same. Both social policy and the civilizing mission have focused on transforming the domestic life of those defined as the poor at home and abroad. The conventional recipe involved: (1) creating the conditions and attitudes of cleanliness so that all matter, beings, and bodies are in their proper place; (2) reforming sexuality by encouraging legal, Christian marriages, monogamy, and the creation of nuclear families; (3) spreading the ideal of private property, beginning with the family home; and (4) reconstructing gender relations and the social division of labor (see Comaroff and Comaroff 1992). Perhaps this project's spectacular lack of success on the savanna and its many unintended outcomes are due to the simple fact that nomadic pastoralists in East Africa and the poor in Europe have very little in common other than being subjects of a the misplaced discourse.

Whatever the case, the first dislocation from Turkanaland to Isiolo town was a result of this project of sedentarization. In this context, there is a cruel irony to an administrative report about the second relocation from town "back" to the savanna. Having been thrown out of town for *not* conforming to norms of colonial civility, when reaching the destination in the South Turkana desert the hope was voiced that deportees would become a modernizing force. Reporting from the final journey of "repatriation", the Isiolo office concluded that the immigrants would do wonders for Turkana District. The immigration created "a real chance of the Isiolo Turkana establishing shops in Turkana which would begin to teach the Turkana the value of money and encourage attendance at stock Sales".[44] The D.O. welcoming the Isiolo Turkana noted that local Turkana were not thrilled by the arrival of a group that from their perspective consisted of mostly poor people without much cattle. On his own account, however, he joined in the chorus of aspiring developers, adding a few points in the report to his own D.C. in Lodwar:

> The effect of their return on the local Turkana is more incalculable. Given the premise of a new "forward" policy in Turkana their greater sophistication should be an asset, particularly in the encouragement of stock auctions, greater trade and disease control, they should also help in the acceptance of the l6/- personal tax.[45]

Tribes, bodies, portraits

At the core of the Western significance ascribed to sedentism is the great attraction of cultivation as a summarizing symbol for civilization—of nature,

---

[44] G.W.L Pryer and M. Wasilewski, "Turkana Repatriation, Isiolo-Kangetet, 6 July–31 August 1958".

[45] 15–9–1958, ADM.15/28/55, DC/ISL/3/1/77.

bodies, and minds. And it is precisely through the idiom of gardening and cultivation that the European colonialists in Africa made sense of their diverse experiences with different types of African "tribes", their own communities, and the whole colonial enterprise. In her book *Colonial Inscriptions* (1995), Shaw elaborates on how the flower garden was a focus of great attention and pride for colonial householders in Kenya. The cultivation of ornamental flowers was an effort to root familiar transplants in African soil; a signposting of European high culture and its capacity to civilize foreign wilderness. Improving African plant species was analogous to the colonial project of improving African humanity. The discourse on gardens, as Shaw demonstrates, brought together European aestethic appreciation; the ideology of social hierarchy based on class, cultural refinement, and race; practical knowledge about improving the race and species; and the desire to make something European out of Africa.

In the poetic economy of colonial Kenya, gardening was a symbolic device deployed in social distinctions within the settler community itself. In *Flame Trees of Thika* first published in 1959, Elsbeth Huxley succinctly conjured the lack of flowers among Boer immigrants to Kenya from South Africa as a sure sign of relative destitution and crudeness, of being from a lower class. In *Out of Africa* (Dineson 1939), Karen Blixen cast the cultivation imagery more widely, including in the poetic equation distinctions between different kinds of local peoples in ethic and occupational terms. Those in spatial proximity to the colonialist, domestic and agricultural workers, were themselves like plants—capable of being cultivated, pruned, and improved. The Kikujyu in their capacity as farmers, laborers and servants came to embody these characteristics. Those spatially removed from the fence of the farm—living on the huge savanna—took on in Blixen's romantic eye a noble savageness. They came to embody the features of the landscape itself—her prime example being the Maasai.

The East African pastoralist groups represent a very interesting case, exemplifying how unstable and shifting the images of poverty and prosperity have been in European conceptualizations of Africa. In the early days of colonialism, an influx of aristocratic settlers from Europe—of which Karen Blixen was a part—constructed nomadic pastoralists like the Maasai and Turkana as their counterpart, although of a more distant form in the evolution of humanity. The romantic image of the nomad as a melancholic outcast, wandering through a doomed landscape existing outside history in which all movements are circular, is captured in this quotation from another Huxley text:

> [T]hese obstinately conservative nomads, wandering with their enormous herds from pasture to pasture, seem like dinosaurs, survivors from a past age with a dying set of values ... aristocratic, manly, free, doomed. (1948:89)

Most significantly, these nostalgic values were central in the larger project of creating tribal constituencies, a project of which the Isiolo Turkana repatriation was a part. At the core of the tribal model, as with the romantic nationalism that inspired it, was a ferocious enchantment with the mystique of a particular place and its people. In this part-for-whole relationship, body and landscape were embraced in a spiritual unity—the soul of a place—the place of a soul. This metonymic construction of a body-scape (Broch-Due 1990, 1993) relies heavily for its evocative force on ideas of tradition and authenticity. The construction of a national tradition in Northern Europe clearly depended on a poetic economy. Growing up in Norway, for instance, I was early made aware of how the molding of each "folk" was not only a matter of language and landscape but of costume, cuisine, music, poetry, and painting. The Norwegian national costume (*bunad*) perfectly exemplifies how particular versions of this poetics were "frozen in time", conserved and declared for posterity as "the tradition", and zealously guarded by national museums. In the contours of the literary examples sketched out above, we can sense that tribe making in Kenya had its own poetic economy focusing on body morphology. Thus, just as the body surface was a powerful medium for forging "folk" identities in the nationalistic project in nineteenth-century Europe, body morphology was turned into an arena in which colonial relations were molded, enacted, and contested in print, paint, and clothing.

A good example is embodied by the couple Joy and George Adamson, more famous for their roles in "repatriating" lions to their "proper" habitat than people, he employed as a game warden responsible for the routing of the Isiolo Turkana move, she accompanying him in search for proper tribal sitters for a portrait collection commissioned by the government. From slightly different angles, they both seemed to share the belief that town living and modernity had a corrupting influence on the noble nomads, albeit only skin deep. It was only a matter of removing their "gaudy modern garments" and returning them to their natural habitat to revive the spirit that bound them to nature. Frustrated by their "poaching", but keenly aware that this was the only survival strategy open to them after being victimized by government raiding and ill-founded projects, George Adamson was among the first to air the idea of repatriating the Turkana, seemingly with the best intentions. For Joy Adamson, the repatriation of the Isiolo Turkana formed part of the larger canvas of romantic representations of native people and their natural habitats. Here we see how the colonial project of defining landscapes and that of defining bodies work in tandem.

In *The Peoples of Kenya*, Joy Adamson displayed a series of striking portraits of subjects colorfully dressed in feathers, furs, beads, and weaponry, their faces and bodies tattooed and painted. Each distinctive portrait was offered for public consumption as a metonym for a tribal whole; in the plural mode the collection as a whole was intended as a metonym for an authentic federal Kenya before the advent of the colonizers. The paintings, commissioned by the colonial government in 1949, took five years to complete, but this was, as Adamson's biographer wrote, "a small price to pay for a uniquely privileged view of this savage and esoteric way of life, doomed to change like much of the spectacular landscape that supported it" (House 1995:166). The problems encountered were very much a matter of finding the proper "primitive" model representative of each cultural environs. Apparently, she kept "detailed notes on her subjects, and unless a ritual or costume was authenticated by at least two elders, she refused to paint it"(167). This pursuit of authenticity was clearly problematic. For a while

> she found some superb subjects among the Nandi above the Rift valley and the Turkana to the west of Lake Rudolf. After that she ran into serious trouble. Anywhere near a town the old traditions were fading or forgotten; sitters turned up in gaudy modern fabrics and vulgar beadwork. (166)

While Joy Adamson lamented the difficulties of finding the right sitter according to her preconceived ideas about the authentic tribal body, plenty of "inauthentic" models seem to have come forth in their "contaminated" costumes, eagerly offering themselves as sitters. This brings us to a very crucial point. Joy Adamson's obsessive interest in tribal bodies collided in interesting ways with an equally overriding local concern with a competitive body politics. The colorful costumes, paint, and tattoos worn by pastoralist peoples in the Northern Province were certainly *intended* to be received by target audiences as an indication of specific motivations and identities but *not* primarily the ones attributed by colonial spectators like Adamson. In sharp contrast to the conserving image of a bounded collective, be it local or national, that European folk costumes typically are tailored to represent, the spectacular pastoral body display is a deployment of personalized power in social relationships. Perhaps it was this energetic agency reverberating through body performances that obliviously enticed her to the portrait painting in the first place.

In other words, the minute attention paid to amplifying body morphology is part of an ongoing political game of projecting fame, deeds, and desire. By accentuating the body appearances of self and animal through adornment and scarification, a Turkana person achieves the momentary image he or she desires, visually as well as verbally declaring socially individual capacity and deeds. This entails a behavioral ethos of assertiveness, one that potentially applies equally well to women and men (Broch-Due 1990). Most significantly, the cartography of bodies and the cartography of animal bellies are united in Turkana expansionist political economy, the projection of a particular forms of social power, embodied by specifically styled peoples and herds, into the unfolding landscape. The Turkana idea of bodyscape is a mapping of the aims and effects of these efforts in the real world, penetrating all its layers physically, socially, and symbolically.

Ebei, the famous war leader in the resistance against the British, the brave *ekile angikilok*, husband of men-wives, the leader of the famous age set *Ngiruru*, illustrates this cultural logic perfectly. My notes from an interview in 1985 with elders who recalled him with passion, contain vivid descriptions of his stature and those of his cattle, the account being constantly abbreviated by their chanting of Ebei's ox name, Lokwaris. Here is an edited version.

> Ebei took the liberty of wearing the leather insignia of both clan clusters and those of intermediate clans. Thus, sometimes he wore gazelle and calfskins and sometimes goat and sheepskins. He always wore a baboon fur hanging from his back, and the tail of a giraffe dangling from his wrists. His right hand gripped firmly the shaft of a long spear with a large blade, and in his left hand he carried an equally impressive shield, square shaped and made from hippo leather. His long hair was styled in a chignon (*emedot*), painted with greenish mud (*apus*), and interwoven with ancestral hair so it reached to his waist; the long black feathers he stuck into the coiffeur would blow in the wind. He is portrayed as garbing himself in so many layers of leather that his body looked even larger and stronger than his already imposing and solid physique.

The point of this corporeal game of aggrandisement is to impress your image on others while simultaneously trying to eclipse the effects the image of others are having on you. It is simply about influence and staying influential. The person, like Ebei, who sees his or her acts echoed and enlarged in something else (cattle, cosmos, clan, age group, or wild beasts) has a better chance to come out a winner than those who for the moment are stripped of these powers. In the words of Strathern "the hidden sociological dimension is that the collective relations aggrandise individual acts" (1988:328).

Clearly, to be effectual in such a bodily game each person's register of regalia is in constant need of replenishment, whether in the form of garments or surface surgery on the skin. Thus, in this continual search for additions, innovations, and surprise—marked places in savanna towns are searched for exotica, for example, in the form of a pair of plastic sunglasses, a pink nylon bra, a woollen hat, and the like. These are precisely the things that in the eyes of the colonial beholder appear as "gaudy modern garments and vulgar beadwork". The search of the unique signature, of authorship, of agency, is accomplished through the medium of an extraordinarily rich bodily morphology, carved and cast to simultaneously display a diversity of social identities. Alongside the biographical information of age, sex, and status, as well as memorable events and impressive acts, the costumes give details of membership in clan, moiety, and other relevant statuses. Many of these are transitory "identities on the move", biographically, socially, geographically, and historically.

If we compare this fluid, dramatic, evolving bodily style with Joy Adamson's "tribal portraits", we find that it precisely echoes the contrasts I evoked earlier between Turkana and colonial maps. Adamsons's delineation of bodily boundaries as claims to a specific ethnic and unitary identity clearly found resonance in the colonial, map-obsessed audience. This was also an audience, as indicated earlier, obsessed with the essence, the "spirit", of the peoples it dominated. In a very telling statement, Adamson wrote of her efforts in this regard:

> I tried to achieve this by holding the attention of my sitters with my eyes and thus automatically, but unconsciously, I believe I was able to convey the spirit of my models. In our "silent eye language" we both communicated in a far more direct way than words could have achieved. (quoted in House 1993:168)

This explanation is a textbook example of an event in which, given that speech is impossible, the encounter becomes a "semiotic event in which a visual language of bodily forms is especially critical" (Hendrickson 1996:15). Unchecked by an interlocutor, this kind of "semiotic event", constituted by physical contiguity and physical commonalties, is particular productive of stereotypes of the other. When picking her sitters, Joy Adamson was clearly influenced by ideas widely in circulation about the exotic primitive. The great irony in all this is that in her quest for the authentic and representative her intuition tended to draw her in exactly the opposite direction. She often ended up with eccentric individuals whose unrestrained individualism spilled over into the bizarre and excessive body styles for which she was so hungry. Often they were prophets, called witchdoctors by the colonialists, who from any perspective were the least representative figures on the scene!

Most significantly, prophets are often antisocial and cast as the other by their own pastoral constituencies. The most successful among them have managed to turn exclusion into influence precisely through their ability to manipulate images of otherness and to play on fears and uncertainties (Waller 1995). Joy Adamson, for example, wrote about the troubling effects of the mesmerizing eyes ringed in white paint of a Maasai *laibon*. Another equally uncanny encounter she described was most probably with a Turkana *emuron* due to the outfit and the greetings given:

> One day, after completing the picture of an elder wearing a large head-dress of ostrich plumes, a monkey coat and a spear, I took a stroll up a lonely valley. I must have come close to his home for suddenly he appeared at the top of the hill dressed in the clothes in which he had posed for me. As soon as he spotted me he came tearing down the slope, shouting and waving his spear. I thought he was doing it for fun. When he came closer I realised he was going to charge me and I had to run as fast as I could to escape. (1967:165)

Prophets typically live apart from ordinary people, conducting most of their affairs in secrecy, and are excessively competitive. In other words, prophets are metaphorically "outsiders" because they operate outside the moral and social boundaries of the community. Their otherness is magnified not only by ambiguous and abnormal behavior but in their dress. The monkey coat is the prime index among Turkana prophets' regalia of their foreign origin.

They not only claim to have come from somewhere else but they even claim to come from another species. This betwixt and between status, the transgressing of boundaries—moral, social and physical, indeed, their prominent role as *cross-dressers*, mixing garments across species, genders, and generations—align prophets with other transforming figures, like the transvestite, who "performs a necessary critique of binary thinking" (Garber 1992:10). Perhaps Joy Adamson's intuitive attraction to prophets was not only formed by stereotypes of the "primitive" in circulation but also motivated by more personal impulses. For in her "exchange of eye-contacts" she conjured up a moment perfectly exemplifying how "discovering ourselves in others" and "discovering others in ourselves" are parts of the same cognitive process.

Joy Adamson was a cross-dresser herself, and her ways of "going native" seemingly caused great scorn and anxiety in the white colonial society. Referring to one of the reconnaissance trips undertaken to find the best route for the Isiolo Turkana move, which Joy joined in her search for sitters, this letter from D.C. Shirreff, Samburu District, offers a case in point:

> The job of supervising the move was given to Don Stone, DO at Isiolo, an ex-fighter pilot with one eye, and George Adamson, who was accompanied by Joy. By the time they reached my district, Stone, George and Joy were travelling in three separate camps and hardly speaking to each other. Joy had the rather disconcerting habit of walking about naked except for a very brief pair of pants, not the sort of behaviour which the elder at Baragoi expected of a white woman. I remember I told her to put a shirt on when attending a stock sale and she thought I was rather stuffy. (quoted in House 1995:145)

This crossing of bodily boundaries set up to keep the "civilized" and the "primitive" apart was of course much more threatening when the culprit was a woman who should, according to the dominant gendered expectation of the time, serve as gatekeeper.

Whatever the source of Joy Adamson's fascination, it is interesting how, in her efforts to portray the essence of the tribal communities she sought, she fixed on precisely the least emblematic of all characters. Perhaps prophets, in their freaky costumes and opportunistic roles as entrepreneurs, have always generated an uneasy tension between "preserving traditions"—a necessary ingredient in the construction of the "imagined community" of the tribal nation—and the civilizing mission based on sedentarization and commodification. Their eccentric appearance, from the perspective of both their home audiences and foreigners, has turned them into a creative source for image making, figuring as icon for both the authentic tribal body in its benign version and for the ultimate "primitive", "filthy", unruly, and uncivilized body of the figure of the "nomad".

Conclusion

As has been demonstrated throughout this essay, there are obvious parallels among the endeavor to reconfigure pastoral bodies, the colonial redefinition of territoriality, and the creation of bounded territories on pastoral land. I hope to have elucidated some of those parallels and how, rather than producing clarity and order on the ground, these body-space projects, in their articulation with local models and practices, generated unending contradiction, confusion, and unintended, tragic outcomes. From one colonial perspective, the Isiolo "repatriation" was painted as move toward authenticity and the preservation of "tradition"—a return to leather and beads; from the opposing perspective, the repatriation was constructed as a modernizing force bringing cash, decent clothing, and change to the backwaters of development.

However, both these rationales were only skin deep, so to speak, and moving beneath the surface of ideology were more pressing political concerns—simply to get the Isiolo Turkana out of the way as soon as possible. Correspondence linked to the case makes clear the understanding among some officers that the result of the move would not be a decorated body dressed in leather walking in lush pastureland, nor a cleansed body dressed in cotton behind the busy counter of a shop, but most likely a starving body huddled in a famine camp in Turkanaland. They were right. The events of the march were a foretaste of what was to come.

The long trek "home" took its heavy toll; many had to be hospitalized en route because of bronchitis, exhaustion, dehydration, and festering sores on their feet. Their goats did not fare better; many turned lame and died in their thousands. Again the wardens expressed their great surprise:

> The immigrant Turkana seemed remarkably unprepared for safari conditions and none really carried adequate water containers. Had water not been found so frequently along the route [added in pen] movement might have suffered serious difficulties. The Turkana women carried an open saucepan of water on their heads from Kipsing to Kangetet![46]

Once again, the traumatized voice of dislocation is heard loud and clear beneath the bizarre tones of the colonial reports.

> Old Turkana showed extreme reluctance to be singled out for lorry party believing they would never see their families again. Some expressed a wish rather to die en route than travel by lorry, claiming that Government would throw them into the nearest river! In some cases they Insisted on going on foot rather than entrust their stock to anyone else.[47]

---

[46] G.W.L Pryer and M. Wasilewski, "Turkana Repatriation, Isiolo-Kangetet, 6 July–31 August 1958".

[47] Ibid.

And, while the victims' voices are erased from the records, their desperation cries out also from beneath the page of reporting from the more sympathetic officer in charge of the so-called lorry party transporting the elderly, nursing mothers, and toddlers, those too weak to march on foot for five weeks.

> I reached Nanyuki ... it was very cold and the semi-naked Turkana began to suffer greatly ... I pressed on and on the really high ground after Thompson Falls it became very cold indeed. The Turkana were in very bad straits and several of the really senile had lapsed into semi-consciousness. I was very worried indeed and considered it essential to get down into the Rift (Valley). ... I was told several times that old people had died and once I had to resort to pushing an old man's head between his knees to revive him. ...The mother's milk had dried up and the babies found the Tetrapak milk provided by the Provincial Commissioner invaluable. ...The convoy got through by great good fortune.[48]

References

Adamson, G. 1968. *Bwana game*. London: HarperCollins.

Adamson, J. 1967. *The peoples of Kenya*. London: HarperCollins.

Anderson, B.R. 1983. *Imagined communities: Reflections on the origin and spread of nationalism*. New York and London: Verso.

Anderson, D.M. 1988. Cultivating pastoralists: Ecology and economy among the Il Chamus of Baringo, 1840–1980. In *The ecology of survival: Case studies from northeast African history*, ed. D.H. Johnson and D.M. Anderson. London: Lester Crook.

Appadurai, A. 1991. Global ethnoscapes: Notes and queries for a transnational anthropology. In *Recapturing anthropology: Working in the present*, ed. R.G. Fox. Santa Fe: School of American Research Press.

Archer, T. 1865. *The pauper, the thief, and the convict: Sketches of some of their homes, haunts, and habits*. London: Groombridge.

Ashcroft, B., G. Griffiths, and H. Tiffin 1995. *The post-colonial studies reader*. London and New York: Routledge.

Barth, F. 1969. *Ethnic groups and boundaries*. Oslo: Universitetsforlaget.

——. 1996. Enduring and emerging issues in the analysis of ethnicity. In *The anthropology of ethnicity: Beyond "ethnic groups and boundaries"*, ed. H. Vermeulen and C. Govers. Amsterdam: Het Spinhuis.

Bhabha, H.K., ed. 1990. *Nation and narration*. London: Routledge.

Bianco, B.A. 1991. Women and things: Pokot motherhood as political destiny. *American Ethnologist* 18:770–85.

Broch-Due, V. 1990. *The bodies within the body: Journeys in Turkana thought and practice*. Ph.D. thesis, University of Bergen.

——. 1993 . Making meaning out of matter: Turkana perceptions about gendered bodies. In *Carved flesh/cast selves: Gendered symbols and social practices*, ed. V. Broch-Due, I. Rudie and T. Bleie. Oxford and Providence: Berg Press.

——. 1995. Poverty and prosperity in Africa: Local and global perspectives. Occasional Papers, no. 1, *Poverty & Prosperity*. Uppsala: Nordiska Afrikainstitutet.

——. 1996. The "poor" and the "primitive": Discursive and social transformations. Occasional Papers, no. 5, *Poverty & Prosperity*. Uppsala: Nordiska Afrikainstitutet.

---

[48] Sd. Deverell, "Report on Turkana Move Vehicle Party, 23 August–26 August '58, DC/ISL/3/1/77".

———. 1999. Remembered cattle, forgotten people: The morality of exchange and the exclusion of the Turkana poor. In *The poor are not us: Poverty and pastoralism in Eastern Africa*, ed. D.M. Anderson and V. Broch-Due. Oxford: James Curry.

Comaroff, J., and J. Comaroff. 1992. *Ethnography and the historical imagination*. Boulder: Westview.

Deleuze, G., and F. Guattari. 1987. *A thousand plateaus: Capitalism and schizophrenia*. Minneapolis: University of Minnesota Press.

Dinesen, I. [Karen Blixen]. 1937. *Out of Africa*. New York: Random House.

Garber, M. 1992. *Vested Interests: Cross-dressing and cultural anxiety*. New York: Routledge.

Foucault, M. 1979. *Discipline and punishment: The birth of the prison*. London: Alan Lane.

Greenwood, J. 1874. *The wilds of London*. London: Chatto and Windus.

Gutkind, P. 1970. *The passing of tribal man in Africa*. Leiden: Brill.

Hendrickson, H., ed. 1996. *Clothing and difference: Embodied identities in colonial and post-colonial Africa*. Durham, NC: Duke University Press.

Hjort, A. 1979. *Savanna town: Rural ties and urban opportunities in northern Kenya*. Stockholm: Department of Social Anthroplogy, University of Stockholm.

House, A. 1995. *The great safari: The lives of George and Joy Adamson*. London: HarperCollins.

Huxley, E. 1948. *The sorcerer's apprentice*. London: Chatto and Windus.

———. 1959. *Flame trees of Thika*. New York: Viking-Penguin.

Knowles, J.N., and D.P. Collett. 1989. Nature as myth, symbol and action: Notes towards a historical understanding of development and conservation in Kenyan Maasailand. *Africa* 59(4):433–60.

Lamphere, J. 1992. *The scattering time: Turkana responses to colonial rule*. Oxford: Clarendon.

Little, P.D. 1998 . Maasai identity on the periphery. *American Anthropologist* 2(100):444–57.

MacKenzie, J. 1988. *The empire of nature: Hunting, conservation and British imperialism*. New York and Manchester: Manchester University Press.

Mayhew, H. 1861. *London labour and the London poor: A cyclopedia of the condition and earnings of those that will work, those that cannot work, and those that will not work*. London: Griffin, Bohn.

Otter, E. von. 1930. *Som officer och storviltjägare i Turkana*. Stockholm: Bonniers.

Rabinow, P. 1989. *French modern: Norms and forms of the social environment*. Cambridge and London: MIT Press.

Said, E.W. 1978. *Orientalism*. New York: Vintage.

Schama, S. 1995. *Landscape and memory*. London: HarperCollins.

Schlee, G. 1989. *Identities on the move: Clanship and pastoralism in northern Kenya*. Manchester: Manchester University Press.

Shaw, C.M. 1995. *Colonial inscriptions: Race, sex, and class in Kenya*. Minneapolis: University of Minnesota Press.

Sobania, N. 1988. Pastoralist migration and colonial policy: A case study from northern Kenya. In *The ecology of survival: Case studies from northeast African history*, ed. D. Johnson and D. Anderson. London: Lester Crook.

Strathern, M. 1988. *The gender of the gift: Problems with women and problems with society in Melanesia*. Berkeley: University of California Press.

Waller, R. 1985. Ecology, migration and expansion in East Africa. *African Affairs* 84:347–70.

———. 1995. Kindongoi's kin: Prophecy and power in Maasailand. In *Revealing prophets*, ed. D.M. Anderson and D.H. Johnson. London: James Curry.

Conservation in the Sahel: Policies and People in Mali, 1900–1998

Tor A. Benjaminsen

Introduction

European conservation ideology and its image of Africa as a special Garden of Eden that should be preserved has, since colonization, influenced environmental policies on the continent. In contrast to the domesticated landscapes and polluted cities of Europe, Africa offered wilderness and what was thought to be untouched nature. It was in Africa that European conservationists were able to influence policies long before they could do the same in Europe (Anderson and Grove 1987; Neumann 1996). The European interest in preservation and control was most pronounced in parts of East and southern Africa, which possessed abundant wildlife, picturesque landscapes, and general natural beauty and diversity. This was also due to a strong presence of European settlers, who managed to influence policies toward local African populations (Anderson 1984; Beinart 1984). Examples of local consequences of such policies are the exclusion of the Maasai from the dry season pastures of Western Kilimanjaro in Tanzania in order to establish settler farms and the eviction of pastoralists and other local people from the many game reserves and conservation areas created in Kenya and Tanzania during the colonial period (e.g., Serengeti, Maasai Mara, Amboseli). Still today, influential international conservation organizations are advocating wildlife preservation at the expense of local people in East and southern Africa (Bonner 1993)

However, also in the drier and more resource poor parts of Africa, for example, in the Sahel, the European conservationist influence has been noticeable. In West Africa, French scientists such as the botanists Chevalier, Adam, and Aubréville, who also worked as civil servants in the colonial Forest Service, supported a vision of severe deforestation caused by local destructive land use (Fairhead and Leach 1996). This included a transformation of the forest into savanna and of the savanna into desert. Aubréville (1949) was in fact the first to use the term *desertification* in an African context. The image presented by these scientists of a general process of environmental degradation and the linking of this image to local resource management

I thank Richard Schroeder, Hanne Svarstad, and Arne Kalland for helpful comments on this essay.

were institutionalized within the French West African Forest Service and even incorporated into the colonial education system at the primary and secondary levels (Fairhead and Leach 1996).

This preservationist approach to environmental conservation has, during the last two decades, lost ground to a new ideology based on "sustainable use" and a participatory approach involving local people in natural resources management. It implies building on the capacity and competence of local institutions and it also converges with a recent international trend toward privatization or enclosure of resources, decentralization, and the disengagement of the state (Benjaminsen 1997a). Using the example of Mali, this essay presents the contrast between colonial conservationist ideas and later national forest policies in the Sahel and current knowledge about environmental change and local agency in the region's environmental management. Furthermore, recent political developments in Mali and the new democratic government's decentralization policy are discussed. The perspective taken is inspired by environmental history and political ecology. A historical approach to environmental change and management is combined with an analysis of forest policies and their social and environmental implications. Political ecology is a rather broad field with many directions (see Peet and Watts 1996). However, there is a common reaction against Malthusian linear thinking about people-environment linkages and a focus on the most powerful political and economic actors influencing environmental change and management.

Forest policies (1900–1991) and local implications

The forest policies that have been enforced in the Francophone Sahel until recently are the direct result of forest laws[1] issued by the French colonial administration. The first law relating to the use of forests and forest products in Senegal and the "Dependencies" was the *décret* of 20 July 1900,[2] which allocated restricted use rights to "indigenous" populations to collect nuts and wood, hunt, and use the forest areas as pastures. However, if the use of the resources jeopardized "the richness of the forest", the *gouverneur général* would take the necessary protective and prohibitive action (art. 23). Apart from these restricted use rights, the *décret* introduced permits for commercial extraction of forest products and fines for infraction of the rules.

---

[1] According to Elbow and Rochegude (1990:1), laws can, in the French system, be divided into *loi* (legislation enacted by vote of the National Assembly), *ordonnance* (legislation enacted by the head of state, with authority equal to that of a *loi*), *décret* (a legal enactment of the executive, often used to clarify a *loi* or to provide the guidelines for its application), and *arrêté* (legislation that may be formulated and promulgated at the ministerial or even the regional level). *Décrets* and *arrêtés* are relevant for the discussion in this essay.

[2] Décret relative au régime forestier du Sénégal et Dépendances (20 July 1900), Archives Nationales, Bamako, Fonds anciens, 3R14.

Then, in 1918, an *arrêté* was signed by the governor of Haut-Sénégal and Niger.[3] This law was primarily concerned with an alleged overexploitation of economic species. In a guide to the law distributed to colonial officers, the governor stated that it was important to preserve certain tree species in the colony and to reserve these species for particular usage.[4] Seven species were protected from cutting. Five of these could, however, be cut upon the payment of a fee to receive a special permit from the district commissioner (*commandant de cercle*). One of the main purposes of this *arrêté* was that hardwood species should be reserved for the wood industry and not used as fuelwood for the "indigenous" populations.

In addition, no trees could be cut within six meters of roads, five hundred meters of small villages (of less than one thousand inhabitants), or two kilometers of large villages (of more than one thousand inhabitants).[5] The fines were the same as the ones defined in the 1900 *décret* (varying from ten to one thousand francs).

The Sahelian forest legislation most frequently referred to is the forest *décret* of 1935.[6] This law is more developed than the two former pieces of legislation. The influence of article 539 of the French Code Civil, or Code Napoléon, from 1800 to 1804, is also more pronounced. The article states that: "Tous les biens vacants et sans maitre ... appartiennent au domaine public." (All vacant assets without a master ... belong to the public domain). This implied that the only way to establish legal ownership of the land was through productive use (*mise en valeur*), which in practical terms meant farming. Thus, pastoralism, the collection of wood and gathering of wild grains, fruits, and medicinal plants, fell outside the European notion of property that was used. The result was that large areas of fallow and silvo-pastoral land were annexed by the state. This led to antagonism between local rules and regulations and the regulations and laws issued by the state. Local communal use rights were regarded as being inconsistent with rational forest management and were restricted or suspended. The forest *décret* of 1935 defined the authority of the state to protect forests from overuse and to protect and restore forest areas that had become degraded (Elbow and Rochegude 1990).

The result of the passage of these laws was a policy that was primarily concerned with conservation. It was generally assumed that local Sahelian populations were causing degradation through overuse of the resource base. Therefore, the system of permits for use and fines for rules violations were created. Research reports (e.g., Aubréville 1949), travel accounts (e.g., Bovill

---

[3] Arrêté no. 722 bis, Archives Nationales, Bamako, Fonds anciens, 3R13.

[4] Circulaire du Lieutenant-Gouverneur au sujet de l'application de l'arrêté du 14 juin 1918 réglementant la coupe et la circulation des bois dans le Haut-Sénégal-Niger. Archives Nationales, Bamako, Fonds anciens, 3R39.

[5] Ibid.

[6] *Journal Officiel de l'Afrique Occidentale Française*, 3 August 1935, 611–18.

1921), and reports from the colonial administration alleged an encroaching Sahara, which was attributed to local resource damage. Similarly, Stebbing (1938) brought widespread attention to the "advancing desert", citing a French colonial officer in West Africa who claimed on the basis of just three years of observation that the desert had advanced southward at a rate of one kilometer per year.

The Forest Service (Le Service des Eaux et Forêts), which was created under the 1935 legislation to implement the forest policy, recruited its agents from the military and the police. The role of the foresters as policemen rather than extension agents continued through the colonial period and persisted after independence. However, in 1981 the functions and tasks of the Forest Service were expanded beyond environmental protection to include forest management. The mandate then shifted to also comprise extension work and technical support to peasants, but still the background of most forest agents and the culture within the Forest Service were more compatible with the police function than with extension work (Brinkerhoff and Gage 1993).

The 1935 forest law has been revised twice in Mali; in 1968 (with only minor modifications) and 1986. Former president Moussa Traoré (1968–91) became a concerned "environmentalist" in response to the increased international focus on environmental issues of the 1980s. To impress donors, the revised forest law of 1986 was made even more severe, with extremely high fines compared to the income level in Mali. "Oppressive legislation during the eighties, such as high fines on cutting branches and on forest fires, was instituted because the president wanted to appear to be a staunch environmentalist to appeal to foreign aid for project funding" (Ribot 1995:32; see also Ribot 1996).

The fact that Malian environmental policy became stricter during the 1980s must again be seen in relation to the revival of the desertification narrative. In the Sahel, this narrative, claiming that increasing population pressure leads to general degradation of the natural resource base, has had a huge impact on the policies of donors and national governments. The desertification narrative, similar to other Malthusian crisis narratives of people-environment relations in Africa, has proven persistent, even in the face of strong empirical evidence against its storyline (Roe 1991, 1995). In fact, Swift (1996) attributes the robustness of the narrative to the fact that it was created by and was convenient for the interests of three main groups of actors: colonial and national governments, international aid donors, and some scientists. These actors are the winners, while the losers emerging from the desertification narrative are Sahelian farmers and pastoralists.

In Mali, the revised forest law of 1986 completely banned bush fires and made wood-saving stoves compulsory. Villages had to pay fines of 300,000 FCFA (francs of the Communauté Financière Africaine) (6,000 French francs [FrF] at the time) if forest agents discovered that there had been a fire on village land. However, fires are customarily set at the end of the rainy season in the Sudanian and Sahelian zones. People consider burning during this period as favorable to bushes and perennial grass. The moisture left in the soil together with the last showers of the season make the bushes send out new leaves and the perennial grass to sprout again. However, for pastures of annual grass burning does not make sense and may even be disastrous if important pastures are lost. Annual grass will not come up again before the next rainy season. Fires in the dry season are generally considered harmful and are not set on purpose. These fires may kill both trees and bushes. Hence, farmers and pastoralists use fire as part of a natural resources management strategy.

Fires change the landscape and contribute to the creation of a man-made environment. However, this environment is beneficial to the natural resources, that the people depend upon. If managed correctly, fires improve these resources. However, this was neglected when the Malian forest law was revised in 1986. The new law[7] created a general campaign against all use of bush fires, and President Traoré made the fight against bush fires one of his favorite themes. Ironically, the hill where the presidential palace is situated (la colline de Koulouba) burned every year at the end of the rainy season, a protest against Traoré's environmental policy that could be observed throughout the capital.

Sanogo (1990:2) defends the banning of bush fires and calls the fact that rural people "consider bushfires the ideal way to restore nature and safeguard the security of its human inhabitants" as "the sad reality which the technical departments of government know only too well, but for which a remedy still has to be found."[8] However, what used to be more of a sad reality was that forest agents often set fire to the land themselves in order to collect fines from nearby villages. In fact, Sanogo (1990) reports of a case where a forest agent was observed by villagers starting a fire.

The 1986 forest law also made wood-saving stoves compulsory. Households without such stoves were fined five thousand FCFA (one hundred FrF) beginning in March 1987. This policy was based on the assumption that widespread deforestation is caused by household fuelwood consumption. Such a perceived fuelwood crisis is often presented as one of the main environmental problems in Africa. Besides being considered a serious problem in the vicinity of urban centers, it is seen as particularly severe in rural arid and semiarid areas, where the demand for fuelwood is believed to be a major contributing factor to the reduction in tree cover (deforestation).

Wood is also frequently presented as the critical and scarcest resource in the Sahelian and Sudanian zones of Africa, with a carrying capacity that has

---

[7] Loi no. 86–42/AN–RM, Portant Code Forestier.

[8] Sanogo (1990:2) describes the attitude of the peasants vis-à-vis the use of fires in the following way: "Without bushfires the karité tree does not produce nuts, the animals graze poorly and grow thin, the villages are invaded by wild creatures such as snakes, and at the slightest false step all land is liable to catch fire and get burned out."

already been exceeded (Gorse and Steeds 1987; Pieri 1994; World Bank 1996).[9] I have earlier examined whether these generalizations, which the 1986 law was based on, are valid when confronted with local data from rural Mali (Benjaminsen 1993). The area chosen for this purpose was the Gourma region, which is situated in the northern part of the country. This is an area often referred to as an extreme case of general and widespread deforestation caused by rural household fuelwood consumption. However, no relationship between deforestation and the local use of fuelwood in the area could be found. The fuelwood was generally dry wood collected from dead trees. The observed deforestation was caused by drought. The fuelwood crisis in the Gourma is therefore related to economic rather than physical scarcity. Collection distances are increasing, as is money spent on fuelwood, but for the region as a whole there is no physical scarcity of fuelwood.

Later, a follow-up study was carried out in southern Mali (Benjaminsen 1997b). This time, the Koutiala area was selected, known to be the part of the country with perhaps the highest population densities and the most intensive land use. By comparing estimated wood production in the area with data on wood consumption, it was suggested that local use of fuelwood does not exceed forest regeneration in the area. However, where there is external pressure on the forest represented by commercial exploitation of wood for sale in the urban centers, fuelwood depletion might occur.

The fact that people collected only dead wood in the Gourma case and that there was virtually no cutting of wood for fuel in the area was well known to the local foresters. However, they abused the power given them by the national policy to fine people. According to the forest law of 1986, it was illegal to cut trees or collect dry wood for sale without permission from the Forest Service. Farmers were even required to secure permits to cut or use trees they had planted themselves on their own land. It was permissable to collect dead wood for one's own consumption, but local forest agents profited from their positions as controllers by arbitrarily imposing fines, even on people who had collected wood for their own consumption, and pocketing a portion of them. During my earlier study of fuelwood management in the Gourma region (Benjaminsen 1993), many of the women inter-

---

[9] Comparing sustainable and actual population densities in the Sahelian and Sudanian zones of West Africa, Gorse and Steeds (1987:13) claim that "the carrying capacity of the natural forest cover is lower than that of crops and livestock with traditional production techniques. The natural forest cover is therefore the most vulnerable part of the ecosystem; ... [and] the actual population already exceeds the sustainable population. The natural forest cover is not just vulnerable, it is already being severely over-exploited." This is said to be caused by household fuelwood consumption. Pieri (1994:94) concludes that "the Sudano-Sahelian zone has already exceeded its capacity to support the people and the Sudan is about to reach that point." This is again explained by excessive fuelwood demands compared to available resources. The World Bank (1996:24), discussing the Sudano-Sahelian belt, says that it "features one of the most rapid annual population growth rates of the continent, despite the fact that in many areas the mainly rural population (about 80 percent in the land-locked countries) is already beyond the carrying capacity at current technological levels. This growth has resulted in a downward spiral of extensive land degradation and fuelwood shortage."

viewed said that they did not collect wood for their own consumption for fear of being fined. This meant that many more people bought wood than would be the case in a less restricted situation. Because of the eagerness of the forest agents to impose fines, an important clandestine wood market emerged in the Gourma. Wood was bought either on the market or from wood traders selling at home or moving around with their merchandise. Some of these persons would sell legally, but most traders selling outside the markets did not have a permit. In the Gourma villages, up to 90 percent of the fuelwood trade was hidden when the survey was carried out in 1990. This rate varied from one local situation to another, depending on the attitude and eagerness of the local forest agent. Fines levied on consumers and traders varied between 5,000 and 25,000 FCFA (100 to 500 FrF or 20 to 100 US dollars in 1990). If the person could not pay, livestock could be confiscated or the "offender" imprisoned.

Because of the policy of harassment and the lack of dialogue with the local communities, the relationship between the Forest Service and rural Malians grew increasingly antagonistic during the Traoré regime. However, since the democratization process commenced in March 1991 the authorities have tried to attenuate this antagonism.

After 1991: decentralisation and *gestion de terroir*

In March 1991, the Traoré regime was overthrown following demonstrations and popular unrest that culminated in a coup d'état. A transitional government was established, and the process of democratisation began. During spring 1992, presidential and legislative elections were organized and the new government headed by President Alpha Oumar Konaré was inaugurated. One of the main aims of the new government has been to implement a decentralization reform.

Many of the implications of this reform regarding land tenure, natural resource management, and the redefinition of the role of the state are influenced by the policies of the major donors in Mali, for example, the World Bank and the French Caisse Centrale de Coopération Economique (CCCE) (Hesseling 1994). The donors undoubtedly have used the occasion to encourage the Malian government to undertake a major reorganization. In fact, these changes are not only encouraged but may sometimes be imposed by foreign aid donors (Mathieu 1994).

In the field of natural resources management, donor policies encouraging decentralization are based on the *gestion de terroir* (GT) approach, which has become popular in project planning in the Sahel. *Gestion de terroir* refers to the actual management by the rural population of its *terroir*, the village land area, including all resources, which is considered to belong to a given village.[10] The GT approach may concern one village with its *terroir* or a group of villages with corresponding *terroirs*. The main aim of the approach is improved environmental management, and it may be seen as a consequence of international donors' concern for environmental degradation during the 1980s as well as a result of the push toward privatization and enclosure of resources that swept Africa during the same period.

In agreement with the Common Property School,[11] the belief is that open access to pastures and forests as a result of state policy is the cause of environmental degradation. By enclosing the land, restricting access, and establishing formal rules regarding resource use, that is, by creating local commons, management is thought to be improved.

The GT approach may also be connected to a general change of conservation ideology within the Western environmental lobby during the 1980s as well as the new focus on people's participation in development thinking during the same decade. Important environmental organizations such as IUCN and World Wildlife Fund (WWF) moved during this period from "preservation" to "sustainable use" as the guiding principle in environmental conservation interventions. While conservation of the environment was still the main aim, the involvement of local people in conservation was acknowledged as necessary to achieve the objective. Earlier, an antagonism between people and interventions to conserve environmental resources had been created through the preservationist approach, as in the case of Malian forest policy. When the Forest Service lost its power after 1991, people reportedly, when they could not punish a forester, reacted by cutting down trees. A policy based on people's participation and making people responsible for their immediate environment would, on the other hand, prevent such antagonism and hence in the long run be more beneficial to the environment. The GT approach, which emerged in Francophone West Africa in the late 1980s, can be regarded as a result of these two new converging trends; sustainable use and people's participation.

Toward a new forest policy?

The new decentralization law, which produces the general framework for decentralization in Mali, points toward some kind of cooperative management (comanagement)[12] between the state and local communities. However, in this and other recent laws pertinent to the decentralization reform, it still

---

[10] However, this does not mean that the *terroir* is exclusive village property. Other groups (e.g., pastoralists) may also have claims on resources on the *terroir* based on established practice.

[11] *Common Property School* is the term used on the bulk of the literature criticising the tragedy of the commons model.

[12] According to Berkes (1997), in comanagement local people must have a stake in conservation and management and there must be a partnership between government agencies, local communities, and other resource users. The idea is to provide local incentives for sustainable use and to share power and responsibility for resource management and conservation.

remains unclear what the role of the local communities vis-à-vis the state, and the Forest Service in particular, will be. There appears to be some room for manoeuver here, and a lot will probably depend on the negotiation power of individual local institutions.

In some respects, the new forest law of 1995,[13] implies a more liberal policy. There is no mention of wood-saving stoves being compulsory, bush fires are allowed under certain conditions, local people have explicit use rights to dry wood for fuel, and only fallows older than ten years are defined as forest (earlier it was five years). However, fines are still high for the cutting of green wood without a permit from the Forest Service, and eleven species are now protected, as opposed to ten in the 1986 law. Several of these species are found on agricultural land and are traditionally protected by peasants because they are beneficial to crops (e.g., *Acacia albida*) or because of the economic importance of their fruits (e.g., the sheanut tree [*Butyrospermum paradoxum*] and the *néré* [*Parkia biglobosa*]). These trees can only be cut with special authorization from the Forest Service. The law does not, however, proscribe the pruning of farm trees, an important concession to farmers.

According to the decentralization reform, each commune elects a council headed by a mayor who is responsible for the administration of the area. Regarding environmental management, the state is only represented by a technical adviser in the commune. This adviser is employed by the government and not by the commune, and it appears that his or her role will be to control permits for use and sanctions for misuse of natural resources in addition to giving technical advice. However, some land belongs to the communes and some to the state, and it is not clear whether the function of the technical adviser will differ on the two types of land.

Following the introduction of the GT approach and the plans for decentralization, a number of donor-funded projects and programs were established with the aim of improving environmental management while building on local institutions. If they succeed, such initiatives will represent a new management option in Mali based on comanagement.

Village conventions in southern Mali

In the cotton zone in southern Mali, two French-funded programs Projet Gestion de Terroir-Développement Local (PGT-DL) and Cellule d'Aménagement de Territoire-Gestion de Ressources Naturelles (CAT-GRN), have been working with villages to establish formal agreements regulating natural resource access in the village *terroirs* that are approved by the authorities. Another initiative in this direction is the Siwaa project (Coulibaly and Joldersma 1991; Joldersma and Fané 1994; Hilhorst and Coulibaly 1996). In 1988, some villages close to Koutiala town complained that outsiders, basically

---

[13] Loi no. 95–004, Fixant les conditions de gestion des ressources forestières.

wood traders from the town, were exploiting their village *terroirs*. The villagers were unable to exclude the wood traders because of the 1986 law. A Dutch-funded erosion control program under the cotton company Compagnie Malienne pour le Développement des Textiles (CMDT) then proposed the establishment of a *terroir* test project. The Siwaa Committee is now a formalized cooperative body involving six villages. Each village has three representatives on the committee. However, the sharing of resources through formal cooperation between villages is a new idea in Mali, and it was not established in the Siwaa case without internal conflict. It appeared that five of the six villages had fuelwood scarcity, while the sixth, M'Péresso, had a surplus on its *terroir*. All the villages experienced problems with woodcutters from Koutiala town. However, M'Péresso people had the impression that the other five villages wanted to use the convention to get formal access to their *terroir*. The other villages, for example, suggested that M'Péresso only should sell wood in the Siwaa zone and not in Koutiala. In view of the higher prices in Koutiala, this was not acceptable to the people of M'Péresso, and also they had an interest in a local convention because it was made clear by the Forest Service that an enclosure of its *terroir* alone would not be acceptable (Hilhorst and Coulibaly 1996).

In 1993, the Siwaa Committee began to work out a common convention, and in March 1995 it was approved by all the villages.[14] The agreement was then presented to local Forest Service authorities for their approval. Similar agreements from the other donor-funded *terroir* programs mentioned above have also been submitted to the Forest Service. In addition, a number of individual villages (eleven according to Sissoko 1995) without external support have undertaken such initiatives themselves. These attempts at converting de facto open access areas into local commons were stopped by the Forest Service in 1995. The argument was that it could not approve conventions before the implementation of the decentralization reform (Hilhorst and Coulibaly 1996). The follow-up now depends on the local elections of 1999 and not least on the strength and negotiation power of the village institutions involved.

The ogokana experience

The Dogon people inhabit the central parts of Mali bordering Burkina Faso. The *ogokana* is an old village institution responsible for natural resources management among the Dogon. The word *ogokana* comes from *ogo* (power) and *kannu* (punishment). Hence, it is an institution with the mandate of enforcing local rules through sanctions. The scope of the traditional *ogokana*, however, goes beyond mere natural resources management. Its purpose is to

---

[14] It seems, however, that the leaders of M'Péresso are not satisfied with the agreement because during the delimitation of communes in 1996 they opted to join with another commune than the five other Siwaa villages (Hilhorst and Coulibaly 1996).

maintain order, peace, and mutual understanding through the management of village land and the resolution of social conflicts.

As already mentioned, the old forest law stated that all land not being cultivated (forests, pastures, and fallows older than five years) belonged to the state. The application of this law through the Forest Service weakened the role of local institutions in environmental management. With the political change that took place in Mali in 1991 and the subsequent government commitment to decentralization, local institutions have resurfaced as potentially important actors in natural resources management. In the Koro District (Cercle) in central Mali, there has been a spectacular revival of the *ogokanas* since 1991. This process has been facilitated by CARE, an international NGO intervening in the area.

In 1984, CARE launched an agroforestry project in the Koro District. As the project developed, CARE found a certain capacity, competence, and knowledge relating to agroforestry among local people in the Koro area and noted that the real constraint in this field was the forest policy practiced at the time. While people had at one time protected seedlings of *Acacia albida* and other species in their fields, the forest policy, at least during the 1970s and 1980s, had put and end to this. Even though the old forest law allowed the pruning of trees in one's own fields, the policy practiced by the foresters included fining people who did this, which led people to neglect the seedlings. This was identified as a major obstacle to natural regeneration in the fields. CARE then, in 1990, managed to convince the Forest Service to allow farmers to use their own trees. The decision considerably increased the participation of farmers in project activities.

Then, in 1992, two conventions were established in the Koro area between the Forest Service and the *ogokanas* of two villages (Youdiou and Guéourou) with CARE as a facilitator. In these agreements, the areas of competence of the *ogokana* are defined as: the protection of trees on the village *terroir*, management of the collection of dead wood for fuel, management of the cutting of green wood for local use, the pruning of trees, and the monitoring of bush fires and promotion of reforestation and other actions to protect the environment. Furthermore, the rights and obligations of the different signatory parties are stated as follows: the administrative authorities should guarantee the efficient application of the convention; the Forest Service has the role of support and control; and the *ogokana* should, in consultation with the village council, guarantee better management of human and natural resources. Finally, CARE's role was to permanently monitor the agreement and inform the authorities on the progress of the test.

Hence, since 1991 the NGO has played a central role promoting the *ogokana* as a competent and unavoidable partner in natural resources management. To the villagers, having a convention is an advantage because it provides an assurance against outsiders with *permis de coupe* (permits to cut trees, which can be bought from the Forest Service). However, in 1993 the two above-mentioned conventions were canceled by the Forest Service. According to the conventions, the two *ogokanas* could collect fines for infractions of the rules, but it seems that in practice few such fines had been collected. However, the Forest Service soon regretted the conventions, especially the mandate allowing the *ogokanas* to levy fines. This was perceived as threatening one of its major sources of revenue. In addition, a study carried out for CARE in 1993 questioned the legality of the conventions because the *ogokana* is apparently not a legal entity that can enter into legally binding contracts (Touré et al. 1993). According to the law, in order for an *ogokana* to become a party in a convention it must first be officially recognized and registered. Because of these two problems, CARE decided to suspend its support of the officialization of the *ogokanas*. It was also concluded, following recommendations from Touré et al. (ibid.), that the decentralization process would solve this problem because village authorities would play important roles in the management of the forthcoming communes. The canceling of the conventions and the lack of officialization of the *ogokanas* do not, however, seem to have had much impact on their activities. They continue to thrive and manage village land with or without a convention.

Conclusion

Studies within environmental history have demonstrated how European conservation ideology has influenced colonial, and later national, environmental policies in Africa. This influence has been particularly obvious in resource-rich East and southern Africa. However, as this essay seeks to demonstrate, also in the drier and more resource-poor parts of the continent, such as in the West African Sahel, the European conservationist influence has been manifest.

The colonial powers needed to justify their conquest of land and resources, and, as Neumann (1996:95) puts it: "The control over nature, either for aesthetic consumption or for production, must be recognised as an integral part of the geography and history of (the colonial power). ... Conquest ... is reflected in and reinforced by social and cultural constructions of property, aesthetics, and nature."

Malthusian ideas of environmental damage caused by local overexploitation had a central place in the colonial discourse on the African environment. This discourse was later inherited by national governments and used by groups within the state bureaucracy to maintain privileges and control over vital resources. The rise of the international environmental movement in the 1970s and 1980s and its discourse on population-induced "degradation" also represented a *coup de pouce* for these groups. The losers of the resulting policies have obviously been African farmers and pastoralists who depend on the natural resource base for their survival.

In Francophone West Africa, a preservationist approach to nature conservation was installed by the colonial power. This policy was based on the claim of serious environmental degradation taking place due to local over-exploitation. The history of forest legislation in Mali is traced in this essay. Social and ecological effects of various forest laws and regulations from 1900 to the present are analyzed. A system of permits for use and fines for rules violations was created by the French colonial government, and a paramilitary Forest Service was made responsible for implementing the policy. This oppressive policy, which persisted after independence and was further encouraged by the increased environmental "awareness" of the 1980s, implied extra costs and obstacles to local resource management. However, empirical studies of the use of forest products in Mali indicate that local management is not as destructive as the forest policy asserted it to be.

After the political change in 1991, a decentralization reform was introduced in Mali. This reform has been encouraged and partly pushed by international trends of privatization, enclosure of land and resources, and the disengagement of the state. However, the fact that decentralization to a large degree has been externally initiated in Mali does not mean that it cannot result in the improvement of local conditions. In contrast to the colonial experience, this external influence may lead to the devolution of more power in natural resources management to the local level. Local people are demanding more control over resources. On the other hand, influential groups within the state administration are seeking to conserve their privileges and power. Hence, it remains to be seen whether real decision-making power will be allocated to the local level or whether natural resources management will continue to be controlled and sanctioned by state officials.

The attempts to establish village conventions in southern Mali and the reemerging *ogokanas* are examples of local institutions demanding more control over land. Forest policy in Mali has already changed for the better in the sense that it implicitly acknowledges the competence of rural people in managing trees and therefore opens up for local institutions an increasing role in natural resources management. But, since the result of the decentralization process is being shaped by the actors involved in the process, the role finally played by local institutions will depend on the strength and negotiating power of each institution. It is therefore too early to say whether the process will lead to comanagement or deconcentrated state management.

References

Anderson, D.M. 1984. Depression, dust bowl, demography, and drought: The colonial state and soil conservation in East Africa during the 1930s. *African Affairs* 83(332):321–43.

Anderson, D., and R. Grove. 1987. *Conservation in Africa: People, policies, and practice.* Cambridge: Cambridge University Press.

Aubréville, A. 1949. *Climats, forêts, et désertification de l'Afrique Tropicale.* Paris: Société d'Editions Géographiques, Maritimes, et Coloniales.

Beinart, W. 1984. Soil erosion, conservationism, and ideas about development in Southern Africa. *Journal of Southern African Studies* 11(2):52–83.

Benjaminsen, T.A. 1993. Fuelwood and desertification: Sahel orthodoxies discussed on the basis of field data from the Gourma region in Mali. *Geoforum* 24(4):397–409.

———. 1997a. Natural resource management, paradigm shifts, and the decentralisation reform in Mali. *Human Ecology* 25(1):121–43.

———. 1997b. Is there a fuelwood crisis in rural Mali? *GeoJournal* 43(3):163–74.

Berkes, F. 1997. New and not-so-new directions in the use of the commons: Co-management. *Common Property Resource Digest*, July, 5–7.

Bonner, R. 1993. *At the hand of man: Peril and hope for Africa's wildlife.* London: Simon and Schuster.

Bovill, E. 1921. The encroachment of the Sahara on the Sudan. *Journal of the Royal African Society* 20:175–85, 259–69.

Brinkerhoff, D.W., and J.D. Gage. 1993. *Forestry policy reform in Mali: An analysis of implementation issues.* Washington DC: USAID.

Coulibaly, N., and R. Joldersma. 1991. *Réglementation de l'utilisation des ressources naturelles. Cas des 6 villages de la zone Siwaa de Koutiala.* Sikasso: Ministère du Développement Rural et de l'Environnement.

Elbow, K., and A. Rochegude. 1990. *A layperson's guide to the forest codes of Mali, Niger, and Senegal.* LTC Papers, no. 139. Madison: Land Tenure Center, University of Wisconsin-Madison.

Fairhead, J., and M. Leach. 1996. *Misreading the African landscape: Society and ecology in a forest-savanna mosaic.* Cambridge: Cambridge University Press.

Gorse, J.E., and D.R. Steeds. 1987. *Desertification in the Sahelian and Sudanian zones of West Africa.* World Bank Technical Papers, no. 61. Washington DC: World Bank.

Hesseling, G. 1994. Legal and institutional conditions for local management of natural resources: Mali. In *Land tenure and sustainable land use,* ed. R.J. Bakema. Issues in Environmental Management KIT-Agricultural Development, bulletin 332. Amsterdam: Royal Tropical Institute.

Hilhorst, T., and A. Coulibaly. 1996. *L'élaboration d'une convention locale au Mali-Sud.* Sikasso: Institut d'Economie Rurale.

Joldersma, R., and N. Fané. 1994. The Siwaa experience: Village land management in southern Mali. *ILEIA Newsletter* 10(2):22–23.

Mathieu, P. 1994. *Attentes et propositions des bailleurs de fonds pour la conférence régionale de Praia sur la problématique foncière et la décentralisation.* Documents de Travail, SAH/D(94)431. Paris: OECD/CILSS/Club du Sahel.

Neumann, R. 1996. Dukes, earls, and ersatz Edens: Aristocratic nature preservationists in colonial Africa. *Environment and Planning D: Society and Space* 14:79–98.

Peet, R., and M. Watts, eds. 1996. *Liberation ecologies: Environment, development, and social movements.* London and New York: Routledge.

Pieri, C.J.M.G. 1994. *Fertility of soils: A future for farming in the West African savannah.* Berlin: Springer-Verlag.

Ribot, J. 1995. *Review of policies in the traditional energy sector: Forestry sector policy report, Mali.* Washington DC: Africa Technical Division, World Bank.

———. 1996. Participation without representation: Chiefs, councils, and forestry law in the West African Sahel. *Cultural Survival Quarterly,* (fall):40–44.

Roe, E. 1991. Development narratives, or making the best of blueprint development. *World Development* 19(4):287–300.

———. 1995. Except-Africa: Postscript to a special section on development narratives. *World Development* 23(6):1065–69.

Sanogo, Y. 1990. *Zooforé: Friend or enemy of the forests? The viewpoint of the son of a Malian peasant.* Issues Papers, no. 15. London: International Institute for Environment and Development.

Sissoko, A.K. 1995. *CMDT et Comité Locale de Développement de Koutiala: Apercu du bilan et des perspectives des interventions pour résoudre les problèmes de développement au niveau du Cercle de Koutiala (1975–2000).* In Rapport PSS no. 10. Wageningen: AB-DLO and DAN-UAW.

Stebbing, E.P. 1938. The encroaching Sahara: The threat to the West African colonies. *Geographical Journal* 85:506–24.

Swift, J. 1996. Desertification: Narratives, winners, and losers. In *The lie of the land: Challenging received wisdom on the African environment,* ed. M. Leach and R. Mearns. London: James Currey and Heinemann.

Touré, M.D., A. Kanouté, and I. Sangaré. 1993. *Prise de responsabilité des institutions paysannes dans la gestion des ressources naturelles renouvelables.* Bamako: CARE International-Mali.

World Bank. 1996. *Toward environmentally sustainable development in sub-Saharan Africa: A World Bank agenda.* Development in Practice Series. Washington DC: World Bank.

Knowledge Claims, Landscape, and the Fuelwood-Degradation Nexus in Dryland Nigeria

Reginald Cline-Cole

The conceptual landscape

A landscape is a hybrid conception (cf. Richards 1995) and, following Mitchell's (1994:10) recent theorization, I understand it as "a certain kind of produced, lived, and represented space constructed out of the struggles, compromises, and temporarily settled relations of competing and cooperating social actors: it is both a thing and a social 'process', at once solidly material and ever-changing". I use the closely related concept of fuelscape to refer to an *unseen* landscape of scarcity and abundance of woodfuel,[1] whose contours are the product of cultural and socioeconomic factors (including institutional arrangements and their implications for effective legitimate command over fuel resources) interacting with demography and ecology (Mearns 1995; Wisner 1988).[2] Fuelscapes occur within and in part define and structure (agro-silvi-pastoral) landscapes. Together, they constitute what this essay refers to as (regional) forestry landscapes, which are understood to represent sites of contestation and cooperation for human agents and state agencies engaged in constructing, maintaining, and modifying wood-fuel- and other forestry-related discourses (cf. Sivaramakrishnan 1995).

Regional forestry landscapes thus evoke a multiplicity of meanings, which are commonly expressed in distinct but related or cross-cutting discourses or underlying sets of practices involving speech, writing, and action (Blaikie 1995). As particular areas of language use related to certain sets of institutions and expressing particular standpoints (Peet and Watts 1993:228), discourses are "frameworks that embrace particular combinations of narratives, concepts, ideologies and signifying practices, each relevant to a partic-

This essay is based on an article entitled "Knowledge Claims and Landscape: Alternative View of the Fuelwood-Degradation Nexus in Northern Nigeria", first published in *Environment and Planning D: Society and Space*, 1998, 16, pp. 311–46, Pion Limited, London.

[1] I use this term to refer collectively to fuelwood, charcoal, and other wood-based (or biomass) fuels; individual fuels are identified by name whenever necessary.

[2] But, while Wisner stresses only the importance of structural location in society in explaining variations in scarcity and abundance, there is little question that the capability and knowledge-ability of the actors in the "fuel game" are also important variables here.

ular realm of social action". And because discourses "vary among what are often competing, even conflicting, cultural, racial, gender, class, regional, and other differing interests" (228), this leads, following Foucault, to the existence of "contradictory" discourses within particular discursive strategies (Mackenzie 1992).

Blaikie (1995) suggests oral testimonies, religious ceremony and ritual, agricultural practice, scientific research papers, multilateral projects, the activities of nongovernmental organizations (NGOs), and World Bank documents as some of the more common "texts" of environment and development discourse. One could, of course, add the role of forests, plantations, and trees as both text and *con*text (cf. Rocheleau and Ross 1995). Yet discourse involves not just such symbols and their meanings, as well as material transformations of society and environment, but also the amalgam/infrastructure of ideas, and so on, from which such texts are composed. Forestry discourses, then, to paraphrase Sivaramakrishnan (1995), are historically contingent constructions through which relations of power and knowledge are articulated and which cannot be understood without reference to contemporary politics and power. Furthermore, the (discursive) practices that express such knowledge/power both produce discourse and "are embedded in technical processes, in institutions, in forms of transmission and diffusion" (4, citing Foucault 1977; cf. also Rocheleau and Ross 1995).

Thus, while the priorities of individual (agro-)foresters interact with those of the state in the production of regional forestry landscapes, the perspectives of the state exist *over*, as well as *alongside*, those of individuals in shaping and controlling landscape form and structure, usually through measures of regulation (cf. Yeoh and Hui 1995). But such hegemony is not self-securing; "it is constructed, maintained, and modified by human agents and state agencies through contest and cooperation" on regional forestry spaces (Sivaramakrishnan 1995:4).

To take one example, Adams (1996:167) argues that "truths" of official—in this case scientific forestry—discourse are "the means through which ignorant but enthusiastic outsiders make sense of complexity, the source of their confidence (often of course misplaced) that they can diagnose problems and prescribe useful solutions, and (of course) the standards by which short-term 'success' and 'failure', and hence career prospects, are judged" (see also Cline-Cole et al. 1990a; and Cline-Cole 1996). They permit, in other words, enthusiastic bureaucrats, scientists, conservation groups, forestry administrators, and managers to interpret regional forestry "with the experiential, technical, cultural and value-laden means" at their disposal (Blaikie 1995:203). More importantly, perhaps, they are implicated in state-constructed measures to shape and control regional landscape and "rationalize" forestry spaces, which are ordered, to varying degrees, according to the logic of prevailing systems of power. But there exist, too, other nonstate (and highly differentiated) discourses, which speak to local cultures, economies, social structures, politics, and so on, and which, like their "official" counterparts, are mobilized through rhetoric and action, multiply over time and in space, and are just as situated, partial, and/or misguided (cf. Schaefer 1995).

Put differently, forestry space is political and strategic, and

> is construed as a site of control and resistance which is simultaneously drawn upon by, on the one hand, dominant groups to secure conceptual or instrumental control, and on the other, subordinate groups to resist exclusionary definitions or tactics and to advance their own claims. (Yeoh and Hui 1995:185)

In the final analysis, however, the (re)production of Nigerian dryland forestry landscapes combines "formal [scientific] ideas and substantive local [indigenous] knowledge" in "interactive and contextualised" ways that make their uncritical separation distinctly unhelpful (Sivaramakrishnan 1996:147–48). Thus, I attempt in this essay, through examining fuelwood-degradation discourses within semiarid Nigeria, to show how dryland forestry, as a site of discourse, delineates the intersections of conceptions of science, visions of progress, and views of nature, among other things (cf. Schaefer 1995). I try to demonstrate how colonial and postcolonial forestry discourses change with shifts in power relations, as forestry activity (e.g., planning, protection, poaching, and regeneration) reconstitutes or elaborates on them and as they reshape each other (cf. Yeoh and Hui 1995). Equally, too, I try to show how discourses (and their associated instruments of power) demonstrate remarkable continuity, frequently spanning both the colonial and postcolonial periods and regional discursive formations.

In addition, given that the driving forces behind landscape (re)production are differentiation and struggle within cultures (Mitchell 1994), the essay also aims to highlight the social interests behind different forestry discourses and to demonstrate that, although several "valid" interpretations or plural "rationalities" of regional forestry landscape production, function, and "change" can and do exist,[3] these need to be seen much less as objective facts, hard and immutable, than as "successfully packaged and promoted 'knowledge claims'" (Blaikie 1995:206; see also Cline-Cole et al. 1990a; and Cline-Cole 1996).

Silviconsult: text and context

The essay is based largely on data from a survey of fuelwood use and farm tree management in northern Nigeria. Using the Silviconsult (1991) report as the principal (but by no means the only) "text" of dryland forestry discourse provides an opportunity for highlighting the way different discourses emphasize particular concepts at the expense of others (Peet and Watts 1993). It also offers possibilities for illustrating how "the construction of landscape is

---

[3] I use "change" here to refer to the monitoring and analysis of the landscape, including its "physical systems" base (cf. Agnew 1995).

inextricably bound to the exercise of control, to the willing closing off of perspective, and to a rationalization and naturalization of what is seen in the
view as a landscape" (Mitchell 1994:10). Above all, perhaps, the report is
ambiguous and (deliberately?) neither smooths out incongruities and incon-
sistencies nor erases oppositional voices and spaces of dissent (cf. Crush
1995). The social locations and historical context shaping its production thus
need to be clarified.

Social locations

The report is the product of an international consultancy. It is based on the
results of the largest and most comprehensive sampling of fuelwood consumption and patterns of use as well as of firewood resources (both standing wood volume and sustainable yield) and farm fuelwood production
ever undertaken in northern Nigeria. The socioeconomic component of the
survey, which is the focus of this essay, was designed and administered by a
multidisciplinary team of four consultants (of three different nationalities)
and seven group leaders (all geographers and all but one Nigerian), all of
whom had prior experience in woodfuel consumption surveys or other
household- and interview-based surveys. The team was dominated by uni-
versity-based researchers and teachers.[4]

One of the consultants (who also acted as lead author of the survey report) and two of the group leaders had previously been involved in what
Silviconsult (1991:20) describes as "a detailed, two year study of fuelwood
production and consumption in the Kano region, including an examination
of the urban energy system (household and non-household consumption of
fuelwood and other energies), the wood trade (supply and distribution aspects), the ecology of wood fuel in the Kano close-settled zone, and the
management of wood resources at the farm and the household level" (see
Cline-Cole et al. 1990a, 1990b). In the absence of competing narratives, their
results, thinking and conclusions seem to underwrite much of the socioeconomic analysis in the Silviconsult report, in addition to providing (some)
data and (a few) assumptions for the models used in estimating existing
wood resources and projecting their availability into the future.[5]

The Silviconsult survey itself covered all of Bauchi, Borno (now Borno
and Yobe), Kano (now Kano and Jigawa), Kaduna and Katsina and Sokoto
(now Sokoto and Kebbi) States as well as the northern part of Plateau State.

---

[4] I supervised data collection in two states, was involved in the design of both the survey questionnaire and sampling frame, and analyzed much of the data on the structure and functioning
of the fuelwood market. I would like to thank Francis Odoom and Mike Mortimore, who coor-
dinated the project, and the Afforestation Project Coordinating Unit (Kano) and Forest Management Coordinating Unit (Ibadan), the commissioning agents, for permission to use information from the study.

[5] The estimation of standing wood resources and analysis of regional land cover and use was
carried out under contract by a separate group of consultants.

Information was provided by an all-male sample of respondents comprising more than 2,100 household heads. Sample households were 76 percent rural (reflecting an assumed regional rural-urban population split) and were stratified on the basis of estimated size and distribution according to ecozones. Household fuel consumption was monitored, with supporting information on all relevant aspects of fuelwood use and management being obtained by means of questionnaire interviews and monitoring schedules. The structure and functioning of the urban fuelwood market was also investigated.

The survey's objectives, terms of reference, and expected output were quite specific and very circumscribed in nature. The instructions were to conduct a detailed analysis of the household fuelwood demand and supply situation as part of a wider program to arrest deforestation and desertification and to suggest policies for meeting this demand into the future as well as for reducing adverse environmental impacts linked to the satisfaction of this demand. Consequently, the project output was expected to report concisely on current household energy consumption patterns and preferences relative to the cost and availability of supplies; review both the potential for and means of promoting energy substitution away from wood and increasing efficiency in energy use; and, finally, recommend a strategy (to include policy measures, investment requirements, and institutional changes) for satisfying long-term household energy needs that would guide federal and state authorities, as well as donors, in planning fuelwood-related projects. It is thus crucial for the survey report to be situated in its historical context.

Historical context

Prior to the survey, forestry bureaucrats and managers, conservationists, rural development practitioners, NGOs, the media, environmental and development journalists, and so on, perceived fuelwood as being increasingly difficult to secure by northern Nigerian households, industries, and institutions. Rural villagers were reportedly walking longer distances to collect firewood. The nominal price of wood, where it is bought and sold, was said to be increasing. Urban supply radii, already wide in several cases, was reportedly expanding rapidly. In the absence of reliable estimates of fuelwood demand or supply, it was assumed that regional demand was outstripping supply and that this was causing environmental degradation, particularly "deforestation" and "desertification".

Although Silviconsult's own starting assumption was that the main causes of deforestation were land use conversions to agriculture and uncontrolled grazing, with problems of fuelwood supply being both a cause and an effect of wood resource degradation, it was the contrasting perceptions outlined above that dominated fuelwood-degradation discourses within state, federal, and international bureaucracies. During the 1980s, therefore, the links between forestry and environmental degradation in the drylands were the subject of national reports produced by the National Committee on

Arid Zone Afforestation (NCAZA) and the Consultative Committee on
Desert Encroachment (CCDE); both committees had, in formulating policy
priorities, depended heavily on data collected from state forestry departments.

At about the same time, the economist Dennis Anderson (1987), in a book
on the economics of dryland afforestation that uses Nigerian case study
material, evoked images of a marked decline in farm tree stocks, a threatened decline in soil fertility, the harvesting of tree stocks without replenishment, and significant encroachment on and degradation of forest and game
reserves. Citing a reported two-thirds decline in regional farm tree densities
since the 1950s for support, he concludes that "the overall picture is that the
loss of trees is as widespread on farms as in forests and woodlands" (30).

Furthermore, and according to one state government source, fuelwood-
induced degradation had by 1989 become "a very well-recognised problem
posing serious threats to the lives and well-being of crops and forests as well
as animals and man himself", because of its "massive and wanton destruction of trees and vegetation" (KSCASE 1989:1). And, with seeming support
in the form of Food and Agricultural Organization (FAO) claims that about a
thousand hectares of dryland forests are cleared annually to provide firewood and charcoal (cited in Osemeobo 1990), it was but a short step from
this to the observation, this time credited to the Federal Department of
Forestry, that some 12.5 million ha of the drylands were affected by
"desertification" (FDFALR 1989, cited in Hyman 1993).

This "desertification jump", as Mortimore (1989a) so evocatively describes it, was just as neatly executed by conservation organizations, politicians, civil servants, development practitioners, the press, and NGOs. NEST,
the Nigerian Environmental Study/Action Team, an indigenous environmental NGO, for instance, speaks of "extensive desertification" caused by
intensive agriculture, heavy grazing, bush burning, and firewood and
browse collection and identifies the "overexploitation for firewood" as a
major cause of deforestation:

> [S]hortages of fuelwood ... are already critical in the ... Sudano-Sahelian States ...
> where nearly 75 percent of the total cooking fuel is derived from plants [where]
> people have resorted to burning cow dung and farm residues, which should
> normally be recycled to the soil as badly-needed manure [and where] scarcity of
> fuelwood is also leading to the modification of cultural practices. (1991:164).

Mortimore (1989a), the consultant social scientist on the Silviconsult survey,
had earlier noted the existence of "a general consensus amongst government
officers [including state governors], and *especially among forestry officers*, that
general environmental degradation is occuring" (emphasis added, p. 41).[6] I,

---

[6] According to Mortimore, the proceedings of the deliberations of the Consultative Committee
on Desert Encroachment, in which these various actors participated, cite the following indicators of desertification in the drylands, in diminishing order of frequency: declining rainfall,

too, recall that starting in the early 1980s, for example, it became fashionable for senior dryland forest administrators to emphasize that existing forest reserves were with few exceptions more accurately described as "sand reserves". The "desertification consensus", of which such attitudes were an integral part, and which holds that between 55 percent and 75 percent of northern Sokoto, Katsina, Kano, Jigawa, Yobe, Borno, and Bauchi States, along with smaller proportions (30 to 45 percent) of southern Bauchi and Plateau States, are "desertified", was much influenced, in Mortimore's opinion, "by the forestry profession's view of environmental degradation, and the difficulties experienced in afforestation programmes" (1989a:9).[7]

But there were other texts, too. It is worth recognizing, for instance, that for dryland inhabitants the years since the 1970s have coincided with changes in rainfall, the most critical climatic parameter in the region, whose totals have diminished and monthly distribution altered at the same time as drought has become persistent and that this period has consequently seen the environment of cultivation, livestock, and forestry systems become fundamentally altered (Kimmage 1990; Lockwood 1991; Mortimore 1989a; Mortimore et al. 1990; Stock 1978). Consequently, villagers have referred variously to the land "dying" (Stock 1978) and the increasing "cost" of drought, erosion, and desertification "finding their way into villages" (Anon. 1989b). "[R]esponsible community leaders down to the village level and ... many individual resource managers report increasing shortages of firewood, construction timber, farmland and grazing land; movement of surface soil; localised gully erosion; and increased mobility of the human and livestock populations" (Mortimore 1989a:7).

By the mid-1980s, and excluding Nigerian government responses such as the federally funded Arid Zone Afforestation Programme, which began in 1978–79, such expressions of concern with fuelwood and degradation in the drylands had elicited significant international responses-as-text, notably from Washington. The extension services of regional agricultural and rural development projects, which were cofunded by the World Bank, for instance, promoted tree planting during this period. Furthermore, in 1987 the bank funded a social forestry program, which was designed to propagate and distribute to individuals, institutions (secondary schools), and NGOs for planting some five million forest and fruit tree seedlings annually. But a

---

with persistent drought and associated dust storms and low humidity; fluvial and aeolian erosion, loose or fragile soils, and active sand dunes; deforestation, scanty vegetation, tree mortality, the invasion of dryland species, and fodder scarcity; diminishing groundwater and the drying of surface ponds and streams; falling farm yields; southward migration of livestock producers and/or farmers; and food scarcity and increased human mortality. And yet, as he argues (1989a:42), any such consensus "rests on an inadequate empirical base [and] should be accepted with caution" because wind and water erosion have not been "measured in enough places ... to provide a basis for regional assessments of soil loss". Overall, he concludes, available data do not allow conclusions to be drawn on soil degradation at the regional level.

[7] Field foresters and forest guards in particular complain about poor seedling survival rates, slow seedling growth, unauthorized grazing and firing of plantation undergrowth, and so on.

considerably more substantial response, this time in collaboration with the federal and state governments, was the initiation of the Afforestation II Project, a multi-million-dollar enterprise designed to stimulate the expansion of afforestation activities across the drylands. Its specific objectives included strengthening the strategic base of the forestry sector; stabilizing soil conditions in threatened areas and improving the supply of fuelwood, poles, and fodder by supporting farm forestry and shelter-belt activities; and increasing the supply of industrial wood through improving the management of existing high-priority plantations and establishing new ones.

Indeed, it would appear that the principal text analyzed in this essay, the Silviconsult report, owed its existence to the birth of Afforestation II, for it was to provide desperately needed input into the program's principal goal of expanding afforestation, that the survey that eventually became Silviconsult was commissioned. Nonetheless, it is worth noting, as a close reading of the Silviconsult report clearly reveals, that the commissioning of the survey was but one of a number of environment-development initiatives of the time. These included the adoption of a national energy policy in 1987; the promulgation of countrywide Environmental Protection and National Resource Conservation Decrees in 1988 and 1989, respectively; and the drafting of a National Conservation Strategy. In addition, and within the drylands themselves, other initiatives such as the Katsina Afforestation Project and the forestry/tree-planting components of the North East Arid Zone Development Project (NEAZDP) were funded separately by the (then) European Economic Community.

The wider discursive context

Much of the foregoing merely reaffirms that forestry discourse, like other forms of environment and development discourse, "permeates and justifies very real interventions and practices with very real consequences" (Crush 1995a:6). At the same time, the variety of means and methods used to achieve these ends, "the forms in which [discourse] makes its arguments and establishes its authority, the manner in which it constructs the world" (3), all need to be brought into focus.

For instance, I still recall a senior forestry administrator's very public reaction to suggestions by a younger (and recently trained) field forester for greater decentralization of control over, and management of, regional woodlands to include a truly meaningful role for local community participation. "I must apologize on my colleague's behalf", the administrator said to participants at the national workshop where this drama was played out, shaking his head somewhat wearily; "he is a young man who does not know that forestry is ultimately about power and control". In the event, the young forester must have understood who held power, and who dominated whom; he did not contradict his *oga kwata kwata*,[8] despite the very public nature of this crushing reprimand (cf. Crush 1995a).

I recall, too, conversations with regional forestry managers, administrators, and technicians, which took place both before and as part of the Silviconsult survey and during which the view of the existence of an acute and worsening regional fuelwood scarcity and desertification situation was repeatedly expressed. Apart from the evidence for the continuing conversion of woodland/forest into agricultural land and an increasing inability to effectively police reserves and profitably manage plantations, little more than references to United Nations and FAO maps (or a variety of derivations) of the assumed extent of desertification, and the spatial distribution of fuelwood demand and availability, were adduced in support of these claims.

It was significant that these claims were presented as either incontrovertible "proof" or seemingly uncontested and uncontestable "facts", for the "data" that would have gone into the construction of such texts had to have come initially from poorly resourced, local-level bureaucracies like those in the drylands, which employed the foresters in question, and which, like these Nigerian institutions, would have been (most) unlikely to have had adequate and reliable information to report in the first place. That the commissioning of the Silviconsult survey was, in itself, a recognition of such a shortcoming was readily and openly acknowledged. But this realization was not in any way allowed to interfere with the ultimate institutional goal of attracting (major) new investment into dryland forestry. Indeed, the irreplaceable value of gap and desertification discourses to the achievement of this goal was both readily appreciated and single mindedly, yet extremely skillfully, pursued. Ambiguity and contradiction of this kind bear testimony to the embeddedness of the local dramas of fuelwood and dryland forestry.

Consider woodfuel gap discourse. While there is a long history of attempts to assess, measure and predict levels of fuelwood consumption and supply in given localities in the drylands (Barrott 1972; Davies 1976; Nash 1941; Trappes-Lomax 1952; Trevallion 1966), regionwide surveys of savanna fuelwood consumption and regional supply are rarer and (sometimes considerably) more expensive undertakings (Grut 1972; Silviconsult 1991; Thulin 1970). Nonetheless, regardless of the scale of geographical coverage, the results of fuelwood surveys have always possessed tremendous potential, as "scientific evidence", for use within discourses of "rational" forestry and fuel efficiency.

Not surprisingly, these exercises in supply-demand analysis have, like other interventionist projects designed to (provide justification for) restrict(ing) or prevent(ing) firewood collection, or expand(ing) supplies, tended to take place during periods of heightened (pan-African and global)

---

[8] The expression is Nigerian pidgin, which loosely translates as "overall boss".

environmental awareness and development consciousness (Anderson and Millington 1987). During such periods, extra resources have been made available, specifically to fund them. A major objective of the Silviconsult survey, for example, was "to assess current and future household fuelwood demand and supply, and to suggest policies for meeting this demand and reducing adverse environmental impacts" (1991:15).

Historically, surveys like Silviconsult have tended to emphasize actual or potential shortages, and the latter's perceived negative social and environmental consequences (see, inter alia, Barrott 1972; Davies 1976; Grut 1972; Nash 1941; Thulin 1970; and Trevallion 1966). This is not merely coincidental. Such surveys have consistently been informed by woodfuel gap thinking, which, according to Mearns (1995:104),

> compare[s] total woodfuel demand, based on estimates of average per capita consumption, with standing stocks and annual growth of trees. A woodfuel supply shortfall [gap] is typically identified. ... Since consumption has to be met from somewhere, it is assumed that it is made up by cutting into tree stocks. The shortfall is then projected into the future, usually in direct proportion to population growth. As consumption rises and trees are felled, the annual growth falls, the shortfall grows and tree stocks are inexorably depleted. Woodfuel demand is therefore assumed to be the prime cause of deforestation ... in turn contributing to accelerated soil erosion and "downward spirals" of environmental degradation.

The study that arguably did the most to legitimize gap theory in Africa, *Fuelwood Consumption and Deforestation in African Countries* (1984), was co-authored by Dennis Anderson and Robert Fishwick, both of whom have Nigerian dryland forestry connections.[9] Not surprisingly, therefore, the classic recent example of the application of gap theory to the drylands remains the former's previously mentioned *The Economics of Afforestation: A Case Study in Africa* (1987), which builds on ideas in the earlier collaborative study. In summarizing gap thinking with reference to the Nigerian drylands, this text highlights perfectly the remarkable degree of "rationality" that permeates discourse strategies of this kind:

> The area has a harsh climate with a long dry season and periodic droughts. ... The estimated population is 20 million, and there is a dense livestock population as well. The destruction of trees in woodlands and on farms because of the demand for fuelwood, agricultural land, and livestock fodder has made the soils more vulnerable to dessicating winds. There is a large fuelwood deficit and a risk of a further decline in the already low productivity of agricultural land. (29)

Accordingly, Anderson continues, fuelwood use and degrading soils interact with cultivation, livestock rearing, and desiccating winds in such an evident, logical, common sense, and mutually reinforcing way that "Many offi-

---

[9] Fishwick worked for several years in the regional forestry service, while Anderson acted as a World Bank consultant on a mission to the area in the 1980s.

cials and community leaders [have come to] recognize the severity of the situation, [so that] afforestation in the arid zone is now regarded as having a high economic priority" (29). Certainly, for the village head of Babamutum in northern Katsina, one such community leader, "persistent and merciless" forest destruction during the 1950s, 1960s and 1970s meant that "my major pre-occupation when I climbed the throne [some twenty years ago] was to create awareness of the situation [of the high cost of firewood, lack of rainfall, wind/rainstorms, soil/wind erosion, drought and desertification, etc.]" (Anon. 1989b:29). Within gap discourse, closing perceived supply-demand gaps to bring fuelwood consumption and tree resources (back) into balance requires tree planting, frequently on a vast scale (Leach and Mearns 1989). Consequently, the village head claims, he "mobilise[d his] people to begin massive tree planting and enact[ed] a by-law against the cutting down of trees" (Anon. 1989b:29).

But to return to Anderson for the moment, his conclusions regarding the existence of a large and growing regional fuelwood deficit were in fact based on "a very broad quantitative estimate of the rate of tree loss in the region because neither an inventory of stocks nor measurements of their growth ... have been undertaken recently". And the main source of this estimate? Field reports by foresters that were "not comprehensive", were "highly approximate", and relied "more on the foresters' judgment and experience than on precise measurements". But this mattered little, according to Anderson, for "the defects in the information ... do not obscure the changes taking place. ... Only the precise magnitudes and locations of the losses in tree stocks are in doubt, not the general trends" (all quotations are from 1987:31). In effect, his fuelwood gap was largely *assumed* rather than convincingly *demonstrated*.

This particular discursive device was widely used during the period of heightened environmental awareness spanning the decades of the 1970s and 1980s. Indeed, by the mid-1970s, one retired provincial forestry officer was confident that "it hardly needs surveys to show the increasing shortage of timber, poles and fuel in most parts of the north [of Nigeria]" (Horsman 1975:70). It was apparently enough to reiterate, preferably authoritatively, the potentially calamitous consequences, that would result from a failure to apply (re)afforestation remedies (cf. Leach and Mearns 1989) or, better still, to "demonstrate" the benefits of tree planting (here claimed in the late 1980s by the Babamutum village head in a discussion with representatives of a locally based but internationally funded afforestation project):

> [Now that] the whole environment [of Babamutum] is covered with trees ... flood and erosion have been reduced, there is now no frequent destruction of houses by wind and rainstorms. Effects of drought and desertification have been on a diminishing pattern which is a most striking achievement, and we have had a bumper harvest this year [1989], while other communities ... have been severely battered by drought and have had a poor harvest. Right now ... the price of fire-

wood has gone down drastically. The natural environment is stable and my people are very happy about this. (Anon. 1990b:30)

And yet the conceptual basis of gap theory is far from unproblematic. Contradictory discourses summarized by Leach and Mearns (1989:6) hold, for example, that serious practical flaws in this theory lead to exaggeration of the size of woodfuel gaps where these actually exist and, by the same token, to exaggeration also of the need for planned forestry interventions. Yet, they note, "this numbers game is played with weak numbers [and,] while this fault is widely acknowledged, the game continues and its conclusions continue to be taken with great seriousness".

This assumes particular significance in the Nigerian case, for the first officially accepted census conducted since 1963 took place in 1991 and revealed a national population of 88.5 million people. According to Gina Porter (1992:371), this was a "considerable surprise as population was expected to have reached about 120 million". Indeed, policy making during the intercensal years when the Silviconsult survey took place was, an International Labour Organization report observed in an apposite metaphor, akin to "trying to run through the forest in the dark without a torchlight" (371). It is significant, therefore, that Silviconsult itself acknowledges that the population estimates used for distributing sample households during its survey were not only outdated but were probably also widely off the mark.

In any case, Leach and Mearns (1989) argue further that it is commonly the case that fuelwood collection is not the principal cause of deforestation; and that the assumption of gap thinking, that consumption increases in line with population, even as supplies dwindle to vanishing point, is unrealistic, for "as scarcity worsens and wood prices or the labour costs of gathering fuel increase, many new coping strategies [fuel switching, tree planting, seedling protection, etc.] would come into play" (13). Yet, while Anderson (1989:13) recognizes the probability of such fuel switching ("as fuelwood becomes scarce, real costs and prices rise, and people turn to substitutes or otherwise reduce consumption") and acknowledges the role of agriculture and grazing in vegetation clearance, he maintains that the firewood "consumption rate is thought to increase *exponentially* with population growth" (emphasis added).[10]

Such neo-Malthusianism clearly serves to establish the conceptual foundations of his case for the existence of a fuelwood gap, to illustrate how landscape functions ideologically "to present, in a selective way, social interaction to the viewer" (Mitchell 1994:10) and to highlight the social interests behind gap discourse in this case. In particular, it is worth noting that the numerous forestry interventions of the 1970s, 1980s, and 1990s, many of which were premised on a need for accelerated farm forestry and shelter-

---

[10] See Cline-Cole et al. 1990a for a demonstration of how and why Anderson's position on this issue is manifestly untenable.

belt and watershed planting, would have been extremely difficult to justify without the demonstration (and, in its absence, the assumption) of the existence of a (growing) regional fuelwood gap and its perceived environmental and economic implications (frequently expressed as a welfare issue). Thus, while it is possible, it is extremely unlikely that bank policy toward, and investment decisions about, Nigerian dryland forestry were not influenced by the work of gap theorists. Indeed, in the case of Anderson and Fishwick, such work was carried out under bank auspices and preceded (possibly unrelated?) cycles of bank investment in dryland forestry.

But Nigerian social interests were served, too, by gap discourse. From the point of view of both federal and state governments, international intervention in dryland forestry was a reflection of official success in attracting development funds into both the national and regional economies at a time when the country was experiencing serious financial problems. In turn, this was "sold" to the public as a manifestation of official support for, and commitment to, the country's forestry (and wider natural resources) sector and as a continuing reaffirmation of the country's "green" credentials. Thus, from a governmental point of view, green capital of this kind also had tremendous value as political capital.

But what about individual interests? Here cursory examination of the social composition of Nigerian foresters who (have) administer(d)/ manage(d) the programs and projects established during and since the 1980s suggests that these were (or would have been) the same professionals who, as federal and state forestry employees, (would have) provided local insights for the analyses, usually based on gap thinking, that established the feasibility of the projects in question. Clearly, the transition from underfunded state and federal forestry services into these more attractively remunerated parastatal sector positions could not but have enhanced professional and personal careers, prospects, and goals.

Similarly, too, after positioning himself as a defender of green land use practices in the passage previously cited, and reinforcing this by underlining the valuable demonstration effect of his village's "success story",[11] Alhaji Usman Uban Dawaki, the Babamutum village head, appealed for—and succeeded in obtaining—greater rural development assistance, particularly investment that would provide support for ongoing forestry initiatives in his village and thereby reward his people's "cooperation [in] and dedication [to]" their "collective" project of greening the local landscape (Anon. 1990b:30).

Taken together, the various discursive strategies deployed above illustrate how "in a regional discursive formation even competing notions often use the same metaphors, interpret in similar ways, perhaps even think with

---

[11] He claimed, for example, that he dispensed forestry advice to other (very impressed) community heads on request.

similar logics", and act as reminders that "discursive formations grounded in material, political, or ideological power supremacies demonstrate a continual tendency to extend over spaces with greatly different characteristics and discursive traditions" (Peet and Watts 1993:231). At the same time, they resonate with the Foucauldian position that power circulates, "is employed and exercised through a net like organisation", and may be conceived of as one of "a multiplicity of discursive elements that can come into play in various strategies" (cited in Mackenzie 1992:693, 694).

It is worth emphasizing, in concluding this section, that neither the discourses above nor the thinking that underwrites them owe their existence to an identifiable cast of actors, or interests and aims, that can be described as *purely* local, regional, or national. This is hardly surprising, for as Crush (1995a: 6) notes,

> the immediate institutional or broader historical and geographic context within which [dryland forestry] texts are produced ... is global in reach, encompassing departments and bureaucracies in colonial and post-colonial states ..., Western aid agencies, multilateral organisations, the sprawling network of NGOs, experts and private consultants, [and] institutes of learning.

Nonetheless, Silviconsult notes that, although dryland forests and forestry now receive more attention from both state and federal governments than previously, forestry is still assigned a much lower financial priority than agriculture and other social needs and that some rural and regional development policies, for example, favor grazing and agricultural land at the expense of forest and woodland. There might be a good case, then, for seeing both the genesis of Silviconsult and the nature and dynamics of the wider discursive context described as responses to such challenges and attempts by a variety of social interests to be formative in outlook—that is, to employ the power of discourse in setting up and framing fuelwood-degradation-desertification linkages as "problems", and to draw selectively from (sometimes fiercely) contested notions in support of a central role for forestry and (agro-)foresters in the resolution of them.

The institutional context of dryland forestry

The history of (in)direct state intervention in regional life dates to precolonial times when a sometimes bewildering array of locally sensitive and community-based institutions and mechanisms regulated land, water, and forest resource access and use, notably in the densely populated close-settled zones of what was overall a highly stratified and inequitable society (Cline-Cole 1994, 1996). Yet, although this did not imply the existence of sweeping and overriding state power over forestry resources, both the colonial and postcolonial states have used such precolonial antecedents to justify intervention, frequently of a qualitatively different kind, in both land and its associated resources (Lugard 1970; Egboh 1985).

Consequently, even though the circumstances surrounding the establishment of a regional forestry service were heavily influenced by the personal political agenda of Frederick Lugard, the region's first high commisioner, he also drew selectively on prevailing elements of imperial environmental and economic concern to justify forestry intervention (Egboh 1985). Indeed, his entire forestry project was underpinned by the clearly expressed beliefs that, first, only the colonial state was capable of both appreciating the full extent and significance of the indirect benefits of forests and of planning effectively for their sustained management, and, second, that local populations were both ignorant of and apathetic toward environmentally sustainable forestry practices (Lugard 1970). Such "beliefs" were widespread within British colonial forestry (Rajan 1994).

Nonetheless, unlike both southern Nigeria and British India, where forestry was perceived as (potentially) capable of making significant direct (export-earning) contributions to regional or wider imperial wealth, early forestry intervention in northern Nigeria was justified largely on the grounds that perceived uncontrolled deforestation (and resulting soil erosion, falling groundwater tables, and desiccation) posed an implied threat to the economic basis of colonial rule. In particular, and as the early years of colonial occupation coincided with a series of poor rainfall years, both the establishment of a regional forestry service and earlier (more sporadic) government forestry interventions were ostensibly designed to arrest deforestation (particularly that caused by shifting cultivation) across the region; counter desiccation in the northern margins of the drylands; and secure future supplies of fuelwood, both for towns and cities and for rural populations in densely populated areas of permanent cultivation (Cline-Cole 1996; Egboh 1985).

To my mind, such an overwhelmingly utilitarian view of regional forestry informed interventions designed more to enhance the physical environment and prevent regional and local timber and fuel shortages than to improve or secure livelihoods directly. Areola (1987:281) appears to share this assessment. Taking the long (colonial and postcolonial) view, he concludes that "it is in the field of forestry that Nigeria has made the most conscious and discernible efforts at conservation". In contrast, however, Morgan believes that livelihood concerns were paramount; he is of the impression that environmental protection during the colonial era may "have been subordinate to the aim of managing the forested or wooded areas to maintain and improve the supply of firewood, poles and timber" (1983:53). Nonetheless, I suspect that we would all agree that a conservationist rather than preservationist "ethic" has historically dominated Nigerian dryland forestry thinking and practice (cf. Rajan 1994).

Such contradictory assessments of the agenda and concerns of regional forestry are due, in part at least, to wide geographical variations in policy design and implementation, in the fine detail of management practice, and in the changing mechanisms used in the pursuit of policy goals. In part, too, such differing interpretations reflect contradictory "readings" of state forestry projects and, equally importantly, of the latter's self-(re)presentation(s), which is characterized by ambiguity and contradiction.

Thus, although Lugard acknowledged early that there were few local forest products of export significance, he also argued forcefully that this shortcoming was offset by regional forestry's crucial import-substitution role, stating that "the comparative value of forest produce cannot be measured solely by the value of exports" (1970:437). In satisfying a "native" demand for fuel, building materials, and so on, and in supplying the government with timber for office, house and railway construction as well as furniture making, he observed, forestry was reducing the region's import bill and in the process keeping taxation levels low (437). Lugard also noted that forestry's "value to the Natives [wa]s very great, and increase[d] as their standard of life improve[d]"(436) and acknowledged the existence of pre-colonial examples of desirable (because "intensive") agriculture combined with tree management as well as flexible community-based forest access control mechanisms.

Yet in a bid to justify the introduction of "modern" forestry and distinguish it from the "primitive techniques" it was supposed to replace (cf. Rajan 1994) he was obdurately insistent on the need to "educate" dryland inhabitants about the benefits of forestry, particularly forest conservation and to "impress [the value of forestry] on the Chiefs and the more intelligent members of the community, in order to gain their willing co-operation in forestry work" (Lugard 1970:436). To complicate matters further still, in practice management appears to have been influenced (contra Lugard) by a perception that local "people generally seemed to appreciate the practical effects of woodland destruction, and to have recognized more readily than people in [the forest-rich] South the need to protect the woodlands" (Egboh 1979:9).

Nonetheless, forestry's "modernizing intent" has persisted. Well into the colonial era, for instance, a forestry official still identified the principal aim of regional policy as "the establishment and maintenance of a balance between forestry and agriculture as integral parts of the rural economy ... and to persuade people to a wise use of land" (Kerr 1940:22). And currently forestry policy "aims at achieving self-sufficiency in forest products through the employment of sound management principles and techniques as well as mobilising human and material resources" (Oduwaiye 1995). Along with the other components of state forestry, policy has emerged as a tool for restructuring livelihoods, reconfiguring landscapes, and reinscribing the relations of power among state forestry institutions and agents and between them and dryland inhabitants (cf. Rocheleau and Ross 1995).

Consequently, new approaches to forestry management have been introduced. Over time, there has evolved an administrative structure, a legal code, and a body of scientific practice as instruments of power (cf. Sivara-makrishnan 1995). A series of ordinances, laws, and acts established and consolidated the framework for the management of the forest estate and the protection of trees on unreserved land. Here, as elsewhere, these legal instruments not only divided regional landscapes into reserved, protected, and communal or village forests, but they also gave power to foresters "to determine how [these and other] forests were to be managed; provided for control, not only of state-owned lands but over forests and lands not belonging to the state"; and (re)defined and restricted tenurial rights in new and often complex ways (Rajan 1994:141).[12] And, as in British India, "at the time of demarcation and protection, grazing, fire, shifting cultivation and the collection of firewood rapidly became stock items on the standard list of ills to be curbed" (Sivaramakrishnan 1996:150).

Significantly, however, "commodification of [forestry] time and space" (Jewitt 1995:72), so closely associated with the introduction of British forestry into India in the 1860s, long predated colonial occupation in the case of the drylands (see Cline-Cole 1994). Consequently, as Jewitt (1995) concludes in the case of India, at no time under colonial or neocolonial rule have "other" discourses been incapable of representing themselves, and of offering their own critiques of state forestry projects. The regional geography of the forest estate offers a richly suggestive "text" here, with reserved and protected areas occupying spaces or sites that could be easily appropriated at the time of the estate's establishment rather than those that were (and in some cases remain) in greatest need of "environmental" protection. As Howard (1976) observes, forestry administrators recognized early that the existence of vast areas of little or no human population presented opportunities for rapid and unproblematic forest and woodland reservation.[13] In another, more recent example, Anderson (1987:36) describes how many Kano farmers "acquired free [tree] seedlings [for afforestation in order] to use the potting soil as fertilizer for their millet (the potting soils were specially prepared at the nurseries and included manure)".

Nonetheless, the regularized layout and regimented management of plantations and reserved and protected forests did add new and alien dimensions to a regional forestry landscape earmarked for transformation according to Western scientific forestry principles: first, by creating new spaces of work, production, and, it was hoped, recreation; and, second, through a concerted (and ultimately futile) attempt to create discrete spaces or areas for farming, forestry, and livestock herding (cf. Jewitt 1995). Of the

[12] A Katsina indigene remembers, for example, the Nabagudus and Mallam Iro Inkos as "veritable forest [reserve] guards and supervisors [during the 1950s] at [which] time it was Iro Inko and Nabagudu; not the hyenas and leopards; that were the 'terror of the forests'" (Katsina 1989:3).

[13] As the forest reservation program in the South had had to contend with sometimes violent protest by and sustained resistance from local communities and their leaders, the Northern Region forestry establishment was desperate to avoid similar confrontations.

new modern forestry landscape components (industrial-type plantations and village forests and fuel areas managed according to specially designed working/management plans), the most alien was the even-aged, exotic tree, monoculture plantation. Within scientific forestry, it was iconographic, an important landscape symbol, concrete evidence of the diffusion of biological Fordism as "modernisation" (cf. Chambers 1988), a concept that still has not entered the local lexicon.[14]

Thus, for a Nigerian colonial forest assistant who came originally from the so-called Middle Belt to the south of the drylands but was stationed in Kano in the 1940s, dryland inhabitants may have practiced "a kind of rotation" and "appl[ied] volume increment methods" prior to and during colonial rule, but only in ways that fell short of the methods used in "our well organised plantations" (Adelodun 1940:30). The methods in question were based on the "scientific" principles of minimum diversity, sustained yield, and the balance sheet, as outlined in working plans, the new forestry management texts (Rajan 1994; Lugard 1970).[15]

But a "plantation lobby" within colonial forestry did not command universal support. Kerr (1940:21), a rare early dissenting voice, for example, was of the opinion that too much valuable time was spent on establishing small plantations for local fuel supply, whose cost "place[d some of] them in the category of skeletons in the departmental records". In his view, the

> subconscious motive for starting many of these plantations was rather to provide a form of "window dressing" than to meet any real economic need for woodlands raised by artificial means. (21).

Support for this view was very long in coming. It would be three decades before Grut (1972:15), a forestry consultant in an independent Nigeria, would reiterate: "Scarce resources of development capital c[ould] be better employed than to be invested at 3% in firewood plantations in the dry Sudan zone". Twenty years further on, at the time of the Silviconsult survey, an inordinate share of forestry research resources was still being devoted to plantation forestry, even though official rhetoric and symbolism had long and consistently highlighted the merits of extension, social, and farm forestry (Silviconsult 1991).

But long before this it had become clear that the motivation for, if not the nature of, "rational", "wise", or "scientific" interventions in northern Nigeria differed significantly from those in southern Nigeria in important respects, not least in its insistence (as a cost-saving measure) on *indirect*

---

[14] There still does not exist, to the best of my knowledge, an equivalent Hausa word or description for tree plantation.

[15] However, it is worth noting that mismatches between policy as outlined in working plans and actual practice "on the ground" were commonplace. Furthermore, projections and expectations were frequently not borne out by actual performance. The predicted performance of the Kano Fuel Plantation, for example, was never realized.

landscape management through local administrations, and using locally generated funds, for the production of "minor" forest produce (fuel, building materials, etc.) for local and regional consumption and use (Egboh 1979; Cline-Cole 1998). In southern Nigeria, where forests were perceived as having imperial commercial significance, forestry management was direct, more intensive, and targeted export timber production (Egboh 1985).

In the interim, and under the 1938 Law for the Protection and Control of Forests, which remains largely unchanged to the present, postindependence states (successors to colonial regions and provinces) inherited responsibility for managing reserved and protected forests as well as communal forest areas. However, and rather confusingly, local government reforms in the 1970s, which confirmed the rights of states to manage and protect reserved forests, also vested the *ownership* of the forest estate in local government councils (LGCs) (Silviconsult 1991; also see Cline-Cole 1998). Currently, therefore, LGCs have full rights to the revenue from fuelwood cutting in state forest reserves (and natural forests and woodlands) located in their "territories", even though they share administrative and management responsibility for reserves with state forestry departments (Hyman 1993).

Not surprisingly, therefore, while much recent agitation for reform of forestry policy, administration, and practice has tended to be formulated as part of wider environment-development discourses of decentralization, community participation, and empowerment, powerful murmurings within some sectors of federal and regional state forestry coalesce around a perceived need to break away from a colonial heritage of sectoral underfunding; to "toughen up" former (sometimes oppressive) colonial forestry laws; to recentralize management and fiscal responsibility in the hands of state forestry bureaucracies and away from local-level administrations; and to increase sectoral policing powers and capacities which often generated tremendous animosity during colonial rule. Indeed, in 1997 one of the longest serving and most respected foresters in the region confided that, in his opinion, the greatest threat to the dryland agro-forestry programs and projects favored by the World Bank and the European Union remains an entrenched belief, shared by many of the most senior members of national and regional forestry administrations, that the only "real" forestry is plantation forestry.

Nonetheless, while Peet and Watts's (1993:231) observation that "certain modes of thought, logics, themes, styles of expression, and typical metaphors run through the discursive history of a region, appearing in a variety of forms, disappearing occasionally, only to reappear with even greater intensity in new guises" is undoubtedly instructive here, its uncritical application would carry with it the risk of downplaying the significance and extent of the *discontinuities* that have characterized both the substance and the rhetoric of state forestry discourses over the long term. For example, while much rhetoric still privileges notions of environmental protection, conserva-

tion, and sustainable forestry management (thereby demonstrating noticeable continuity in scientific forestry's "modernizing" intent into the postcolonial period), actual forestry practice has become inordinately dependent
on the redefinition of (forestry) space for extending its influence and authority onto "private" land and into the everyday lives of the regional population (cf. Rocheleau and Ross 1995).

The direct intervention in land management practices beyond the boundaries of legally constituted spaces in reserves and protected areas implied
here has been achieved, initially through the (only moderately successful)
introduction of ordinances, laws, and rules banning or limiting bush burning and tree cutting in unreserved areas, but subsequently also, and considerably more effectively, via a plethora of community, farm, social, and other
forms of "decentralized" forestry initiatives that promote particular (so-
called scientific) ways of "seeing" and "doing" forestry among dryland inhabitants, which remain remarkably reminiscent of (scaled-down) plantation
models, methods, and ethics (Anderson 1987; Anon. 1990a; Cline-Cole 1997;
Hyman 1993). Nonetheless, these processes of "extending" forests/
plantations to farms have not only seen the state/parastatal monopoly over
the funding and practice of "modern" forestry broken (e.g., NGOs now participate actively in regional forestry) but they also offer opportunities for
exploring how the complex interaction between formal ideas and substantive local knowledge intervenes in the (re)production of regional fuelscapes
and forestry landscapes (for which, see Cline-Cole 1997).

In its contribution to and continuation of such processes, Silviconsult
should perhaps have highlighted the lack of gender perspectives within
dominant regional forestry discourses. This represents a "loud" silence,
given that all (re)productive spaces are routinely (en)gendered in the drylands and consequently that interventions in regional forestry landscapes
(including both the Silviconsult survey and its policy/program/project recommendations) impact on gendered patterns of resource access and control
(Schroeder 1997). However, the European Economic Community/Federal
Government of Nigeria EEC/FGN Katsina Afforestation Project has initiated
a women's program, run by a woman extension officer, with the aim of
"involving women in project activities [such as] introducing energy efficient
wood burning stoves to rural households, promoting the planting of trees in
compounds and farms and encouraging the setting-up of small nurseries by
women's groups" (Anon. 1990a:19).[16] More recently, there have been
attempts to elicit female forestry "texts" through the medium of community
drama/dance/song (Frances Harding, personal communication).

Nonetheless, overall, forestry discourse has not been unduly preoccupied
with questions of female (farm) tree ownership and (indirect) use and man-

---

[16] It is not clear whether these were priorities identified *by, with,* or *for* women in the project
area.

agement, even of the valuable fruit trees that are being actively promoted by farm forestry programs and projects, and has been little interested in shrubs, which can represent significant female forestry assets and interests. In contrast to their southern Nigerian counterparts, dryland female voices have not been prominent in contesting dominant forestry discourses (Egboh 1979; Cline-Cole 1996, 1997).

Imagining fuelscapes

Silviconsult estimates that some 13 to 18 million tons (19 to 24 million cubic meters per year) of wood, slightly more than 1.7 to 2.2 million tons of dried sorghum stalk (*kara*) and 0.1 million tons of charcoal are consumed annually in the study area. Most (some 85 percent) of this is accounted for by rural consumption, mostly for domestic cooking, which consumes more than 80 percent of regional wood and at least two-thirds of available *kara* supplies. Overall, more than 80 percent of households (rural and urban combined) reported the use of wood for cooking, although other domestic cooking fuels like *kara*, kerosene, gas and charcoal had reportedly been tried as substitutes for wood by 64 percent, 26 percent, 6 percent and 2 percent of households, respectively. Many rural household consumers purchase at least some of their fuelwood.

Although average per capita daily consumption of wood was roughly the same as that for *kara* (1.45 and 1.52 kg, respectively), on a heat-equivalent basis wood generates 87.3 percent of the gross calorific value of domestic energy consumed by households, compared to *kara*'s 7.7 percent contribution. Household wood consumption increases during the cold season when space heating becomes necessary; significant quantities of charcoal are also used at this time of year for heating.

When analyzed by ecozone, the greatest concentrations of rural wood demand are to be found in the high population density zones of Kano-Jigawa-Katsina and Sokoto-Kebbi. Rural households in Sokoto and Kebbi combined account for 30.4 percent of regional rural consumption; together, Kano and Jigawa consumers account for a further 28.9 percent, with Katsina households using 17 percent of the remainder.

Rural household per capita consumption of both wood and *kara* increase northward with increasing aridity and diminishing biological productivity, with wood consumption peaking where average distance traveled to collect wood is least and *kara* consumption in the main sorghum-producing zone.

Prior to Silviconsult, it was widely believed that most fuelwood consumed by rural households was collected from farms or nearby bush. Survey results provide confirmation of just such a heavy dependence on self-provisioning, with only 43 percent of households collecting less than a tenth of their wood. "Own" land remains the most important source of fuelwood in rural dryland Nigeria; it represents the primary source of wood for two-thirds of households. But with about half of fuelwood-collecting households exploiting more than one source, supplies are obtained from "other" lands are too. Also, even though distance traveled does not seem to significantly affect the quantity of fuel collected or used, actual distances covered during collection trips remain reasonable, even for the 40 percent of household members who now travel further than in the past. More than half of collecting households travel no further than two kilometers to obtain wood, with almost four-fifths of collectors traveling three or less.

Across much of the drylands, where the practice of female seclusion restricts most Muslim women of reproductive age to their compounds between the hours of sunrise and sunset and men and women occupy separate public and private spaces, rural fuelwood collection and/or purchase, that assumed most archetypal of female tasks, is carried out overwhelmingly by men (frequently aided by children). Thus, although postmenopausal women can collect medicinal and edible plant food (roots, leaves, fruits, bark) and firewood from source areas frequented by men, female and male fuelwood collectors/gatherers (are expected to) "work" separate gendered spaces. At least one district head in Kano extended this principle to Local Government Council (LGC) reserves, woodlots, and Communal Forest Areas (CFAs) by designating separate collection areas and periods for men and women within such areas.[17] Also, because stark wasteland, wilderness, or open grassland is dreaded by Muslim Hausa in general, but particularly by their women, such areas are not particularly favored by female wood collectors.

As with those for rural consumers and consumption centers, firewood catchment areas of state capitals overlap widely. The city of Kano has the widest supply radius of all. Wood originating in every state in the region is sold in its fuelwood markets, and about 40 percent of its wood imports originate in supply zones more than two hundred kilometers distant. In contrast, Maiduguri, Zaria, and Sokoto town depend on supplies that are wholly or largely obtained from rural sources within their respective state boundaries. However, Jos, like Kano, is a very heavily "import dependent" regional capital. Overall, some three-quarters of regional urban supplies travel between fifty-one and two hundred kilometers from source to site of consumption, with the lion's share coming from distant woodland and fallows.

Negotiating and maintaining access to a wide range of wood sources requires sometimes considerable investment in the institutions (land and tree tenure rules, family and kinship systems, community organization, gender relations, local government forestry rules, etc.) that ensure effective legitimate command over the wood reserves that they contain (Mearns 1995). With reference to rural collectors, land and tree tenure (or type of access right), in particular, emerges as an important institutional variable here: tree owners frequently prevent other people from cutting them, sometimes by

---

[17] Interview with Madakin Kano, Emir's Palace, Kano, 11 July 1985.

fencing them. Thus, 84 percent of people who borrow or hire land report that they collect less than 10 percent of their fuelwood needs, while 78 percent of this stratum of the population buy more than 95 percent of their total supplies. In contrast, 67 percent of those who claim that they "own" farmland report that they collect 10 percent or more of their fuelwood free, with only 37 percent of this latter group buying up to 95 percent to meet their consumption needs.

But physical accessibility is only one of a number of defining characteristics of the morphology of regional fuelscapes. Thus, tree and shrub species within these gendered spaces are commonly carefully evaluated and ranked in terms of their fuel characteristics and value. And, although regionally there exist in excess of seventy fuelwood trees, dryland consumers generally express clear preferences for specific species (*Anogeissus, Parkia, Piliostigma,* etc.) or for fuelwood with particular (and varying) combinations of fuel characteristics (ignition and combustion properties, smoke production). That consumers routinely settle for what is available or accessible rather than what is preferred in no way alters the contours of the mental maps that are fuelscapes. Indeed, preferred species sometimes command higher prices in the market, and some dealers and transporters indicate that (reported) availability of preferred species does influence their choice of supply sources. Thus, two of the three species that were best represented in shipments were *Anogeissus leiocarpus* and *Combretum micranthum,* two of the three top-ranked firewood trees. The eight most frequently cited firewood species stocked by urban dealers are also some of the most highly rated firewood trees.

When exploited by rural consumers, live as well as dead branches of fuelwood species (both available and preferred) are cut and whole trees may be felled. More people reportedly cut live branches (58 percent) than do not (41.7 percent), and significantly more people (almost two and a half times as many) reportedly fell dead trees than do live ones. The incidence of single-tree management (pruning, lopping, pollarding, and coppicing) appears to increase with the size of the farm tree population managed by a smallholder.

In contrast, urban supplies tend to be for the most part the product of the systematic exploitation of woodland/fallows in areas where increasing distances have to be walked to secure fuel supplies for domestic consumption at the same time that the need for supplementary income induces the cutting of fallow/natural woodland for sale to urban fuel traders (Mortimore 1989). In these and other areas, the significance of fuelwood-related activity for rural livelihoods is well established, and local populations participate actively both in production (providing paid labor for felling/cutting) and commerce (selling wood to urban-based transporters and traders as well as consumers). Not surprisingly, dealers report that more than four-fifths of all firewood subsequently traded is bought rather than collected free; they claim, perhaps not entirely surprisingly, that such wood is obtained less from live whole trees and twigs/branches than from deadwood.

Regional standing wood resources are estimated at some 335 million cubic meters, with farmland, which accounts for about half of regional land use, supporting a fractionally larger proportion of this total than woodland (much smaller in extent but with a much higher volume per unit area). However, sustainable regional wood offtake is only an estimated 6.8 million m^3, with the most important sources being woodland and shrubland and, to a lesser extent, land under cultivation. Ecozones of peak aggregate demand and those of greatest available supply do not represent a perfect spatial match. Currently, therefore, regional fuelwood demand is estimated to exceed sustainable production by a factor of 2.9 to 3.7, producing an annual deficit of 13 to 18 million m^3 per year.

This deficit represents, according to both Hyman (1993) and Silviconsult, an excess annual offtake of between 3.8 percent and 5.4 percent of regional growing stock. But it is worth emphasizing that this fuelwood gap exists *only when supplies imported into the region from further south are discounted.* In reality, imports actually *do* make up a significant proportion of this regional fuelwood "gap", for the increasingly frequent and sometimes prolonged shortages of refined petroleum products, for example, are not characteristic of the fuelwood market. The range of excess annual offtake values specified is thus based, at best, on informed "guesstimates", and needs to be handled/used in a sensitive and discerning way, particularly as Silviconsult repeatedly notes (and occasionally laments) a prior lack of information on vegetation productivity, growth rates, standing wood volume, and so on, all of which are needed for calculating the existence and size of the regional fuelwood (im)balance.[18]

The significance of this caveat cannot be overstated, for prior to Silviconsult the nature of regional fuelwood exploitation was widely described as "indiscriminate", and its impact on regional vegetation and environment was regularly represented as consistently and overwhelmingly destructive. Indeed, the perceived modification of *agri*-cultural practices as a consequence of fuel overexploitation in the zone around the city of Kano had been *pre*-scribed in graphic detail by Eckholm et al. (1984), whose imaginary geography mapped configurations of rising fuelwood demand, sale of biological capital, farmland stripped of trees, and collapsing agricultural systems.

As the most densely populated rural sector in all northern Nigeria, and possibly the area with the highest potential fuelwood demand in the whole of the Sudan-Sahel belt, this Kano close-settled zone has long been almost irresistible as a candidate for "dissection" by people interested in theorizing

---

[18] It is interesting to note that both Anderson (1987) and Silviconsult identify the need for obtaining comprehensive, reliable data on literally all major areas of regional forestry as a consideration of prime importance.

the relationships between population dynamics, fuelwood consumption, deforestation, and land degradation, commonly without the benefit of first-hand field experience and/or data. It was such a fuelscape imaginary that Anderson (1987) adopted as the cornerstone of his application of gap discourse to the drylands as a whole.

Nonetheless, field foresters I spoke with readily acknowledge, in support of Mortimore's (1989a) observation, that it is in the more distant source areas of urban fuelwood supplies, those characterized by low but rising population density where land clearance for agriculture (sometimes using mechanical means) is proceeding apace and intensive agro-forestry systems with livestock cannot be supported by the family labour available, that the greatest threat to rural woodstock is found (see also Cline-Cole et al. 1990a). Several saw no reason to argue with the observation that in (some) peri-urban and near-farmland areas, smallholder farm tree management appears to be consistent with sustainable management, with tree densities remaining stable or even increasing (Mortimore et al. 1990) "as urban [fuelwood] demand is deflected from valuable multi-purpose farm [trees] to distant natural woodland and fallows" (Mortimore 1989a:11). However, almost all point out, admittedly under pressure, that information of this kind possessed the potential for undermining efforts to attract external funds because it did not provide support for the widespread claim, favored by donor/funding agencies, that farmland trees and communal resources in close-settled zones were those most at risk of "overexploitation".

(Re)Producing regional forestry landscapes

Landscape components

According to Silviconsult, current trends in land use are dominated by the transfer of woodland, shrubland, and shrub/grassland to cultivation/agriculture, which accounts for nearly half of regional land use (48.3 percent). Mixed shrub/grassland (20.3 percent), woodland (17.8 percent), shrubland (11.1 percent), grassland (2.4 percent) and water (0.1 percent) make up the remaining land use categories. The regional forest estate of some 42,000 ha of plantations, 4,000 to 15,000 ha of woodlots, and 1,000 to 2,000 km of shelter-belts is subsumed within these land use categories.

Symbols, meaning, and material transformation of regional environments and societies are inextricably entwined in the (re)production of (the components of) regional forestry landscapes (cf. Blaikie 1995; Cline-Cole 1994). Read in all their complexity, landscapes function "as activity, site, morphology, symbol, and mediation in systems of surplus value production" (Mitchell 1994:14). Consequently, they are "structured" in a wide variety of ways, including on the basis of age, race, religion, and gender. As table 1 illustrates, different categories of dryland inhabitants engage with forestry spaces, products, and processes in a variety of ways, as agri-silvi-pastoral

Table 1. *Landscapes and fuelscapes as text and context*

Karkara

Dominates rain-fed cultivated upland farmed at over 60% intensity.

Supports human settlement. Mature(ing) economic tress and shrubs, and selected grasses, on annually cultivated farmland, which is grazed by domestic and transhumant cattle after the annual harvests.

Trees and shrubs are private property, which are:

(1) inheritable, mortgageable, and marketable;
(2) sometimes owned separately from land;
(3) usually belong to (overwhelmingly male) smallholders, but women have full control over the disposal of the products of *Parkia* trees, which they can inherit, and of henna (*Lawsonia inermis*) and zogale (*Moringa oleifera*) shrubs, which are widely considered women's plants.

The landscape configuration of plants along field boundaries, around compounds and settlements, interspersed with crops, and along cattle droves reflects the goals of agroforestry, whereas species composition and plant density are heavily influenced by the mutual compatibility of land uses and familiarity with, and availability of, "appropriate" planting material.

Saura

Areas of immature parkland or patches of fallow (*fako kekuwa*) shrubland or grassland in *karkara*, which are farmed at 30–60% intensity and in which farm trees are only just emerging from the sapling or shrub stage.

Shrubland (6.6%), and shrub grassland (17.1%) contain significant quantities of the total dryland standing-wood volume.

Grazing, burning, and woodcutting are important land use features, and land is sequentially or concomitantly grazed, fallowed, and cultivated.

Although species composition may be similar to that in *karkara*, trees and shrubs here need to be more fire resistant than are *karkara* plants if they are to survive regular controlled burning of undergrowth by hunters and livestock rearers.

Where land productivity is low and agricultural opportunities limited, the collection and sale of fuelwood to urban suppliers; the manufacture of mats, rope, and baskets from the leaves of the *dum* palm; and beekeeping may become important remunerative activities.

Daji

Natural woodlands occupy some 17% of the drylands: either *daji dawa*—open, uninhabited bush managed as common property resources: or stark wasteland, wilderness, or open grassland (*dokar daji*), which function as de facto open access areas.

"Wilderness" areas, which are either completely unfarmed or, where they contain remote farmlands, farmed at less than 30% intensity, are also important sources both of timber and of nontimber forest products:

(1)   building poles, fuelwood, and construction timber for local and regional use;
(2)   invaluable dry season grazing and fodder resources for livestock;
(3)   raw material for craft occupations such as carving; and
(4)   "bush meat" (mammals and large birds), which constitutes an essential source of animal protein in locally wooded or forested areas.

Most extensive in areas of low population density, higher rainfall, or favorable edaphic conditions, these include a diversity of habitats: uncultivable shrubland in *saura* and *karkara*, often devoid of trees; extensive areas of rangeland or tree savanna; riparian woodland (*kurame*); hill forest blocks; forest outliers; and so on.

Customary tenure and statutory land law do not provide title to grazing, water, and fodder resources, hence herder access to these resources depends on the tolerance of sedentary communities.

Fadama

Seasonally flooded wetlands containing well-drained *saura* and *daji*, as well as irrigable and uncultivable marshland, sometimes in complex spatial and temporal mixes.

Double-cropped *fadama* areas can be as intensively exploited as *karkara*, and where irrigated bunded plots are interspersed with economic trees and shrubs a characteristic farm parkland is sometimes identifiable, although the tree species are likely to be exotic fruit trees planted for the commercial value of their products.

Like *daji*, these wetlands are increasingly contested and bear the expanding (and intensifying) imprint of human and livestock activity, with floodplain parkland and uncultivated areas producing fodder, fuel, (wild) food, medicines, grasses for grazing, roofing material, and so on.

Makiyaya and burtali

Makiyaya are communal grazing lands and watering points (often *daji* with some *fadama*) reserved primarily for sedentary livestock. *Burtali* are drove routes or tracks lined by living hedges, which link *makiyaya* to each other, to local settlements, and sometimes to regional transhumant routes.

The hedge plants that mark the location of *burtali* do not only protect farms against animal incursions (and thereby reduce farmer-herder conflicts); they also yield useful medicines, latex, and so on.

Burtali have fallen into disrepair in some areas and, like increasing areas of *makiyaya*, have been encroached upon by farmers in others.

practice, gathering activities, and religious ritual are all enacted as discourse. But such discourses are permeated, as always, by ambivalence and contradiction.

Wall (1988:141) notes, for example, that the "polarity between the civilised village community of Islam and the uncertain, wild domain of the bush is a basic component of Hausa thought ...":

> Islam, the domestic world of the family, and the cultivated fields are *gari*, "the village". It is the place of life. In stark contrast to this is the wild, uncultivated bush (*daji*) surrounding the village, with its lack of habitations, wild animals, uncontrolled forces, and malevolent spirits; and the pagan Maguzawa with their blood sacrifices, witchcraft, and sorcery. Outside the village is the place of death. The polarity between the civilised village community of Islam and the uncertain, wild domain of the bush is a basic component of Hausa thought. (141)

However, not only are Hausa not all Muslim, but even Muslim Hausa retain pre-Islamic practices and beliefs more commonly associated with non-Muslim Hausa. Landscape discourses of non-Muslim Hausa, the Maguzawa, emphasize a proverbial "aversion to the crowding and strangeness of the Muslim village: 'They can't really live there!' said the pagan Maguzawa man [when he saw the village]" (130–31). In the remote rural districts where they are currently (mostly) found, living in kin-based settlements among their farms in the bush, Maguzawa share an association with flora and fauna, which is mediated by a basic dogma of

> the recognition and elevation of iskoki [or spirits], which are everywhere—in the sky, the forest, the hills, in bodies of water and in the cities of men. Those whose names are known, and who have a definite cult, generally have a favourite kind of tree or some other specific locale where they like to stay and where sacrifices are offered to them, each spirit having its appropriate animal. ... It is not the object itself, however, as the naive observer might conclude, which is receiving the sacrifice, but the spirit associated with it. (Fergusson 1973:164–65).

In reality, however, essentialized "humanized" and "wilderness" components of regional forestry landscape imaginaries are linked by complex (re)productive relations/ties over time, across space, and between "cultures", as table 1 makes clear. Fuelwood production is clearly a major beneficiary of such links, which underscore the significance, for livelihood security, of the spatial and temporal integration of access to, control over, and movement between different components of regional forestry landscapes (Cline-Cole 1995).

But fuelwood production, even for insatiable and increasingly distant urban markets, cannot by itself explain the nature, pace, and direction of the kind of regional landscape change identified above. In particular, it fails to account satisfactorily for why virtually all natural woodland shows visible signs of human and livestock interference. In part, this is because *daji* represent important (re)productive spaces in other ways (cf. table 1). For instance, migratory livestock herders, who are overwhelmingly Muslim and Fulani, use them as sites for encampments, farms, and the like. Their women, unencumbered by the restrictions associated with the practice of female seclusion among their sedentary counterparts, traverse these wilderness areas regularly (and for the most part safely) as they travel to and from local markets in and around the settlements where they sell dairy products to members of the "civilised village communities of Islam", among others. Furthermore, livestock animals belonging to sedentary owners that are left in the care of migratory herders alternate between woodland grazings, uncultivated river valley grazings and farmlands in the continued integration of *gari* and *daji* (Mortimore et al. 1990).[19]

In general, these landscape imaginaries appear to privilege notions of sustainable *livelihood diversity and security* over those of sustainable forest/woodland management per se. Unlike the "dedicated" (fuel) reserves and plantations of scientific forestry, *daji* is not simply a source/reserve of woodfuel whose clearance for, or (partial) replacement by, "nonfuel/nonforestry" land uses leads inexorably or inevitably to its "degradation" and/or complete annihilation. Indeed, as the area under cultivation will continue to increase at the expense of woodland and other land uses in the foreseeable future, an appreciation of the process by which *daji* is transformed into a more intensively cultivated space (*saura* and *karkara*) and some of the implications for fuelwood provision/availability at the local level may be instructive.

Landscape production

As Howard (1976:18) describes it, "When agriculture based on a bush fallow cycle is replaced by permanent agriculture, the vegetation develops into a stand of evenly-spaced mature trees preserved for their economic importance". Elsewhere,

> on uncultivated areas, especially where the soil is shallow, shrubland dominated by small thorny trees ... occur. ... [W]here fallowing is practised, shrubland may be interspersed with [permanent cultivation]; when the transition to permanent cultivation takes place, shrubland is eliminated and selected trees are protected to grow to maturity. (Mortimore et al. 1990:45–46)

Farm tree numbers, then, tend to increase as natural forest and woodland decrease in extent. Somewhat paradoxically, therefore, cultivated and recently fallowed fields, which currently occupy about a third of regional space, support some 36 percent of total dryland standing wood volume in the form of a variety of species of trees and shrubs at varying stages of

---

[19] It may be worth recalling the observation made earlier that all regional forest reserves are also de facto grazing reserves.

growth and density.[20] More than half of a sample of 790 landowning respondents in the Silviconsult survey reported up to ten trees growing on
their land and a further 9.5 percent, between eleven and twenty trees, less
than 10 percent of respondents owned farms that were completely treeless.

A widespread belief prior to Silviconsult was that, although protection
and planting of trees, woodland, and forests are desirable practices, they
have either been abandoned by or are alien to present-day northern Nigeri-
ans, who therefore need to be (re)taught these skills. In the event, the survey
revealed that a significant proportion of farmland trees is planted: only 7
percent of landholders do not plant trees. Most of those who reportedly
plant trees (85.2 percent) claim to plant between one and twenty. But wild-
ings are also protected, with only a third of farmers reporting that they do
not provide any protection for seedlings that regenerate spontaneously; of
the two-thirds who do, the commonest form of protection offered is from
grazing animals.

Purchased inputs are insignificant in indigenous forestry, with people
investing labor but little cash in trees, which are nonetheless perceived as an
important component of farmland capital. Farmland tree planting and
tending are considered a "normal" or integral part of the activity of farming,
and I have been told repeatedly that farm tree planting and tending are
skills that are not learned separately from those of crop farming. In some
cases, the value of farm trees has been estimated at levels that are higher
than those of farmland.

That trees are reportedly planted and protected in such numbers is
enormously significant. Indeed, it is doubly significant because such
tree/shrub propagation is occurring not just *despite* but because (and as an
integral part) of contradictory processes of long-term increases in population
totals and densities and intensifying pressure on agricultural land, on the
one hand, and an expansion in extensive farming and the extension of
smallholder cultivation into woodland, pasture, and fallow on the other
(Mortimore 1989a). In other words, the vegetation does not simply and in-
explicably "develop" in the manner suggested by Howard (1976); it is
clearly the product of (sub)conscious processes of landscape (re-)creation.
And, although the end results may vary widely over time, across space, and
with the scale of geographical analysis, they still reflect fusions of particular-
istic and Western scientific knowledges interacting within specific contexts
(cf. Sivaramakrishnan 1995).[21]

A comparison of preferred fuelwood and farm tree rankings is instructive in emphasizing that the hierarchy of species preference for fuelwood

---

[20] This contrasts markedly with the situation in areas that have been cleared by mechanical
means and on which mechanized farming is (to be) undertaken. These are the "completely
bare" farmlands in newly cleared areas referred to by Anderson (1987), for example.

[21] For example, increased tree planting by rural producers in some parts of the drylands have
been attributed (by the producers themselves) to the influence of forestry extension workers.

use differs in significant ways from the hierarchy of species preferred for farmland planting and protection. Thus, although fourteen of the twenty most frequently occurring farm trees also double as the most preferred fuelwood trees, the correlation between the two rankings is very weak. Low-ranking fuelwood trees (can) rate much higher as farm trees because of the value of their edible or industrial (i.e., nonfuel) products, while prominent fuelwood trees (can) appear relatively low in the ranking of farm trees, sometimes because they compete with food crops.

This is hardly surprising, for fuelwood is but one of several products accruing to farm forestry and its practitioners. Thus, in (some) peri-urban and near-farmland areas, smallholder farm tree management appears to be consistent with sustainable management, with tree densities remaining stable or even increasing (Mortimore et al., 1990) "as urban [fuelwood] demand is deflected from valuable multi-purpose farm [trees] to distant natural woodland and fallows" (Mortimore 1989a:11). Fuelwood exploitation can be compatible with other, seemingly competing uses, even when they occupy the same geographical space: pruning and lopping for fuel, for example, can reduce shade on crops, improve fruit yields, and provide cuttings for burning on farms to provide nutrients for agricultural crops (Mortimore et al. 1990).

Yet, although fuelwood can be the most important forestry product by value and volume in areas like the Hadejia-Jama'are floodplain (on the Jigawa-Yobe borderlands) and Gundumi Bush (Sokoto), among others, farm trees actually provide a sometimes bewildering range of services and products. Indeed, I have had the temerity, repeatedly over the years, to scribble marginal additions and updates in my copy of Dalziel's *Hausa Botanical Dictionary*, that indispensible companion of dryland fieldworkers. Of these products and services, those that exert the greatest influence over farm forestry practice are fruit and food production and shade provision; environmental protection and fuelwood production are no more than subsidiary aims of farm forestry. Furthermore, dryland inhabitants perceive that perceptive combinations of (frequently now "nativized") exotic species (e.g., neem for shade and mango for fruit) and indigenous species respond to the achievement of the forestry goals that they themselves consider dominant.

Contesting landscape

Because landscape is like a commodity, Mitchell (1994:10) observes that its "evident (that is, temporarily stabilised) form ... often masks the facts of its production". It is worth exploring, therefore, however briefly, and through the use of anecdotes, how contested and contestable the processes of regional forestry landscape (re)production can be, starting with that old acquaintance the village head from Katsina. All the forestry successes achieved in Babamutum were, according to Alhaji Usman Uban Dawaki, "attribut-[able] to the cooperation and dedication given to me by my people regard-

ing tree planting as a personal commitment to each and every one of them"
(Anon. 1990b:30). However, in doing so, in speaking *for* the community as a
whole, he is careful to employ a discourse of landscape that erases social
conflict from view (cf. Mitchell 1994).

Yet elsewhere, in the Tofa area of neighboring Kano State, I found
"popular participation" in forestry endeavors of the kind described being
expressed in the somewhat problematic terms of "respect". Here

> formal rules governing the felling of farm trees ... used to be effectively enforced
> at the district level by the district head. ... It appears that the effectiveness of the
> system depended on the personality of the district head as well as on the real
> authority he wielded (described ... as the respect he could muster); the most re-
> spected (or authoritarian) district heads were the most successful in enforcing the
> rule, and village heads in such districts had little difficulty in getting villagers to
> cooperate. (Cline-Cole et al. 1990a:91).

One of the most "respected" district heads of Tofa in living memory left me
in no doubt about his conviction that the *masu saurata* (aristocracy) had a
duty to instill environmental awareness and education of this kind in the
talakawa (commoners): "we *were* the torch lights [*sic*] of the North, and it was
our responsibility to enlighten the populace about things that were designed
for their own good".[22] Can it be merely coincidental, then, that Alhaji
Usman's "love of trees" in Katsina was developed during an apprenticeship
served under his then district head (of Baure), "who was also a lover of
trees"? Undoubtedly, the stories about the spatialization of social struggle
that hide beneath the discursive shade (to borrow Rocheleau and Ross's
[1995] imagery) of Babamutum's tree and shrub landscape, with its generous
peopling of indigenous species, await detailed empirical investigation.

In the interim, it is no doubt instructive that when rural opponents of a
winning candidate in gubernatorial elections in Kano State in the 1980s
chose to demonstrate their displeasure, they opted for the symbolic act of
cutting down very valuable *rimi* (silk cotton) trees on their farms. The candidate in question, Alhaji (Mohammed Abubakar) Rimi, shared his name
with the tree. Prior to this, his (self-proclaimed) "populist" ways, and in particular their (perceived) erosive effect on the (supervisory) authority
"traditionally" wielded by village and district heads in matters related to
natural resource management, had been cited as a significant input in local
processes of farm tree loss by at least one village head, who was clearly unhappy at the loss of respect that both accompanied and encapsulated what
he perceived as such reckless empowerment of the *talakawa* (Cline-Cole et al.
1990b).

But, in fact, across the drylands as a whole the importance of space in
social landscape process had long been recognized and was reflected in a
whole range of discourses of this kind. Precolonial dryland rulers were, after

---

[22] Interview with Madakin Kano, Emir's Palace, Kano, 11 July 1985.

all, *sarakunan kasa*—rulers of the land—who supervised the functioning of common property and private institutions regulating access to, control over, and exploitation of forestry resources; territorial authority provided a structure for political control (Cline-Cole 1994). In exercising the latter, they could, and did, redefine forestry space in the "common good", delineating *burtali* and *makiyaya*, for example, while catering to the needs of both the destitute and dispossessed as well as the privileged in precolonial society. British colonial authority assumed this (and much greater) responsibility through the right of (re)conquest, its foremost early local representative insisted, even as he deployed such assumed (in both senses of the term) authority in the constitution and policing of a forest estate (Lugard 1970). The exercise of such institutional rights and responsibilities at all levels needs to be seen as necessary, Mitchell (1994) would no doubt argue, because the production of regional forestry landscapes needs to be continually reinforced and protected.

Indeed, not only were such processes much in evidence during the colonial period as social struggle acquired a quite distinct spatial base, but their contemporary postcolonial resonance was dramatically underscored in May 1992. Early that month, and in the middle of a kerosene and cooking gas shortage, Katsina town fuelwood dealers refused to arrange deliveries of wood to the town. They were protesting, first, that their producer-suppliers in rural Katsina State were being subjected to a sustained campaign of harassment by state forestry agents acting in collaboration with conservation interests, who were trying to promote the idea that producers should plant a replacement seedling for each tree cut to supply wood for fuel, but also that out-of-state supplies were either being subjected to a surcharge at the source or were being prevented from crossing state boundaries altogether. Incredibly, in a revival of forestry rules that were first introduced in 1942, Katsinawa were "not allowed to fell trees or cut live branches on their own land without the permission of an Agricultural or Forestry Officer" (Hyman 1993:35) and were being discouraged from producing fuelwood for nonlocal (predominantly urban) trade.

The protest represented, as far as I am aware, the first "strike" by firewood dealers in the drylands in living memory, although Kano dealers had been privately threatening just such an action for several years. It was a significant but localized manifestation of more widespread contestations over the "place" of the firewood trade in postcolonial northern Nigeria, as the periodic enforcement of administrative restrictions on the cutting and transport of trees, particularly cross-border transport, have increased in frequency (and sometimes duration) since the 1970s. In cities like Kaduna and Kano, therefore, fuelwood dealers have formed unions, which are constructing their own critiques of the representations made of their trade. Thus, in attempting to control Katsina's urban spaces in this way, the firewood dealers were responding in kind to the initial spatialization of the conflict, which the increased policing of production zones and border-crossing points between states represented.

Nonetheless, it would be too simplistic to reduce such contests over space and place to a straightforward conflict between "green" state forestry officials committed to sustainable resource use and fuelwood dealers and transporters who, driven by the profit motive, "mine" regional fuelwood reserves. For these contests are also, and maybe more importantly, about struggles for supremacy and resources between local and state government institutions. Within the context of the "dual mandate" arrangements for the management of the regional forest estate mentioned earlier, locally stationed (but state employed) forest guards and patrols who threaten revenue collection by attempting to prevent perceived unsustainable levels of tree cutting and woodland or forest exploitation have had to contend with the wrath of local government administrators. Historically, poorly resourced local government administrations, which are inadequately equipped to either effectively police or sustainably manage forest estates, have concentrated on generating income from the issuance of exploitation permits and licenses while steadfastly resisting attempts by state forestry administrators and managers to limit such exploitation. Thus, large-scale woodcutting entrepreneurs acquire supplies from thinly populated areas remote from human settlements of any significant size, frequently felling sizable swathes of natural woodland after obtaining cutting permits at nominal cost from LGCs.[23] Under conditions of sparse human population and practically nonexistent policing of the forest estate, violations of permit rules (e.g., overcutting) are commonplace, remain mostly undetected, and go largely unpunished.

But, although state forestry officials are distinctly unhappy about what they perceive as the inability of LGCs to effectively oversee the forest estates under their jurisdiction, even *their* departments lack adequate resources to either protect natural woodland or bring such areas under systematic management. Plagued by serious levels of understaffing, they also have to contend with livestock interests, which constitute a powerful political voice and insist on the right to largely unrestricted access to the grazing and watering resources that reserved and protected forest areas both contain and represent. And, like their LGC counterparts, state forestry officials have also been unable to overcome political and administrative hurdles in the way of securing the conviction of big-time, well-connected wood dealers who either violate permit rules by acquiring supplies outside authorized coupes or ignore commercial exploitation rules altogether by failing to obtain licenses in the first place (Silviconsult 1991).[24] Thus, when state forestry administrators en-

---

[23] License fees are incredibly low and come nowhere near the replacement cost of trees in either natural woodland or plantations.

[24] In any case, penalties for violating forestry rules are too low to act as a deterrent.

list the help of locally stationed members of the national or federal police in their periodic blitzkrieg raids on the so-called uncontrolled and illegal movement of wood within, but particularly *across*, state boundaries, this is as much a statement about the deployment of power and authority to (particularly but not exclusively unlicensed/illegal) dealers/transporters as it is to legally constituted LGCs who issue perfectly lawful exploitation licenses in the exercise of a responsibility that state forestry services would dearly like to see transferred "upward" to them.

To take a final example, (transhumant) spaces of and for mobile livestock rearing have long (and often) been (fiercely) contested by sedentary communities. Indeed, a recognition of the necessity for delimiting *burtali* in precolonial times was in itself an acknowledgment of the seriousness of the frequency, scale, and intensity of such contestations. Currently, in *karkara* and *saura*, nomads have been accused by smallholder cultivators of damaging trees, stealing stored crop residues, and failing to prevent crop damage by their animals (Mortimore et al. 1990). In some *fadama*, invaluable dry season grazing areas, land is sometimes fenced off and nomads charged a fee or "rent" for access; in others, an expansion in state-sponsored small-scale irrigation farming is encroaching on "traditional" grazing and watering spaces and damaging valuable fodder trees and shrubs. Also "[l]ivestock specialists are put at a disadvantage when free fodder is collected from rangeland and sold back to them by farmers—in times of scarcity at prohibitive prices" (Mortimore, 1987:20).

In some instances, then, particularly where *daji* has been expropriated for large-scale farming, long-term pastoral Fulani residents have contested these actions in court (Mortimore 1987). When conflicts between sedentary cultivators and pastoralists arise, two early manifestations of heightening tension include the encroachment on pastoral women's spaces by nonpastoral men, who render them unsafe and threatening by physically denying pastoral women access to local markets, and the deliberate propagation of plants that are known to be poisonous to livestock along farm boundaries adjoining *burtali* that are heavily used by transhumant livestock.

Conclusion

In juxtaposing alternative knowledge claims that have been packaged and promoted around regional fuelwood-deforestation-degradation linkages in an "interactionist" way (Blaikie 1995), this essay has tried to recognize the merit of critically evaluating different dryland forestry discourses. Thus, far from attempting to question the (in)validity, (ir)rationality, or (in)correctness of the viewpoints presented, the essay has tried to highlight, first, the groups, interests, and social forces that are mobilized in (or excluded from) the construction and perpetuation of the various discourses that these viewpoints or "projections" represent and, second, the values, norms, and experiences around which the discourses are articulated.

This is more than mere self-indulgence, for, to paraphrase Paul Richards (1995), regional forestry landscapes are as interpretively flexible as they are diverse, being all at once, or in parts, neglected reserves, effectively policed commons, overgrazed pastures, well-wooded farmland, intensively exploited natural forests and woodland, failed shelterbelts and woodlots, and beautifully manicured live hedges. Forestry discourses, in other words, are seen as struggles or negotiations over meaning (cf. Mackenzie 1992).

Juxtaposing discourses in this way, Richards (1995) argues, creates space for different landscape "visions" to be regarded as "virtual realities", which are sustained by specific social forces and (technologies of) representation. There is a convergence here, it seems, both with Foucault's exhortation (Mackenzie 1992:694–95) "to reconstruct the knowledges or 'reverse discourses' that have been marginalised, to place 'the claims to attention of local discontinuous, disqualified, illegitimate knowledges against the claims of a unitary body of theory which would filter, hierarchise and order them in the name of some true knowledge and some arbitrary idea of what constitutes a science and its object'", and with Zukin's (1991, cited in Mitchell 1994) view of landscapes as contentious, compromised products of society, which are shaped by power, coercion, and collective resistance. In any case, environmental discourses concern the impacts of people on their environment, as well as the impact of the latter on the former, and are "cultural construct[s that] cannot be divorced from ... particular human activity system[s]" (Agnew 1995:148).

References

Adams, W.M. 1996. Irrigation, erosion and famine: Visions of environmental change in Marakwet, Kenya. In Leach and Mearns 1996.
Adelodun, M. 1940. Firewood supply from Kano farms. *Farm and Forest* 5 (2):30.
Agnew, C.T. 1995. Desertification, drought and development in the Sahel. In Binns 1995.
Anderson, D. 1987. *The economics of afforestation: A case study in Africa.* Baltimore: Johns Hopkins University Press.
Anderson, D., and R. Fishwick. 1984. *Fuelwood consumption and deforestation in African countries: A review.* World Bank Staff Working Papers, no. 704. Washington DC: World Bank.
Anderson, D., and R.R. Grove, ed. 1987. *Conservation in Africa: People, policies, and practice.* Cambridge: Cambridge University Press.
Anderson, D., and A. Millington. 1987. Political ecology of soil conservation in Anglophone Africa. In Anderson and Grove 1987.
Anon. 1990a. EEC/FGN Katsina Afforestation Project: Brief information about the project. *Greenlight* 2(2):11–20.
——. 1990b. Re-Babamutum. *Greenlight* 2(2):28–34.
Areola, O. 1987. The political reality of conservation in Nigeria. In Anderson and Grove 1987.
Barrott, H.N. 1972. Firewood supplies in Katsina Province. *Nigerian Journal of Forestry* 2(2):62–66.
Binns, T., ed. 1995. *People and environment in Africa.* Chichester, Sussex: Wiley.
Blaikie, P. 1995. Changing environments or changing views? A political ecology for developing countries. *Geography* 80(3):203–14.

Chambers, R. 1988. Bureaucratic reversals and local diversity. *Institute of Development Studies Bulletin* 19(4):50–56.

Cline-Cole, R. 1994. Political economy, fuelwood relations, and vegetation conservation: Kasar Kano, northern Nigeria, 1850–1915. *Forest and Conservation History* 38(2):67–78.

———. 1995. Livelihood, sustainable development, and indigenous forestry in dryland Nigeria. In Binns 1995.

———. 1996a. Dryland forestry: Manufacturing forests and farming trees in Nigeria. In Leach and Mearns 1996.

———. 1997. 'Promoting (anti-)social forestry in northern Nigeria? *Review of African Political Economy* 74:515–36.

———. 1998. Knowledge claims and landscape: Alternative views of the fuelwood-degradation nexus in northern Nigeria. *Environment and Planning D: Society and Space* 16:311–46.

Cline-Cole, R., J.A. Falola, H.A.C. Main, M.J. Mortimore, J.E. Nichol, and F.D. O'Reilly. 1990b. *Wood fuel in Kano*. Tokyo: United Nations University Press.

Cline-Cole, R., H.A.C. Main, and J.E.N. Nichol. 1990a. On fuelwood consumption, population dynamics and deforestation in Africa. *World Development* 18(4):513–27.

Crush, J. 1995a. Imagining development. In Crush 1995b.

———. ed. 1995b. *Power of development*. London and New York: Routledge.

Daniels, S. 1992. Place and the geographical imagination. *Geography* 77(4):310–22.

Davies, H.R.J. 1976. *Town and country in North Central State of Nigeria*. Samaru Miscellaneous Papers, no. 63. Zaria, Nigeria: Institute for Agricultural Research, Ahmadu Bello University.

Eckholm, E., G. Foley, G. Barnard, and L. Timberlake. 1984. *Firewood: The energy crisis that won't go away*. London: Earthscan.

Egboh, E.E. 1979. The establishment of government-controlled forest reserves in Nigeria, 1897–1940. *Savanna* 8(2):1–18.

———. 1985. *Forestry policy in Nigeria, 1897–1960*. Nsukka: University of Nigeria Press.

FDFALR (Federal Department of Forestry and Agricultural Land Resources). 1989. *Draft Forestry Policy*. Abuja: FDFALR.

Fergusson, D.E. 1973. Nineteenth century Hausaland: Being a description by Iamam Imoru of the land, economy, and society of his people. Unpublished Ph.D. thesis, University of California at Los Angeles.

Fishwick, R.W. 1961. *Some notes on the history of forestry in Northern Nigeria*. Kaduna: Government Printer.

Foucault, M. 1977. History of systems of thought. In *Language, counter-memory, practice: Selected Essays and Interviews*, ed. D.F. Bouchard. Oxford: Oxford University Press.

Grut, M. 1972. *The market for firewood, poles, and sawnwood in the major towns and cities in the savanna region*. Rome: FAO.

Horsman, J. 1975. Production forestry in the Zaria area. *Savanna* 4(1):70–74.

Howard, W.J. 1976. *Land resources of central Nigeria: Forestry*. Surbiton, Surrey: Land Resources Division, Ministry of Overseas Development.

Hyman, E. 1993. Forestry policies and programmes for fuelwood supply in northern Nigeria. *Land Use Policy* 10(1):26–43.

Jewitt, S. 1995. Europe's "Others"? Forestry policy and practices in colonial and postcolonial India. *Environment and Planning D: Society and Space* 13(1):67–90.

Jones, B. 1938. Dessication and the West African colonies. *Geographical Journal* 91:401–23.

Katsina, I.M. 1990. Foreword. *Greenlight Magazine* 2(2):2–4.

Kerr, G.R.G. 1940. Some notes on the forestry situation in northern Nigeria. *The Nigerian Forester* 1(1):20–22.

Kimmage, K. 1990. *North East Arid Zone Development Project (NEAZDP) socio- economic survey: Summary results*. Gashua: NEAZDP.

KSCASE (Kano State Committee on Alternative Sources of Energy). 1989. *Report on alternative sources of energy*. Kano: Ministry of Agriculture and Natural Resources.

Leach, G. and R. Mearns. 1989. *Beyond the woodfuel crisis: People, land and trees in Africa.* Earthscan.

Leach, M. and R. Mearns, eds. 1996. *The lie of the land: Challenging received wisdom on the African environment.* London: International African Institute in association with James Currey and Heinemann.

Lockwood, M. 1991. Farmers' perceptions of population pressure in southern Kano. Cambridge: African Studies Centre. Mimeo.

Lugard, Sir F. [later Lord]. 1970. *Political memoranda, revision of instructions to political officers on subjects chiefly political and administrative, 1913–1919.* Edited by A. Kirk-Greene. London: Frank Cass.

Mackenzie, F. 1992. "The worse it got, the more we laughed": A discourse of resistance among farmers of Eastern Ontario. *Environment and Planning D: Society and Space* 10:691–713.

Mearns, R. 1995. Institutions and natural resource management: access to and control over woodfuel in East Africa. In Binns 1995.

Mitchell, D. 1994. Landscape and surplus value: The making of the ordinary in Brentwood, CA. *Environment and Planning D: Society and Space* 12(1):7–30.

Morgan, W.T.W. 1983. *Nigeria.* Harlow, Essex: Longman.

Mortimore, M.J. 1987. The lands of northern Nigeria: Some urgent issues. In *Perspectives on land administration and development in Northern Nigeria,* ed. M. Mortimore, E.A. Olofin, R. Cline-Cole, and A. Abdulkadir. Kano: Bayero University.

——. 1989a. *The causes, nature, and rate of soil degradation in the northernmost states of Nigeria and an assessment of the role of fertilizer in counteracting the processes of degradation.* Environment Department Working Papers, no. 17. Washington, DC: World Bank.

——. 1989b. *Adapting to drought: Farmers, famines, and desertification in West Africa.* Cambridge: Cambridge University Press.

——. 1993. Population growth and land degradation. *Geojournal* 31(1):15–21.

Mortimore, M.J., E.U. Essiet, S. and Patrick. 1990. *The nature, rate, and effective limits of intensification in the small holder farming system of the Kano Close-Settled Zone.* Ibadan: Federal Agricultural Co-ordinating Unit.

Nash, T.A.M. 1941. Fuel consumption in relation to minimum temperature. *Farm and Forest* 2(1):34–36.

NEST (Nigerian Environmental Study/Action Team. 1991. *Nigeria's threatened environment: A national profile.* Ibadan: NEST.

Oduwaiye, E.A., ed. 1995. *Forestry for urban and rural development in Nigeria.* Proceedings of the 23rd annual conference of the Forestry Association of Nigeria (FAN). Ikeja: FAN.

Peet, R. and M. Watts. 1993. Introduction: Development theory and environment in an age of market triumphalism. *Economic Geography* 69:227–53.

Porter, G. 1992. The Nigerian census surprise. *Geography* 77(4):371–74.

Rajan, S.R. 1994. *Imperial environmentalism: The agendas and ideologies of natural resource management in British colonial forestry, 1800–1950.* Ph.D. thesis, University of Oxford.

Richards, P. 1995. African landscapes as virtual reality: The social shaping of technologies of landscape representation. Paper presented at the workshop African Farmers and Their Environment in Long-Term Perspective, Wageningen Agricultural University, the Netherlands.

Rocheleau, D., and L. Ross. 1995. Trees as tools, trees as text: Struggles over resources in Zambrana-Chacuey, Dominican Republic. *Antipode* 27(4):407–28.

Sagua, V.O., A.U. Ojanuga, E.E. Enabor, P.R.O. Kio, A.E. Kalu, and M.J. Mortimore. 1987. *Ecological disasters in Nigeria: Drought and desertification.* Lagos: Federal Ministry of Science and Technology.

Schaefer, J.T. 1995. Editorial Foreword. *Comparative Studies in Society and History* 37(1):1–2.

Schroeder, R. 1997. "Reclaiming" land in The Gambia: Gendered property rights and environmental intervention. *Annals of the Association of American Geographers* 87(3):487–508.

Silviconsult Ltd. 1991. *Northern Nigeria household energy study.* Bjarred, Sweden: Silviconsult.

Sivaramakrishnan, K. 1995. Colonialism and forestry in India: Imagining the past in present politics. *Comparative Studies in Society and History* 37(1):3–40.

——. 1996. The politics of fire and forest regeneration in colonial Bengal. *Environment and History* 2(2):145–94.

Stock, R.F. 1978. The impact of the decline of the Hadejia River floods in Hadejia Emirate, Kano State. In *The aftermath of the 1972–74 drought in Nigeria*, ed. J. van Apeldoorn. Zaria: Federal Department of Water Resources and Centre for Social and Economic Research, Ahmadu Bello University.

Thulin, S. 1970. *Wood requirements in the savanna region of Nigeria*. Rome: FAO.

Trappes-Lomax, A.F. 1952. Consumption of fuel—Kano. *Forestry Information Bulletin* 10:2.

Trevallion, B.W. 1966. *Metropolitan Kano twenty year development plan, 1963–1983*. London: Newman Neame for Greater Kano Planning Authority.

UNCOD (United Nations Conference on Desertification). 1977. *Desertification: Its causes and consequences*. Oxford: Pergamon.

UNEP (United Nations Environment Programme) 1992. *World atlas of desertification*. London: Edward Arnold.

Wall, L.L. 1988. *Hausa medicine: Illness and well-being in a West African culture*. Durham and London: Duke University Press.

Watts, M. 1983. *Silent violence: Food, famine, and peasantry in northern Nigeria*. Berkeley: University of California Press.

Wisner, B. 1988. *Power and need in Africa: Basic human needs and development policies*. London: Earthscan.

Yeoh, B.S.A., and T.B. Hui. 1995. The politics of space: Changing discourses on Chinese burial grounds in post-war Singapore. *Journal of Historical Geography* 21(2):184–201.

Zukin, S. 1991. *Landscapes of power: From Detroit to Disney World*. Berkeley: University of California Press.

Placemaking, Pastoralism, and Poverty in the Ngorongoro Conservation Area, Tanzania

Nina Johnsen

> Must my people eat dust to get a share in tourism?
>
> *Tepilit ole Saitoti, 1992*

Most of the approximately 240,000 tourists who visit the Ngorongoro Conservation Area (NCA) each year are likely to get only a quick glance at the area's human inhabitants after having visited the world famous Ngorongoro Crater. As they pass in their dust-generating Land Rovers on their way to Serengeti National Park they will probably spot a group of Maasai *ilmurran*, "warriors", waiting at the roadside with their spectacular ceremonial headdresses of ostrich feathers, doing their jumps on the spot, hoping to have their picture taken for a few shillings. After having taken the obligatory photos and hesitantly paid the requested fee, the tourist is likely to comment briefly to his or her traveling companions that it is sad to see how these Maasai prostitute themselves for money. What the tourist usually does not realize, however, is that the Maasai is likely to feel equally embarassed that he is forced to "eat dust" in this way, utterly degrading himself so that his family may avoid eating dust tomorrow.

To the Maasai in Ngorongoro Conservation Area, eating dust is not only a consequence of waiting for tourists at the roadside; it is above all a powerful metaphor for suffering hunger. Not only is it to be expected that every year at the height of the dry season, when milk yields are at their lowest, most people will have to live on *enkurma* (flour), which without milk or fat in it, they say, is like eating dust because "it makes you thin like a stick until your skin looks just like dust". In recent years this condition has become almost permanent due to a highly accelerated poverty process caused by uncontrolled disease in the cattle herds of the pastoral Maasai. Yet for the Maasai living in the park there are not many alternatives to eating dust, one way or the other; the customary subsistence alternative of reverting tem-

Fieldwork took place from May 1992 to May 1993 in and around the administrative village of Endulen in the southwestern part of Ngorongoro Conservation Area, Tanzania. It was financed by the Danish Research Council for Development Research and the Danish Research Council for the Humanities and carried out as part of a doctoral study on Maasai medicine hosted by the Institute of Anthropology, University of Copenhagen. This essay is based on a chapter from my Ph.D. thesis (1998).

porarily to hunting or agriculture in times of crisis in the pastoral mode of production has become illegal with conservation area status. Food security for the human inhabitants has so far not been included among the responsibilities of the park management, and few Maasai have obtained employment with authorities or the booming tourist industry, nor have their communities been granted a share in the revenue from tourism in compensation for imposed restrictions. When milk yields are too low to feed the family, there are therefore few other options than to try to obtain the desperately needed cash directly from the well-fed tourists as they quickly pass by. Thus, among the Maasai, eating dust is feared to be an unavoidable consequence of having become part of world tourism.

In the following, I hope to demonstrate how the process of making Ngorongoro Conservation Area into a distinctive place of international recognition, a sanctuary set aside for the preservation of wildlife for the international community of "global villagers" to enjoy, is rooted in the drawing of conceptual and administrative boundaries around the shared territories of wildlife and human communities in which mobility is the crucial determinant for both. Thus, a failure to realize Ngorongoro Conservation Area as a culturally created landscape, not one of the world's last residues of unspoiled nature, has triggered a human poverty process constituting a for once sadly real tragedy of the commons.

The making of places

Among others, Gupta and Ferguson have called attention to the implicit notions of space in the social sciences (1992). They state that challenging the assumed isomorphism of space, place, and culture "raises the question of understanding social change and cultural transformation as situated within interconnected spaces. The presumption that spaces are autonomous has enabled the power of topography to conceal the topography of power" (8). Rather than taking the autonomy of primeval communities for granted, we must foreground the spatial distribution of power relations in analysis in order to understand the process whereby spaces achieve distinctive identity as places (8). Similarly Hastrup and Olwig (1996) argue that the idea that cultures and people are naturally rooted in particular places has meant that mobility has been regarded as a special and temporary phenomenon, gypsies and nomads providing the exception to prove the rule. After all, nomads move in well-defined patterns within fixed territories. We should, with de Certeau (1986:117), realize that "space is a practiced place" and "practices overlap, intersect and blur the boundaries of place" (Hastrup and Olwig 1996:4). Indeed, territories themselves are overlapping (4). Hastrup and Olwig recommend that the notion of culture be theoretically reconceptualized through detailed ethnographic studies of the siting of culture as a dynamic process (3).

The history of the Serengeti National Park and the adjacent Ngorongoro Conservation Area offers an example of actual place making that challenges the interrelated idea that certain sites are relictual representations of a primeval, natural order and can be discretely preserved as such through the construction of national parks. The demarcation of Serengeti National Park and Ngorongoro Conservation Area as discrete and bounded, if unfenced, units of nature preserved for the benefit of the global community has meant that the pastoral Maasai customary inhabitants, while locked within imaginary but nevertheless real boundaries, in practice have become deterritorialized to the point that not only has their cattle economy broken down but their customary mechanisms of mutual self-help in times of individual destitution have been disrupted. The customary strategies of the Maasai in times of crisis in their pastoral mode of production, that of temporarily reverting to hunting or agriculture, are not compatible with conservation objectives. Hunger is now a chronic condition for the pastoralists in Ngorongoro Conservation Area, and management has frequently voiced the point of view that the only possible long-term solution to the problems of the approximately forty thousand Maasai is relocation outside the park areas. Thanks to tourism, Ngorongoro District is one of the best foreign-exchange-earning districts in all of Tanzania, with an annual income of over US\$ 10 million from visitors' fees alone (Lane 1996:5); yet it has one of the poorest infrastructures in the country. In almost 40 years of managing nature conservation, it has increasingly been ignored that the foresighted legislative construction, of the Ngorongoro Conservation Area as a multiple land use zone was originally made in recognition of the harmonious interaction between man and biosphere (Homewood and Rodgers 1991:2). Mary Leakey found the first homonid skull the year the Ngorongoro Conservation Area was established (33). Ever since, archaeological excavations, especially around Olduvai and Laetoli, have increasingly documented Ngorongoro Conservation Area as one of the cradles of mankind, with a history of some 3.5 million years of homonid occupation (31), and it now seems evident that pastoralists have been present in the area for at least 2,000 to 2,500 years (57).

However, the main objects of conservation, the communities of wildlife, also have been highly affected by this process of place making, and it becomes increasingly apparent to conservationists that the overall development may not have been for the better in the long run. Despite constant surveillance of the crater, the rhinos are almost extinct due to commercial poaching, whereas the wildebeest population has erupted to the point where soil erosion is observable in certain corridors on their migratory routes. Tourist vehicles are likewise responsible for the onset of soil erosion on the crater floor, and further demands for such amenities as bathing facilities at the lodges on the crater rim are likely to restrict vital water resources available not only to Maasai and staff within the area but to the communities outside, as the Northern Crater Highlands form an important water catchment zone for a much wider area. The ban on pastoralist pasture maintenance through range burning has meant an increase in detrimental wildfires and the spread of inedible, coarse grasses, reducing the carrying capacity of highland pastures.

While not wishing to romanticize indigenous cultures as inherently closer to nature or more ecologically sound per se, I nevertheless do propose that a major reason for the present state of affairs in Ngorongoro Conservation Area stems from a failure to comprehend the processual interconnectedness of Maasai territoriality and the spatial distribution of wildlife.

Parking pastoralists

Due to growing concern over big game hunting, the Ngorongoro Crater was declared a closed reserve by the British colonial administration in 1928, and the entire Serengeti-Ngorongoro area was declared a national park in 1940, but before 1951 there was no active enforcement of legislation (Århem 1985a:49). At that time, legislation did not affect the rights of people living within the area; indeed, they were given positive assurance that there would be no interference with their rights to live in the park. Practical conservation measures nevertheless increasingly restricted human activity by forbidding hunting, regulating human settlement and the movement of livestock, and banning the use of fire in range maintenance. In 1954, all cultivation in the area was prohibited and, in the words of Århem, "a single-use concept of conservation, epitomized by the notion of National Park, came to dominate resource management" (49). Restrictions caused political unrest among residents, which in 1959, on the eve of British rule, resulted in the partitioning of the original park into two separate land use units, Serengeti National Park, reserved exclusively for wildlife, and Ngorongoro Conservation Area, in which human inhabitation was not only permitted but was to be actively developed according to the multiple land use designation.

In fact, several sublocations in the conservation area had long been subject to agricultural alongside pastoral activities (Århem 1985a:54). In the Endulen area, commercial farming was practiced by non-Maasai immigrants until the 1970s.[1] In the early part of the century, some Germans had established a couple of farms on the crater floor (Homewood and Rodgers 1991:70). They functioned for some decades, coexisting with the abundant wildlife and the Maasai, but at the time of the establishment of the conservation area, there were only ruins left. Although such ruins could be seen as a

---

[1] Agriculture was formally prohibited in 1954, when the area was still included in the colonial Serengeti National Park construction. However, after the partition into two separate land use units in 1959, cultivation was in some areas allowed as an acceptable form of land use under the multiple land use policy. In the Endulen area, which lay outside the park until the 1959 reorganization, as well as in the Empakai area in the Northern Crater Highlands, cultivation was quite extensive by the mid-1970s. However, feeling that it was now getting out of hand, all cultivation in Ngorongoro Conservation Area was banned in 1975 (Århem 1985a:53–54; 1985b:33).

national heritage, they have gradually been removed. Reportedly, staff
members stationed at the Headquarters just above the crater have utilized
the remains in constructing their living quarters. Less favorably interpreted,
the removal of such testimonies to former settlement helps uphold the illusion that the crater is a natural zoo.[2]

The vast Serengeti Plains had been the customary home area of the
Ilserenget section of the pastoral Maasai, who had for long occupied the area
permanently and utilized the plains as their wet season commons together
with members of other independent Maasai sections coming down from the
surrounding mountains in the rainy seasons. From then on, however, it was
set off as the commons of wildebeest and other migratory ungulates for the
sake of the global community. An international lobby concerned with preserving the unique, migratory routes of the wildebeest had successfully
argued that continued competition for resources between wildebeest and
Maasai cattle would inevitably prove detrimental to the wildebeest. In exchange for improved water facilities and the establishment of veterinary
services, the Ilserenget Maasai in 1959 agreed to vacate the area and move
permanently to their former dry season refuge in the mountains. The
Ngorongoro Highlands were the long-established[3] common dry season
grazing ground of three independent territorial sections of the Maasai, the
Ilserenget, Ilsalei, and Ilkisongo, all coming up from surrounding plains
(Jacobs 1965:124). Today some forty thousand members (Potkanski and Loft
survey, pers. comm.) of these sections live permanently within the area, and
a disease-aggravated poverty process is forcing them to become increasingly
sedentarized in the highlands around the few trading centers in the area.

Seen from a Maasai point of view, the history of conservation management has been one continuos process of land alienation, restrictions imposed
on their pastoral mode of production, and broken promises. The only alternative solution to a future total eviction from the area so far voiced by management is the possibility of a further subzoning of the conservation area, in
which case the Maasai inhabitants will be further concentrated on still less
land. In realization of their precarious political position, the Maasai in
Ngorongoro Conservation Area have in recent years responded to this process of deterritorialization by seeking recognition and reterritorialization.
Internally, this has been articulated around their local leading ritual special-

---

[2] The crater represents a microcosm of the great variety of habitat types and species. At the time
of the expulsion of the Maasai from the crater, it was believed that larger mammals were unable
to climb the crater walls, an assumption later found to be false (Homewood and Rodgers
1991:132). Thus, the Ngorongoro Crater was seen as a naturally bounded zoo.

[3] Following Fosbrooke (1948), it is generally believed that the Maasai around 1850 conquered
the Ngorongoro Highlands from the Barbaig, called by the Maasai *Iltaatwa*, hence their name,
Oldoinyo Laaltaatwa, "the Mountain of *Iltaatwa*" for the Ngorongoro Highlands. An interchange-
able name is *Osupuko Laaltaatwa*, *osupuko* connoting "dry season pasture" or "wetland". However, Homewood and Rodgers find evidence that Maasai occupation may have begun much
earlier, perhaps as early as 1700 A.D. (Anacleti 1991:59). Ngorongoro is a Swahili corruption of
the Maasai name for the crater itself: *Korongoro*.

ist, who had a dispute over ritual payments with his Ilkisongo colleague in the Monduli Mountains and controls the coordination of an important age-set ritual throughout the independent sections of Maasai. Claiming local ritual payments signifies that the Maasai in Ngorongoro should be recognized as an independent section of the new political era. Externally, they are seeking political recognition and empowerment by forming their own Ngorongoro-based nongovernmental organizations (NGOs) in the hope that through this new form of organization they may be recognized as equal counterparts in the process of development.

Disease interactions and the processes of pastoral poverty

That the removal of competitors has been of great advantage to the wildebeest is evidenced by the fact that their numbers grew from an estimated 200,000 in 1959 to a peak of an estimated 1.4 million in the late 1970s. Now their numbers have apparently stabilized at around a million (Potkanski 1994b:49), amounting to an annual population growth rate of close to 15 percent before the stabilization. In comparison, Maasai cattle holdings dropped from an estimated 161,000 in 1960 to an estimated 113,000 in 1987 (Homewood and Rodgers 1991:146). In the same period, the human population has probably doubled. The wildebeest eruption has meant that the wildebeest that come to calve now migrate well beyond the unfenced boundaries into the corner of the Serengeti Plains formally lying within the Ngorongoro Conservation Area. The wildebeest eruption has been further facilitated by forced immunization of Maasai cattle against rinderpest, a disease communicable to wild ungulates. However, malignant catarrh fever, which the calving wildebeest transmit to cattle, practically debars the Maasai from utilizing the crucial, highly nutritious, short-grass associations of the plains, forcing them to stay almost all year in the tall but nutritionally inferior grasses of the highlands. This enforced sedentarization in the relatively moist and well-vegetated mountains again leads to drastically increased levels of tick-borne diseases in cattle, both east coast fever, also known as theileriosis, and within the last fifteen years a new cerebral variety of this disease.

Together malignant catarrh fever and especially the two varieties of theileriosis, none of which are as yet curable but only preventable, are responsible for truly alarming and accelerating cattle mortality rates. Thus, in the period from the dry season of 1991 until, and including, the rainy season of 1992 the altogether 593 Maasai individuals living in an average, arbitrarily chosen locality would have had together 1,574 head of cattle if all could have been kept alive. However, they lost 37 animals to east coast fever, 250 to bovine cerebral theileriosis, and 115 to malignant catarrh fever. These diseases were responsible for a total mortality of more than 25 percent. At the same time, this locality additionally had to sell and slaughter a total of 268 head of cattle in order to fulfill their ceremonial obligations and especially to get money to buy food (primarily maize flour) and medicine. Together they then had a total of 904 head of cattle left, part of which were ill at the time of the survey. For simplicity, I have omitted developments in small-stock, but it does not look much better. The overall tendency is toward more small-stock relative to cattle (Homewood and Rodgers 1991:146–48), in itself a sure sign of growing poverty among pastoralists.

It has been calculated that for a pastoral economy in similar surroundings in which cereals form an important part of the diet the absolute subsistence minimum is five head of cattle per capita (Kjærby 1979). Homewood and Rodgers argue that, although livestock counts in Ngorongoro Conservation Area are for methodological reasons not quite reliable, they are nevertheless revealing of tendencies. Recalculating cattle and small-stock holdings into standard stock units, they document a drop from an average of twelve standard stock units per capita in 1960 to five in 1979 (1991:216). Thus, before the first registered outbreak of the new, cerebral variety of theileriosis, the Maasai economy was balanced on the absolute subsistence minimum. Since that time, cerebral theileriosis has rapidly taken on epidemic proportions, and Potkanski (1994a) has calculated that in 1992 average stock holdings were about three per capita. These data reveal that only since the introduction of cerebral theileriosis has the cattle economy dropped well below the absolute subsistence minimum but also that the development of such conditions has been long term. From this perspective, it is not surprising that the clan-based systems of pooling livestock gifts to help unfortunate, destitute kinsmen reestablish their herds are rapidly breaking down (Potkanski 1994a, 1994b). Few have any cattle left to redistribute.

Facing hunger, most Maasai, much to their own regret, have adopted small-scale hoe cultivation, primarily of maize and beans. Agricultural activities being illegal within the premises of the conservation area, such cultivators were in the beginning imprisoned and prosecuted,[4] but, realizing that the government was unable to provide alternative means of survival on a permanent basis, the ban on cultivation was temporarily lifted in 1992. However, this is only a temporary solution to what seems to be an inherent problem of the park construction.

As relatively wealthy families tend to farm at least as extensively as families with no or few cattle, because such families have more resources for buying seed corn and are less occupied with meeting immediate dietary needs the adoption of agriculture is apt to widen the gap between rich and poor families over time.

As mentioned, when the Maasai were moved into the national park—or, rather, when the park moved into Maasai territory—it was part of the deal that management should develop a veterinary service to assist in developing

---

[4] With a keen sense of irony, some of these lawbreakers pointed out to me that in prison they had actually improved their general knowledge of agricultural techniques, as prisoners are occupied with the raising of crops.

Maasai pastoral production. In the last decade or so, an increasingly smaller part of the total budget has been set aside for this purpose, with salaries taking the greatest toll. In the same period, the Ngorongoro Conservation Area has only for the first time been able to make a profit, despite the fact that each year the area has been visited by well over 100,000 tourists (Homewood and Rodgers 1991:238–41), a number that has steadily risen to almost 250,000 (*Chicago Tribune* 1996). The veterinary staff is almost exclusively non-Maasai and poorly paid. Because of the isolation of the place, there are few possibilities of earning an additional income, which is often necessary to keep children at school. Such conditions favor corruption, among other things, leading to the private sale of veterinary drugs if they reach the area at all.

Thus, in recent years veterinary dip services have become practically defunct. Ideally, cattle should be dipped in acaricides once a week in the dry season and twice a week in the rainy season in order for prophylaxis to be optimally effective. During the year I did fieldwork, the Endulen dip was only working one week, and that happened to be in the dry season when tick control is not as urgently needed. This is not a new development. Homewood and Rodgers, who were doing their research in Ngorongoro around 1981, described what may have been the first outbreak of the new cerebral variety of theileriosis in Olairobi village close to the Crater headquarters of the Ngorongoro Conservation Area Authorities (NCAA) (1991:185). They stated laconically: "Tick control is a major management issue. The NCAA has in the past held responsibility for acaricide dips for livestock: these are largely non-functional" (184).

Many Maasai have wholeheartedly adopted the idea that modern veterinary drugs are superior to their own indigenous remedies. Such informants also say that there are now many more serious cattle diseases in the area, many of them unknown in the past. Thus, they hold that bovine cerebral theileriosis is caused by a new variety of ticks. In general, Maasai are keenly interested in participating in a sensibly planned development aimed at keeping their cattle healthy. The most development-oriented Maasai purchase their own drugs for tick prophylaxis from Arusha or Nairobi; but prices are high when drugs are individually purchased, and the cost for one year of effective prophylaxis for one animal is likely to exceed its market value.

In 1994, a Danish nongovernmental organization conducted a pilot project to demonstrate that it was possible to halt cattle death rates caused by tick-borne diseases completely by relatively small means and efforts simply by performing the dipping program as prescribed and leaving it in the control of the pastoralists themselves. Hardly any cattle died of tick-borne diseases during the experiment (Loft, pers. comm.). However, because of the political sensitivity of the issue of pastoralism in Ngorongoro, and a change of minister for development in Denmark, the main program never materialized. The economic situation in Tanzania leaves no hope that the dipping program will be restored in the foreseeable future.

Maasai traditional veterinary practice has mainly consisted of prophylaxis. One of the crucial components has been seasonal rotations between pastures based on intimate knowledge of disease interaction between wildlife and cattle, insect-borne diseases, high-risk periods, and range burning (cf. Branagan 1962). Thus, disease minimization is an integral aspect of Maasai territoriality.

Applied policies on the East African commons

Working with Maasai in the Monduli Mountains (cf. Ndagala 1991), the Tanzanian sociologist Ndagala has called attention to the enormous impact Garrett Hardin's theory of the tragedy of the commons (Hardin 1968) has had on policies applied to African pastoralists (Ndagala 1990). Hardin's theory has been misunderstood to represent a general model of an in-built mechanism for land degradation in the pastoral mode of production in which the combination of communal ownership of land and individual ownership of cattle inevitably leads to land degradation, which in turn leads to famine, as, according to Hardin's model, each herd owner is supposed to think only about how to obtain a maximum increase of his herds, even at the expense of the future quality of pastures (McCabe 1990). Thus, in Hardin's worldview, communal ownership is a tragedy.

Hardin had no firsthand experience with African pastoralist production, and he did not account for the social mechanisms regulating communal landownership systems of pastoralists. Besides, the model was never intended to be an economic or ecological theory. The image of the communal pasture was presented only as an illustration of Hardin's *demographic* hypothesis that individual self-interest will result in the abuse of a commonly held resource.

Yet, the theory has been widely taken to reflect an economic as well as an ecological dogma about pastoralism. "This belief", Ndagala writes, "gave rise to 'top-bottom' conservation policies and programmes in pastoral areas. Pastoralists have always been told what to do. The 'bottom-up' approach, whereby they would be listened to and their ideas taken into account, has never been a popular one" (1991:176). Based on anthropological and ecological interdisciplinary fieldwork data on land use practice among the Turkana of northern Kenya, McCabe has demonstrated (1990) that the same attitude has prevailed in development projects there;[5] yet Turkana pastoralism was in no way found to degrade land through overstocking beyond the carrying capacity of the environment.

---

[5] See Peters 1994 for a similar observation on applied policies concerning pastoralism in Botswana.

McCabe reminds us that such attitudes toward pastoralism did not originate with Hardin. In the fourteenth century, Ibn Khaldun discussed the same issue (1990:82). Within anthropology Herskovits in 1926 formulated a similar interpretation of pastoralists as economically irrational and automatically driven to overstocking by an obsession with cattle of a basically symbolic and ideological nature, the so-called East African cattle complex. Herskovits's ideas were partly prompted by the 1920s Dust Bowl environmental catastrophe in the United States, which caused an enormous outpouring of writings on soil conservation (Enghoff 1990:10). On a general level, such arguments reflect the mental attitude of the territorially fixed and bounded agri-culture toward the mobile and seemingly unbounded nomads.

Hardin's theory gained support because it underpinned convictions already widespread, and, most importantly, it happened to appear in print just prior to one of the recurrent, extended Sahelian droughts in which hundreds of thousands of head of livestock succumbed (McCabe 1990:82). Hence, Hardin's theory took on the character of a prophecy. Subsequently, desertification of productive rangelands took place on a hitherto unobserved scale, causing donor agencies and national governments on a wide scale to adopt the position that pastoral nomadism was inherently destructive to the environment. Accordingly, there was implementation of development programs aimed at privatizing formerly communal rangelands. McCabe observes that "these projects have met with almost uniform failure, and have in many instances contributed to increased human suffering and the further degradation of the land" (83).

There is some evidence to support the theory that extended droughts are a feature of the climatic regime constraining the long-term resource utilization of East African rangelands. The International Livestock Center for Africa has recorded the occurrence of serious droughts in northeastern Africa with an unstable frequency of about a decade from 1918 to 1984 (McCabe 1990). In the decade following the appearance of Hardin's theory, East Africa actually experienced two such periods immediately upon each other. Probably, this unusual climatic variation has contributed significantly to lending Hardin's commons model explanatory value, especially as it also coincided with an increase in externally donated development programs.

The history of the Ngorongoro Conservation Area and the Serengeti National Park provides an example in which the former commons of pastoralists have been transformed into the commons of the global community along similar lines of reasoning.

Maasai territoriality

Ndagala demonstrates that pastoralists have a system of resource conservation based on territoriality, and in his view land degradation is a reflection of the "erosion" of territoriality in traditional pastoral resource management (1990:176). Much of the misunderstanding of pastoralist utilization of rangelands, he states, stems from a failure to capture precisely which features are important in control over land in pastoral production. In studies of pastoralism, the focus is solely on the land concept, whereas for pastoralists it is not so much land in itself that is seen as a resource; rather, it is the resources of water, salt licks, and types of pastures contained in such a territory that are important. Particularly resources like permanent water and mineral sites may be scarce and scattered. A pastoral territory, Ndagala states, "envisages an area which includes all these components while, at the same time, allowing easy mobility and manoeuvrebility should the need arise. It is an area-unit that encompasses all the spatially-dispersed elements necessary to pastoral production. Territory is, in a way, an ecological notion" (176). Thus, the boundaries of a territory are defined in relation to resource accessibility.

It is precisely because of the failure to understand this crucial notion of territory that it has been possible to alienate large tracts of land from pastoralists in the belief that such areas were free land, not realizing that apparently unused land areas controlled by pastoralists were crucial drought reserves purposely set aside to enable survival in the irregular but frequent years of drought.

The Maasai have a system of free access to pastures but owned rights to water. A territory, as well as the section of people who occupy it, is called *olosho* (pl. *iloshon*). Within each *olosho,* there are several subunits, or localities, *enkutoto* (pl. *inkutot*), and in daily matters of production and political decision making these *inkutot* are the operative units, whereas *iloshon* are mainly operative in ritual matters and in the past in warfare. Usually, *iloshon* take their names from pastures, often plains, whereas *inkutot* generally take their names from characteristics of the landscape in which they are situated. When people move to settle in the area of another *olosho*, however, the name of their *olosho* of origin tends to cling to them, often into the succeeding generations, indicating that *iloshon* are as much a matter of social groupings of people. With large-scale migrations of entire local communities having taken place in the past, original names have tended to become perpetuated, regardless of the fact that they no longer refer to the main area of permanent settlements. This is to be seen, for instance, in the case of present day Ilpurko, who are spread over a much larger area, and most of them in Kenya, whereas the Purko Hills are situated on the Tanzanian side of the border.

It is above all residence within the territory of an *enkutoto* that gives herd owners access to the use of the territory. However, as Ndagala has observed, each territorial unit has a territorial overflow into other such units utilized in times of emergency after consultation with the respective *inkutot*. It is normal for members of one *enkutoto* to move into another for grazing purposes, and in extreme cases cattle can be moved all the way from southern to northern Maasailand (1990:177). It is, moreover, not uncommon to take up permanent residence with an *olosho* other than one's own.

Territorial divisions and subdivisions mainly indicate the units predominantly used by a certain group rather than areas of exclusive use. Territory is, as Ndagala puts it, "a communally-controlled resource subject to a number of corporate rights and obligations" (1990:177). Maasai maintain that in principle they can graze their cattle everywhere in Maasailand (Potkanski 1994b:18), but in practice this happens only in cases of emergency such as widespread drought (ibid.). In the extreme droughts of the 1970s, much livestock was moved between Kenyan and Tanzanian sections in an attempt to relieve the effects (ibid.).

In daily practice, Maasai distinguish between primary and secondary user's rights to pastures and between individual and collective rights in water. Man-made water resources are individually controlled and inherited by a man's eldest son, whereas natural sources are collectively controlled by *inkutot*. Still, a needy herdsman will never be denied water. Likewise, individual controllers of wells will share any surplus with kin and neighbors.

Visitors from other *inkutot* or *iloshon* hold secondary user's rights, and through time and regular seasonal use secondary user's rights become institutionalized (ibid.). When this happens, it is no longer necessary to negotiate use. Secondary user's rights are often reciprocal, and they do not necessarily follow the boundaries of *iloshon* but primarily reflect broad, ecologically continuous and self-sustaining units that may encompass adjacent areas belonging to other *iloshon*.

It is evident from Potkanski's registration of herd movements (1994a, 1994b) that the overall tendency in the system of stock movements is that well-watered mountains are utilized as dry season pasture and for permanent settlement, whereas plains are wet season pastures with temporary camps. The Maasai in Ngorongoro Conservation Area explicitly base their movements on a system of recognized ecological zones of dominant vegetation types, nutritional values, availability of water, and disease interaction patterns.[6]

Mobility is crucial in such a climatically restricted system, as seasonal movements of herds facilitate distribution over wide areas, allowing periodically exhausted pastures time to recover. The gradual alienation of especially high potential pastures accompanied by the imposition of fixed boundaries has restricted mobility everywhere in Maasailand, as in other pastoralist areas of Africa, forcing the pastoralists to sustain their herds on diminishing territories of a relatively poorer quality of grazing.

---

[6] In Johnsen 1996 and 1998, I examine more closely these Maasai notions of ecology and landscape, as they are equally fundamental in Maasai medical practice.

From communal to public land control

Ndagala points out that "when territorial control was taken over by the State, pastoral territory lost its communal character and became public land", and at the local level public land was finally taking on the commons character assumed by Hardin's model to be intrinsic in pastoral grazing systems (1990:182). That most *inkutot* had boundaries overlapping those of others was, Ndagala states, most disturbing to government officials, who thought it unrealistic that such large land tracts were ever effectively utilized, and they used their doubt to prove that the pastoralists did not know their own boundaries. Consequently, supposedly unoccupied land could be readily alienated and firm boundaries drawn around the pastoralists (177). This is nowhere more clearly borne out than in Ngorongoro Conservation Area, where considerations toward the outside public ever since 1959 have increasingly prevailed over considerations toward the communities within.

In Tanzania, this mainly took place in the heyday of socialism. *Ujamaa*, as it is called in Swahili, was a political program of villagization intended to facilitate socialism and self-reliance and adopted after the famous Arusha Declaration of 1967. It aimed at concentrating dispersed rural settlements into manageable, concentrated units according to a ten-cell principle to allow easier access to development facilities as well as easing administration. In the words of Århem:

> The welfare of the people now came to the fore. The elimination of oppression and poverty became the supreme goals of development, increased production the means. ... The ujamaa village was defined as a voluntary association of people living and working on communally owned land. It was conceived as a revitalisation of the traditional concept of *ujamaa*, meaning "communal living" and implying unity and self sufficiency, cooperation and sharing. (1985a:19)

Thus, the *ujamaa* campaign was envisaged as a popular movement designed to mobilize the peasant masses politically, and it aimed at a change from below and within, involving full community participation (20). It is quite striking that this program was actually modeled on concepts closely akin to those assumed to affect the pastoral mode of production negatively in Hardin's model. Contrary to expectations, and several five-year plans, the program never gained popularity among the peasant masses, and only measures of coercion introduced in the early 1970s could secure the program some progression. Despite the fact that the *ujamaa* program was the largest resettlement effort in the history of tropical Africa (22), most *ujamaa* villages were only registered by name in order to receive government benefits and never established communal relations of production or nucleated villages.

The failure to implement the intended program among the country's majority of agriculturalists caused the government to realize that it had to take a different course among pastoralists (Århem 1985a:40), resulting in the launching of the so-called Operation Impernati in Maasailand, *impernati* meaning "permanent settlements" in Maa (41). The operation was launched in 1974–75 with the aim of establishing permanent settled, livestock development villages, and planning teams went out to instruct the pastoralists to form villages according to already existing localities. According to Århem, the process was generally quite smoothly carried out, and the village layouts were flexibly imposed and adapted to local conditions, often modeled on existing land use and settlement patterns. Hence, movements were mostly minor relocations within already occupied territories. But coercion did occur; according to Århem, at least two settlements in Ngorongoro were burned in order to make the inhabitants move, and the general planning was poor. Often people were simply concentrated around preexisting trading centers (42).

Since that time, much voluntary concentration around these villages and trading centers has taken place, and continues to do so, but above all this is an effect of the poverty process. Facing hunger, destitute herdsmen move to where the food is. Concomittant with these efforts toward further sedentarization the Maasai were excluded from the crater and the ban on cultivation was more forcefully enforced.

Århem points out that the Maasai responded tactically and strategically to this operation, displaying initial compliance in the expectation of acquiring rights of occupancy in their land and helping to defend their pastures and water sources against encroachment from agriculturalists, yet at the same time fearing the villagization program as another step toward their subjugation and alienation from their land. And rightly so, because in effect the operation imposed a new, supreme authority structure on the traditional communities by introducing a nationwide system of party-controlled local leaders even in the smallest *enkutoto*. This weakened the traditional political leadership and located the center of authority outside the pastoral community, as Århem noticed (1985a:44). It was also an effort to create more nucleated and sedentary settlement patterns, and it restricted the size of settlements as well as herds, creating among the Maasai a general feeling that their very way of life, the transhumant mode of production, was under threat.

In retrospect, my informants in the Endulen subarea generally agreed that compared to other ethnic groups in Tanzania they had been left largely unaffected by the villagization campaign, as the various traditional *inkutot* had merely registered as ten cells united in forming the various wards of the new villages and few actual movements of people took place there. Likewise, in quite a few instances *balozis*—local, elected representatives—in the new governmental system were in 1992–93 traditional leaders who had become members of the Chama Cha Mapinduzi (CCM) ruling party in order to be able to display the party membership book that determined eligibility. However, Maasai of the elders' generations in Ngorongoro Conservation Area have generally come to realize that the skills of the younger and better educated generations are crucial in dealing with the outside world, and quite a few *balozis* are recruited from their ranks now. In contrast, in internal political matters traditional modes of organization and leadership continue. Ndagala (1990:177) is of the opinion that villagization has meant a radical change in the Maasai concept of *enkutoto*, to the effect that people now identify more with their respective villages "such that boundaries of *inkutot* have become almost identical to those of the villages" and today movement is more limited (177).

People in Endulen definitely identified strongly with their residential *inkutot*, as they have always done, and they did indeed recognize rather uniformly that life nowadays is much more sedentarized than it used to be. However, contrary to Ndagala's findings, most of my informants primarily connected growing sedentarization to the drastic decline in herd sizes and thus their perceived, collective state of poverty. According to such explanations, it is quite simply that when a man's herds dwindle below a certain level moving cattle becomes uneconomic and a waste of energy, no longer a means of facilitating further growth.

Thus, indirectly the goals of Operation Impernati have been achieved, and perhaps the fact that it was partly modeled on concepts not entirely strange to the pastoral Maasai internal organization accounts for its relative success, but the net result has been contrary to intentions. The Maasai cattle economy in Ngorongoro Conservation Area has definitely become highly sedentary and stock levels in cattle have been drastically reduced. However, this does not mean that the Maasai have been able to increase their contribution to the national economy or the supply of beef and dairy products, nor have they received a larger share of the benefits of development. On the contrary, stock reduction is the result of growing levels of disease, not improved breeding and care; pastures appear to have become even more depleted, not because of overstocking beyond carrying capacity but because of undergrazing and (particularly) poor maintenance. The only development benefits established by the colonial and conservation area authorities, manmade water facilities and the veterinary services, have been allowed to deteriorate almost completely. Vis-à-vis the rest of the nation, the Maasai seem to be more underdeveloped than ever.

The process of transforming customary, communal landholdings into public land controlled by the state apparently culminated by the end of 1992, when the Tanzanian government declared that it intended to abolish the judicial principle of customary land rights henceforth. However, coinciding with the announcement of future democratic elections and a change to multipartyism, this decision caused immediate concern among national and international observers, and the final decision on the issue remains to be made. At present, the Maasai citizens of Ngorongoro Conservation Area are recognized to have "deemed rights of occupancy" like other customary land holders in Tanzania, although these have proved as susceptible to alienation as elsewhere in the country. While they are currently unable to acquire title deeds to land even where their houses are built, they have recently seen two foreign hoteliers acquire titles to land and construct tourist hotels on the Crater rim. Consequently, they fear that their rights of occupation in NCA are less than secure than those of outsiders, and they will one day be forced to move out. (Lane 1996:18)

There are still hopes that this situation may improve in the future, however, as the new National Land Policy makes mention of customary titles (18).

Historically, most Maasai territorial sections have taken their names from plains such as the *Ilserenget*, the *Ilkisongo*, and the *Ilsalei*, indicating that it is here they feel at home. Plains names have tended to cling to sections of people even after large-scale, sectional migrations, of which several took place under colonialism, including forced migrations onto reservations. The Maasai in Ngorongoro Conservation Area, however, are now effectively landlocked and sedentarized in the highlands, having increasingly less access to the plains. Reflecting that the highlands are now their home, Maasai inhabitants commonly refer to the conservation area they live in as Oldoinyo Laaltaatwa, or Osupuko Laaltaatwa, that is, the "Mountain" or the "Dry Season Refuge of the Tatoq", thereby reflecting awareness of historical territorial processes. In their efforts to establish their de facto territorial unit as an independent section of the modern era recognized to have its own specific problems, they are currently forming their own NGOs, and attempting to obtain political recognition and influence over decision making through donor-sponsored development projects.[7] These organizations, however, are formed around the internationally recognized name Ngorongoro. Parallel to gaining external political recognition through building such modern institutions around traditional and modern types of community leaders, internally the local leading ritual specialist, *oloiboni* Birikaa, has concomitantly sought

---

[7] Indigenous NGOs are mushrooming in Ngorongoro Conservation Area, as they are all over Tanzania, following new legislation in the early 1990s facilitating the formal acknowledgment of indigenous NGOs. Thus, the Ngorongoro Pastoral Survival Trust (NPST), Ngorongoro Pastoralist Development Organisation (NGOPADEO), and a proposed Ngorongoro Environmental People's Organisation (NGOEPO) are all independent organizations of the 1990s. Although they are all Ngorongoro based, none has an NCA-wide constituency (Lane 1996:27). Besides these, there is the Korongoro Integrated People's Orientation to Conservation (KIPOC), which despite the name is based in Loliondo, the capital of Ngorongoro District, which lies outside of Ngorongoro Conservation Area. KIPOC has some support among NCA residents, but is mainly seen there as representing the Maasai in Loliondo. Since its foundation in the late 1980s, it has been working for wider pastoral cooperation. Thus, following the NAFCO wheat project scandal in Hanang District in 1992–93, in which Amnesty International, among others, documented systematic and severe violations of the human rights of the agro-pastoral Barabaig, whose land had been alienated, a KIPOC local committee was formed there. One Maasai NGO, Inyuat e-Maa, the Maa Development Organisation, formed at the First Maa Conference for Culture and Development held in Arusha in December 1991, has from the beginning seen the need for an umbrella organization for the many subsequent NGOs. This tendency toward very localized and parallel, almost competitive, NGOs seems to reflect to a large extent the strongly egalitarian internal political organization of the Maasai. Maasai NGOs also have been formed in other areas.

independent ritual recognition for their de facto section through a dispute over age-set payments with the supreme *oloiboni* in Monduli.[8] Externally as well as internally, the Maasai in Ngorongoro Conservation Area are currently deeply engaged in a process of reterritorialization.

Poverty, hunger, and the "eating of dust"

P. S. Nestel observes of Maasai dietary practices that:

> Historically, the Maasai have been characterised as tall lean people who subsist on a protein rich diet of milk, meat and blood. There is, however, little quantitative dietary data on the food and nutrient intake of traditional Maasai to substantiate this view. (1989:17)

Nestel did her studies in Kajiado District, Kenya. Compared to the marginalized position of the Maasai in Ngorongoro, the Maasai in Kajiado are much closer to the metropolitan center of their nation-state and they participate much more in the market economy. Money and commercial food items such as maize flour, rice, chapatis, tea, vegetable fat, and sugar are obtained more easily than in Ngorongoro, just as the option of adopting agriculture exists as a legal pursuit in Kajiado. The status of Ngorongoro as a conservation area leaves only the option of traditional economic activities to its residents. Thus, development has left the Maasai in Ngorongoro the sole choice of clinging to their customs, whereas Kenyan development efforts have established group ranches with collective title deeds in Kajiado. As such, Kenyan efforts to transform traditional Maasai land tenure systems into group ranches correspond to the Tanzanian Operation Impernati, reflecting the political goals of a liberal market economy and a socialistic planned economy, respectively. However, the Kenyan development design has not lessened the overall pressure on the pastoral economy, and the effects of population pressure, sedentarization, and poverty are widespread in Kenyan Maasailand as well.

Nestel demonstrates how Kajiado Maasai are not particularly tall but are particularly lean (1989:27). In spite of their easier access to purchased food items and vegetable products, Kajiado Maasai have little cash, and they continue to get most of their calories from milk; however, its availability varies seasonally and is heavily dependent on rainfall and the number of cattle owned. Purchased maize-meal was the alternative staple when insufficient milk was available (19).

Nestel further states that meat consumption is dependent upon the number of dead or dying animals and the timing of ceremonies, and that both types of events are largely seasonal. Thus, according to her, meat cannot be

---

[8] Internally, the political aspects of this dispute were to a large extent discussed as a case of sorcery and retaliation seen to be typical of leading *iloibonok*. For a lengthy analysis of such aspects of Birikaa's dispute, as well as its wider symbolic significance in Maasai discourse, see Johnsen 1998.

classified as a food staple. Blood "was rarely drunk" (20), nor was it in Ngorongoro Conservation Area in 1992–93. Nestel concludes that, although the Maasai in general get a very protein-rich diet, they also get only 65 to 80 percent of the World Health Organization's (WHO's) recommended daily intake of energy. Moreover, they primarily get their calories from milk, not maize. She concludes that increased sedentarization will eventually lead to increased undernutrition among the Maasai, as sedentarization generally reduces the availability of milk (29).

Barbara Grandin (1988) collected data on a group ranch close to where Nestel worked, and she demonstrates how even the diets of relatively wealthy Maasai are remarkably low in caloric content. The Maasai in Kajiado uniformly subsist on an energy intake far below WHO's recommended daily intake, regardless of wealth, and total caloric consumption averages 70 percent of the recommended daily intake (7). There is nothing to suggest that energy intake could in any way be higher in Ngorongoro Conservation Area. Grandin seems to advocate study of the appropriateness of WHO's standards in the case of the Maasai (7).

Although it was to be expected that the custom of giving gifts of milk or loaning lactating animals to poor households would to some degree level out actual intakes between wealth strata, Grandin states, neither this nor milk sales alone can account for the uniformly low energy intake. The reason, she argues, is to be found in the allocation of milk between humans and calves. "Maasai do not speak of milking cows; they speak of milking calves", the terminology thus underscoring the Maasai perception of competition between calves and people (1988:12). The point is that "Maasai perceive milk as a limited good which must be carefully apportioned between human and calf use. As rich households have so many more cattle, they can afford to leave more milk for their calves and still obtain targeted levels of offtake" (16). According to my informants, the Maasai do in fact speak of "milking cows"; the point is that, when one wants to emphasize the actual competition between calves and human beings for milk, one may in this way adopt the metaphorical—almost ecological—notion of "milking calves".

Extreme seasonal and precipitational variations form a constraint on pastoral production; drought and hunger are never far away. As Grandin formulates it: "Savanna pastoral systems are characterized by cycles of boom and bust which result largely from drought and to a lesser extent from disease and political instability" (1988:4). Thus, when relatively well-to-do Maasai maximize the portion of milk left to the calves they are adopting a strategy of maximizing long-term survival by investing in the family's capital and long-term food security in order to anticipate future situations of constraint.

In Ngorongoro Conservation Area, the present poverty levels are primarily caused by diseases, the levels of which are ultimately politically conditioned, yet droughts, periodical or prolonged, are often the immediate cause of starvation, as was the case in 1992–93. Thus, hunger in Maasai experience is intimately linked and hence associated with the dry season. Only a few purchased food items are available in Ngorongoro, mostly from non-Maasai entrepreneurs who have settled as shopkeepers within the area, although a few successful herd owners are setting themselves up as shopkeepers in remote localities. Endulen Hospital runs a maize mill where people come to have their own maize ground for a fee, and it is one of the main local providers of maize, as it sells at the lowest possible price. However, most people in Ngorongoro have serious problems finding cash to purchase food.

If in surplus, milk is manufactured into sour milk and butter. Goats and sheep are sometimes slaughtered "just for food" by better-off Maasai but never cattle, unless the animal is about to die anyway. As dead cattle cannot easily be wasted, carcasses are usually eaten unless the cause is a disease transmittable to humans. Eating cattle is the equivalent of eating one's wealth, hence any slaughter is a sacrifice in a restricted economic as well as a more encompassing social sense. The occasion is invariably ritualized.

Due to poverty, milk is increasingly replaced with *enkurma*, literally "flour", a fluid and drinkable porridge made from maize flour. If possible, milk is added in order to improve the nutritional value; alternatively, butter, animal fat, or purchased vegetable fat is used. However, as poverty in Ngorongoro is extremely severe, many households are forced to cook their *enkurma* only with water. Eating *enkurma* without milk or butter or any type of fat in it is like "eating dust". It may fill the stomach, but it does not "keep the blood" of the body, that is, it does not keep the body fat and healthy but leaves it utterly dry. Drawing on similar imagery, a child feeling hungry may say to its mother, "Mama, I'm having a dry season", thus implying "a widespread lack of milk", as one informant pointed out to me.

Poor women having a little milk frequently try to sell it in order to get cash to buy enough maize flour to feed the family through the day, but there are few buyers. That the change of diet implies an increase in the burden of women was expressed in the following way by an elderly woman who frequently sold milk in order to buy grain: "Now that we do not have enough, cattle it is actually women's responsibility to feed the family, whereas formerly it was the responsibility of their husbands [...] it is not that they do not provide for their families, it is simply that they do not have [the means to do it with]". Thus eating dust seems to be an inevitable consequence of the frequent periods of "bust".

Wild plants are perceived of and eaten only as medicine. A few are added to diluted milk in the dry season, and these are described as "medicines against hunger". Århem states that in the 1980s, only women would drink milk from small stock (1989:3). At present, small stock is often milked, and the milk is drunk by all. As mentioned, the relative ratio of small stock to cattle has been constantly rising through the last decades, reflecting increasing poverty and extreme mortality in the cattle herds.

Merker, writing of the Maasai of almost a century ago, stated that sheep's milk was much relished because of its fat ([1904]1910:32). The present milking of small stock seems to be a reversion to a long-established practice adopted in times of widespread poverty and lack of cattle, frequently triggered by drought or epidemics.

The history of pastoral poverty and the conservation of wildlife

Basing their arguments on regular aerial wildlife censuses conducted by the Kenya Rangeland Ecological Monitoring Unit, Homewood and Rodgers conclude that areas in which Maasai pastoralists are settled generally have a remarkably high concentration of wildlife compared to national parks with no human inhabitation bounded by agricultural communities (1991:197ff.). These censuses suggest that wildlife conservation in a number of national parks in East Africa is actually dependent on Maasai pastoralist rangelands as buffer zones for the survival of the migratory or seasonally dispersing wildlife populations and that pastoralism contributes to the retention of a high wildlife component.

However, few Maasai in Ngorongoro Conservation Area have so far gained any economic benefits from conservation or tourism. Although there have lately been creative Maasai initiatives directed toward getting a share of the economic benefits of tourism, such as crater donkey safaris guided by Maasai "warriors", these are not approved by management. Despite an increasing interest in people and cultural matters among tourists, and the unique position of being the only Tanzanian safari park allowing human habitation, the only tourist activity based on Maasai participation so far approved by management is the so-called cultural village at the crater rim, where Maasai from the nearby *inkutot* regularly display their dances and beadwork.

Facing hunger and the lack of solutions to their poverty, some Maasai in Ngorongoro Conservation Area are for the first time beginning to feel resentment toward wildlife, which could lead to the widespread killing of animals for meat. Maasai are otherwise extremely food selective, preferring to subsist solely on the products of their domestic stock, but they have periodically reverted to hunting in times of poverty and severe crisis in the pastoral mode of production. Primarily eland and buffaloes are likely targets; based on observations of types of excrement and shared diseases, these species are classified as relatives of cattle, just as it is said that gazelles are relatives of goats. There is a myth used to tease women about their economically subordinate position, saying that these groups of wildlife were once the domestic stock of women that escaped and turned wild because the women were too busy feeding their children. Now fathers have lost most of their stock, too, and it has in a sense been turned into wildlife. So far, there is no evidence that Maasai have participated systematically in commercial poaching, but if no solution to their problems is found it may only be a matter of time.

Livestock disease haunts not only the present state of the economy in Maasailand, but it was from the beginning of the era of written history an all-important factor in the encounter between customary occupants and colonizers. Prior to the actual colonization of East Africa, livestock imported into the Sudanic region with the purpose of feeding the colonizing armies had introduced rinderpest, a hitherto unknown cattle disease on the African continent.[9] This was the first of the series of epidemics that the great ritual specialist, *oloiboni* Mbatiany, had prophesied would mark the demise of the Maasai people (cf. Johnsen 1992). Fast as a bush fire it swept ahead of the colonial troops all over East Africa, causing an estimated more than 95 percent of all indigenous cattle to succumb in very short time (Kjekshus 1977:130). This historical period of cataclysmic epidemics is collectively memorized among the Maasai as the Great Disaster, and at present, a century later, many Maasai voice the anxiety that they are now facing another, perhaps final disaster.

Thus, the population and livestock patterns first met by the Europeans were as far from normal as they could be, although the newcomers generally seem to have perceived them to have been in a perpetual state of static, undeveloped primitivity. When the Europeans finally reached the territories of the Maasai, the impact of their presence had already swept ahead of them; the Serengeti Plains were practically devoid of people and domestic herds, having been hit hard by the rinderpest epidemic, which was quickly followed by epidemics of smallpox, cholera, and eventually famine and mass starvation. It was in Ngorongoro that the German explorer and scientist Baumann in the early 1890s met and described some greatly emaciated *"Jammergestalten"* of Maasai refugees from Serengeti barely surviving on hides and carcasses (1894:31). Baumann estimated that of the Maasai, who depended solely on livestock for their food, fully two-thirds had starved to death. In Serengeti, he apparently only found scattered bands of "Wandorobo", though they were clearly of a type closely related to Maasai (167). The majority of these so-called Wandorobo hunters were quite likely destitute Maasai victims of the epidemic who had opted to stay in their previous surroundings and adopt a temporary strategy of survival as hunter-gatherers. The term *Wandorobo* is a Swahili corruption of the Maasai term *Iltorrobo*, meaning a destitute herdsman and with the implication that he is practicing hunting for subsistence until he can acquire new herds.[10]

---

[9] Likewise, sand fleas, or jiggers, were introduced in Africa by the British, arriving in the ballast of the good ship *Thomas Mitchell*, which came from Rio de Janeiro in 1872 (Kjekshus 1977:134).

[10] *Iltorrobo*, or *Wandorobo* in Swahili, and *Ndorobo* in the Anglicized form by which they are more commonly known among Africanist scholars, are one of the riddles of anthropology because the term carries this double connotation of "poor Maasai" and "hunter-gatherers". There has been a long debate as to whether the Ndorobo form distinct ethnic groups of their own, or whether they are a residual category of destitute nomads (cf. inter alia Kratz 1980; Kenny 1981; Blackburn 1982; Chang 1982; Spencer 1973; Galaty 1982; Distefano 1990; and Klumpp and Kratz 1993; but also Baumann 1894).

At the time of the colonial encounter, then, Serengeti seemed practically unoccupied. Additionally, most Maasai were hard up, employing all possible means to build up their herds again. For many of them, this meant temporary association with other ethnic groups practicing agriculture or, as some of the Ilserenget had done, for a time adopting a subsistence strategy as hunter-gatherers.[11] My informants claimed that this was the time when Maasai for the first time began to practice agriculture for themselves, having learned the techniques from the people who gave them refuge, with whom they had previously exchanged grain for dairy products and hides as a precaution against dry season milk shortages. So, in reality, when the European conquerors eventually met the Maasai considerable territorial rearrangements had already occurred throughout the Maasai *iloshon*, and alternative strategies with regard to modes of subsistence were being widely practiced. However, such rearrangements in modes of production and territorial occupancy were largely temporary survival strategies.

Thus, the two major customary strategies for food security during crises in the pastoral mode of production are both incompatible with conservation management objectives, leaving the destitute and severely impoverished majority of the Maasai in the park with no legal alternatives to overcome their present poverty. The parallels between the Great Disaster and the present state of affairs in Ngorongoro Conservation Area are striking and as such are frequently explicated by informants, indicating that Maasai adherence to customary territories is remarkably stable over time despite colonial turmoil and subsequent interventions on behalf of the state. So are the consequences of submitting pastoral forms of production to the principles of agri-cultural hegemony. Thus, the process of making a narrowly bounded, natural place out of Ngorongoro, a sanctuary set aside for the international community to preserve and enjoy, has long historical roots. It is basically a history of politically created problems, the imposition of the values of the agri-cultural mentality on other systems of landscape maintenance. As such, in Ngorongoro Conservation Area the power of the extraordinary topography has steadfastly tended to conceal that what is really at stake is the topography of power, the present state of affairs in the natural communities clearly indicating that something is wrong with the assumption that pastoralist subsistence pursuits are antagonistic to conservation objectives.

The Maasai have intimate knowledge of wildlife, and the few who have found occupations as safari guides are quite popular among tourists. That the Maasai have a great potential as coconservationists is illustrated by the following. In 1992, the Maasai in Osinoni had for some time hosted a pair of

---

[11] The same kind of cataclysmic effects of the epidemics have been reported from all parts of Maasailand, and for some *ilmurran*, mainly from the northern sections of the British Protectorate, it became a popular strategy of economic resurrection to assist the conquerors in their campaigns to pacify other ethnic groups, for which service the British rewarded them with shares in the cattle they had raided from the newly pacified (Waller 1976). Still other Maasai survived by seeking refuge and famine relief with the early missionaries (Meinhof 1904).

hippopotamuses at their only permanent cattle watering point, a man-made dam constructed as part of the agreement to vacate Serengeti. This pair seemed to have materialized out of nowhere, as it were; there had never been hippos in that area before, the closest colonies being some seventy kilometers away. When the hippo pair grew into a family of three, people in Osinoni were extremely proud to display this local attraction to passing visitors, Maasai as well as expatriates. Despite the fact that having hippos at the watering point was not without problems for their herd boys, they expressed much concern when eventually the hippo baby died.

What I am arguing here is that a new approach to conservation based on community cooperation and local empowerment is probably the best way to reverse this long-term development. Maasai communities do in reality have a strong potential as cost effective allies in conservation once it is realized that their pastoral mode of production is not antagonistic to conservation objectives but to a large extent the very means by which this landscape was culturally constructed in the first place.

References

Århem, Kaj. 1985a. *The Maasai and the state: The impact of rural development policies on a pastoral people in Tanzania.* IWGIA Document 52, Copenhagen.

——. 1985b. *Pastoral man in the Garden of Eden.* Uppsala Research Reports in Cultural Anthropology, Uppsala: Department of Cultural Anthropology and Scandinavian Institute of African Studies.

——. 1989. Maasai food symbolism: The cultural connotations of milk, meat, and blood in the Maasai diet. *Anthropos* 84:1–23.

Baumann, Oscar. 1894. *Durch Massailand zur Nilquelle. Reisen und Forschungen der Massai-Expedition des Deutschen Antisklaverei-Komite in den Jahren, 1891–1893.* Berlin: Geographische Verlagsbuchhandlung Dietrich Reimer.

Blackburn, R.H. 1982. In the land of milk and honey: Okiek adaptations to their forests and neighbours. In *Politics and history in band societies,* ed. E. Leacock and R. Lee. Cambridge: Cambridge University Press.

Branagan, Denis. 1962. A discussion of the factors involved in the development of Maasai Land. Manuscript.

Certeau, Michel de. 1986. *The practice of everyday life.* Berkeley: University of California Press.

Chang, Cynthia. 1982. Nomads without cattle: East African foragers in historical perspective. In *Politics and history in band societies,* ed. E. Leacock and R. Lee. Cambridge: Cambridge University Press.

Chicago Tribune. 1996. On safari. Tanzania redirects popular parks' multiplying traffic. *Chicago Tribune,* 17 November.

Distefano, John A. 1990. Hunters or hunted? Towards a history of the Okiek of Kenya. *History in Africa* 17:41–57.

Enghoff, Martin. 1990. Wildlife conservation, ecological strategies, and pastoral communities: A contribution to the understanding of parks and people in East Africa. Paper presented to the workshop Pastoralism and the State in African Arid Lands, organized jointly by the Scandinavian Institute of African Studies and the Department of Social Anthropology, University of Gothenburg, 26–29 April.

Fosbrooke, Henry A. 1948. An administrative survey of the Masai social system. *Tanganyika Notes and Records* 26:1–50.

Galaty, John G. 1982. Being "Maasai", being "People-of-Cattle": Ethnic shifters in East Africa. *American Ethnologist* 9:1–20.

Grandin, Barbara E. 1988. Wealth and pastoral dairy production: A case study from Maasailand. *Human Ecology* 16(1):1–21.

Gupta, A., and J. Ferguson. 1992. Beyond "culture": Space, identity, and the politics of difference. *Cultural Anthropology* 7(1):6–23.

Hardin, G. 1968. The tragedy of the commons. *Science* 162:1243–48.

Hastrup, Kirsten, and Karen Fog Olwig. 1996. Introduction to *Siting culture: The shifting anthropological object*, ed. K. Hastrup and K.F. Olwig. London and New York: Routledge.

Homewood, K.M., and W.A. Rodgers. 1991. *Maasailand ecology: Pastoralist development and wildlife conservation in Ngorongoro, Tanzania*. Cambridge Studies in Applied Ecology and Resource Management, Cambridge: Cambridge University Press.

Jacobs, Alan H. 1965. The traditional political organization of the pastoral Maasai. Ph. D. thesis, Nuffield College.

Johnsen, Nina. 1992. *Pastorale profeter: En undersøgelse af profetbegrebet og dets anvendelighed i forbindelse med Maasai og Nandi Laiboner i Østafrika*. [Pastoral prophets: An investigation of the concept of prophet and its applicability in relation to Maasai and Nandi Laibons in East Africa.] *Magistra scientiarum* thesis in Anthropology, University of Copenhagen 1990. Specialerække nr. 48, 1992, Institute of Anthropology, University of Copenhagen.

——. 1996. The forest of medicines: Maasai medical practice and the anthropological representation of African therapy. *Folk: Journal of the Danish Ethnographic Society* 38:53–82.

——. 1998. *Maasai medicine: Practicing health and therapy in Ngorongoro Conservation Area, Tanzania*. Ph.D. thesis, Institute of Anthropology, University of Copenhagen.

Kenny, Michael G. 1981. Mirror in the forest: Dorobo hunter-gatherers as an image of the other. *Africa* 51:477–95.

Kjærby, Finn. 1979. *The development of agro-pastoralism among the Barbaig in Hanang District*. Research Papers, no. 56, Bureau of Resource Assessment and Land Use Planning, University of Dar es Salaam.

Kjekshus, Helge. 1977. *Ecology control and economic development in East African history: The case of Tanganyika, 1850–1950*. London: Heinemann.

Klumpp, Donna, and Corinne Kratz. 1993. Aesthetics, expertise, and ethnicity: Okiek and Maasai perspectives on personal ornament. In *Being Maasai: Ethnicity and identity in East Africa*, ed. T. Spear and R. Waller. London: James Currey.

Kratz, Corinne. 1980. Are the Okiek really Maasai? Or Kipsigis? Or Kikuyu? *Cahiers d'Etudes Africaines*, 79:355–68.

Lane, Charles. 1996. *Ngorongoro voices: Indigenous Maasai residents of the Ngorongoro Conservation Area in Tanzania give their views on the proposed General Management Plan*. Forests, Trees, and People Program working paper, United Nations, Food and Agriculture Organization.

McCabe, J. Terrence. 1990. Turkana pastoralism: A case against the tragedy of the commons. *Human Ecology* 18(1):81–103.

Meinhof, Carl. 1904. Über M. Merkers "Masai". *Zeitschrift für Ethnologie* 36:735–44.

Merker, M. [1904] 1910. *Die Masai: Ethnographische Monographie eines Ostafrikanischen Semitenvolkes*. 2d ed. Berlin: Verlagsbuchhandlung Dietrich Reimer, 1910. Rpt: New York and London: Johnson Reprint Corporation, 1968.

Ndagala, Daniel Kyaruzi. 1990. Pastoral territoriality and land degradation in Tanzania. In *From water to world-making*, ed. Gísli Pálsson. Uppsala: Scandinavian Institute of African Studies.

——. 1991. *Territory, pastoralists, and livestock: Resource control among the Kisongo Maasai*. Uppsala: Acta Universitatis Upsaliensis/Uppsala Studies in Cultural Anthropology, no. 18.

Nestel, P. S. 1989. Food intake and growth in the Maasai. *Ecology of Food and Nutrition* 23:17–30.

Peters, Pauline E. 1994. *Dividing the commons. Politics, policy, and culture in Botswana*. Charlottesville and London: University Press of Virginia.

Potkanski, Tomasz. 1994a. Livestock as collective vs. individual property: Property rights, pastoral economy, and mutual assistance among the Ngorongoro/Salei Maasai of Tanzania. Ph.D. Department of Ethnology and Cultural Anthropology, University of Warsaw.

———. 1994b. *Property concepts, herding patterns and management of natural resources among the Ngorongoro and Salei Maasai of Tanzania.* Pastoral Land Tenure Series, no. 6. London: International Institute for Environment and Development.

Spencer, Paul. 1973. *Nomads in alliance: Symbiosis and growth among the Rendille and Samburu of Kenya.* London: Oxford University Press.

Waller, Richard. 1976. The Maasai and the British, 1895–1905: The origins of an alliance. *Journal of African History* 17(4):529–53.

Reproducing Locality: A Critical Exploration of the Relationship between Natural Science, Social Science, and Policy in West African Ecological Problems

James Fairhead and Melissa Leach

Introduction

In March 1997, the British politician Tam Dalyell, once again highlighted the plight of West Africa's forests in the British Parliament. To quote:

> Deforestation in countries such as Nigeria, Ghana and Côte d'Ivoire over the past two decades has reached such a pitch that it may have caused the collapse of the West African monsoon system, says Fred Pearce (*New Scientist*). Certainly huge tracts of West Africa have suffered severe droughts over this period. I asked Lynda Chalker, the [then] overseas development minister, what was being done nationally and internationally to stop deforestation. Chalker said that the Overseas Development Administration has invested heavily in forestry programmes through its bilateral development programmes, and that about 200 projects are currently running or are under preparation at a total cost of £182 million. One particular aim is to ensure that local communities are able to manage their natural forests sustainably. The Global Environment Facility (GEF) was established in 1992, said Chalker, to provide grants and concessional funding to countries for projects and programmes specifically in climate change, biological diversity, international waters and stratospheric ozone. ... Britain, said the minister, has committed some £130 million to the GEF so far and intends to play a positive part in negotiating its "replenishment funding" later this year. Both Chalker and I have had the benefit of the sage advice of José Lutzenberger, the great Brazilian ecologist, and once his country's secretary for the environment. He forecast doom if more forests were destroyed. His Cassandra-like warnings seem even more relevant to West Africa than South America. (*New Scientist*, 22 March 1997, 50–51)

Tam Dalyell drew on a review by Fred Pearce headed "Lost forests leave West Africa dry":

> Droughts in West Africa over the last 20 years may have been caused by the destruction of rainforests in countries such as Nigeria, Ghana and Côte d'Ivoire ac-

This essay is the result of our joint and equal coauthorship. It is based on research made possible by the generous support of the Economic and Social Research Council of Great Britain (award to Fairhead, no. L32027313393), and the U.K. Department for International Development via earlier program support to the Institute of Development Studies, University of Sussex (to Leach). The authors would like to thank the many commentators on earlier drafts of these arguments, most especially Christopher Fyfe, Sara Berry, James Mayers, Simon Rietbergen, Eric Lambin, Paul Richards, and François Ruf. Responsibility for errors of fact and interpretation remains, of course, with the authors alone.

cording to a new study [by] the Centre for Global Change Science at the Massachusetts Institute of Technology. ... At the beginning of this century, the West African coastal forests covered around 500,000 square kilometres. Since then, up to 90 percent have disappeared to make way for farms and other kinds of human activity such as mining. Overgrazing, expansion of arable land and the substantial growth of the timber industry are the main culprits. Several studies have predicted that deforestation of the Amazon basin will have a similar impact in Brazil, but Xinyu Zheng and co-author Elfatih Eltahir, also of MIT, say that the effect may already be happening in West Africa. They point out that the proportion of total forest cover that has been cleared is much greater in West Africa than in the Amazon. ... The "worst possible scenario for tropical deforestation in West Africa", the authors say, would see "all the forests replaced by savanna". This according to Zheng's model, "could cause the collapse of the monsoon system". So far that has not happened, and the authors admit that their model is fairly crude. But they point out that since 1970, rainfall over the whole of West Africa has been lower than before, apparently confirming their predictions. (Pearce 1997:15)

The article in *Geophysical Research Letters,* that Pearce reviews, and that through him captures Dalyell's imagination, suggests that:

Early in this century, rain forests covered a large area, about 500,000 square kilometres, along the Atlantic coast. Today, less than 10% of the primary rain forest is left, with credible predictions that hardly any primary forest is likely to be left by the year 2000. ... The fraction of the primary forest that has been cleared in West Africa is significantly larger than the corresponding fraction for the Amazon region (~ 0.9 versus ~ 0.1). ... [D]eforestation in West Africa may introduce a significant change in land cover which is likely to cause substantial changes in energy fluxes from the surface. (Zheng and Eltahir 1997:155–58)

These authors developed a mathematical model that predicts how rainfall might change if all the forests presumed once to have existed in the forest zone are replaced by savanna. They use their results to suggest the need for further research using a more realistic model than their own, "e.g. a 3-D regional model for West Africa, along with a more sophisticated land surface scheme". Nevertheless, their results, once filtered (disseminated) to the *New Scientist* and thence to the British Parliament, are functional to very different ends. They legitimate, indeed motivate, the vast actions taken by international aid agencies "to ensure that local communities are able to manage their natural forests sustainably". The implication is that local communities have not yet achieved this. In this, the globalization of environmental discourse has been producing "locality" in African communities in highly particular ways.

This production of locality has several linked dimensions. It encompasses, first, the attribution to locales of responsibility for ecological problems and presentation of the nature of these locales in conformity with global interpretations of these problems. Second, it refers to particular images of the material relationships between local inhabitants, their land and resource use, and ecological change, linked to administrative responses to "improve" these. And, third, the history and sociality of locales is produced as particular images of community, demography, ethnicity, and so on linked to forest cover change are disseminated. These produced images of locality are not merely external to the lives and livelihoods of West African land users but intersect with them in important ways. The interventions by governments and development agencies linked to them have, we suggest, frequently been severely detrimental in material terms, contributing to impoverishment and hindering farmers' efforts to enrich their landscapes. This reflects the extent to which produced images of locality overlook important processes in ecology and people-environment relations in the region. Yet in articulation with local discourses about social change, modernity, and ethnicity, for example, produced images of locality have sometimes been reproduced and further reinforced.

In seeking to demonstrate and explore further these dimensions of the production and reproduction of locality through global interpretations of ecological problems, this essay pursues three strategies. First, we show how interpretations of forest cover change in West Africa, linked to nefarious effects, including climate change, have been inextricably linked to policy formulation and administrative interventions. Interpretations have, we would argue, been similarly functional throughout the colonial and postcolonial twentieth century: while there have been many different experiments in national policies aimed at achieving "sustainability" over this period, there have nevertheless been strong historical continuities in the science used to frame "the problem".

Second, we attempt to show how depictions of forest history have become deeply embedded within the social-scientific canon of the region.[1] Modern works in the social sciences that suggest the social and economic causes of forest cover loss, and their historical time scale, generally reproduce and reinforce earlier analyses. As such, they support remedial policies in agriculture, forestry, and conservation policy similar to those of the 1930s and 1950s, and the "locality" they depict continues to be framed by colonial conservation policy. Third, we explore how the data used to predict climatic collapse, whether in the present or past, massively exaggerate the rate and extent of recent high forest loss, such that the conclusions they reach are inevitably false. We thus argue that, with a few notable exceptions, historians may not have sufficiently retheorized landscape history. Many works within the social, as well as natural, sciences are now central to the knowledge-power relations that today produce locality in such a way as to remove resources from inhabitant's control.

It is salutary to realize the extent to which recent research on landscape history in Britain and Europe has completely overturned views that were

---

[1] See Fairhead and Leach 1996 for a similar argument presented in detail for the case of the Republic of Guinea.

orthodox as recently as the late 1950s . Hoskins, in his seminal 1955 text on English landscape history, considered that for the greater part of the pre-Roman period in Britain "the population was less than fifty to a hundred thousand", and "the first two thousand years of agriculture from Neolithic times onwards has left little mark upon the landscape" (Hoskins 1955/1988:29). He suggested that in the fifth century A.D. "the great majority of the English [Saxon] settlers faced a virgin country of damp oak-ash forest, or beech forest on and near the chalk" (39).

Subsequent archaeological and historical study has completely transformed this image. As Taylor summarizes:

> There were far more people living in England in Roman and pre-Roman times than has hitherto been realised. England, far from being a largely empty country with vast unpopulated areas until Saxon times ... had far more people within it than at the time of the Norman Conquest. By the time of the Roman conquest in A.D. 43 there were perhaps two million people. .... England was crowded, perhaps over-crowded, with most of its land exploited to a great or lesser extent. The primeval forests had long since gone and what remained, perhaps less than exists today, was the product of two or three phases of clearance and regeneration and was also carefully managed. ... [The Roman era] led to yet another rise in population, perhaps to a level of four million people by the third century A.D. When the Saxons finally arrived in England, they thus came, not to an empty land of forests, marshes and moorland with the insubstantial remains of a few thousand primitive people, but to a crowded, totally exploited country, covered in fields, roads, towns, villages, and farmsteads, all organised into a complex system of land-holding and with political, administrative and religious boundaries not only fixed but of great antiquity. (Taylor 1988 in Hoskins 1955/1988, 15–17).

That such a profound change in British landscape historiography can occur in the space of thirty years indicates the extent to which early analysts had interpreted the evidence available to them within a set of powerful presuppositions, notably that high populations are a feature only of modern history and that past low populations had a relatively inconsequential impact on natural vegetation.

Colonial historians of African population and landscape history, we would argue, shared this interpretative framework. Indeed, the period of "empty forests" is generally considered to have been somewhere between 500 and 100 years ago. We can expect, therefore, that the recent explosion in studies of African landscape history will provide more than "a history of trees" and be linked to a systematic reconsideration of population and economic history in the region. Equally systematic archaeological and social-historical studies will be needed to see this through.

Forest loss and forest-climate theories in science, policy, and administration

The notion of human-induced climatic desiccation in the tropics has an intellectual and institutional history dating back at least four hundred years (Grove 1997), and many of the first modern analysts of West Africa's forest-climate relations had already made up their minds about the impact of forest cover loss on climate. Here we briefly sketch the conceptual frame within which forest and climate history were usually interpreted during the early decades of the twentieth century and consider the links and networks through which such ideas were embedded in forestry administrations.

West Africa's vegetation zones—as represented in the general description that the botanist Chevalier compiled from administrators' vegetation maps—were key to this conceptual framework (Chevalier 1900, 1911, 1920, 1933). Zones were originally based on descriptions of observed vegetation, or "representative" patches, but there nevertheless remained an ambiguity with zonal delineation based on the "potential" vegetation that could exist under given climatic conditions. This ambiguity between actual and potential vegetation gave room for speculative deduction about vegetation history by assuming that a zone once carried its potential vegetation. Early in the century, such speculative vegetation history was most strongly elaborated in the savanna areas on the northern margins of the forest zone. Early foresters and botanists considered that actual savannas were bioclimatically capable of supporting forest, and thus assumed that forest had once existed, having since been savannized through inhabitants' farming and fire-setting practices (e.g., Unwin 1909 for Sierra Leone; Thompson 1910 for Ghana; Chevalier 1910, 1912, for Benin). In 1938 (and 1949) Aubréville helped to formalize these speculations, purporting to give a precise delimitation of this ex-forest zone within the savannas, the zone that became popularly known in Anglophone circles as "derived savanna". Aubréville's calculations included the possibility that climates able to support forest and the forest itself might have disappeared together, implying that forests once existed in areas too dry for them today. He calculated the old forest extent by inflating present rainfall figures by 15 to 20 percent, the supposed amount lost by deforestation according to trials in France, and then estimating past forest area using assumptions correlating this climate with forest potential (1938:88). While more major fluctuations in climate were known to have occurred naturally, such variation was thought to have occurred too long ago to be relevant to understanding present vegetation dynamics.

Vegetation zones provided a basis for defining "natural" vegetation (and its local and regional variation) and hence for conceptualizing subsequent modifications of it. The terms *forêt primaire* and *forêt vierge* were used in Francophone circles, and *primeval forest* in Anglophone ones, to refer to what, in the wider discipline of ecology, soon came to be termed the *climax vegetation*, the climatic climax being the maximum vegetation that a region's climate could support and the equilibrium to which vegetation would return through succession following disturbance (Clements 1916).

Despite ecological debate concerning the suitability or otherwise of equilibrium models, it was on equilibrium notions, climax, and succession that the emerging science of forest ecology was built in West Africa. That there was a balanced, natural vegetation against which present, disturbed vegetation could be compared in turn supported the development of concepts and methods for defining and measuring departures from it. Scientists identified "stages of degradation"—for example, those stages supposed to occur in the stepwise degradation of forest to savanna—so that the occurrence of any particular vegetation form came to indicate both a temporal degradation trend and how far such degradation had got. Thus, by examining the species composition, diversity, and associations (phyto-sociology) of a vegetation form, it became valid to deduce vegetation history. Particular vegetation "indicators" were employed in this deductive process: for instance, the presence of "relic" forest patches and trees and of "relic" oil palms were taken by Francophone and Anglophone botanists to suggest extensive forest cover in the recent past. The "histories" suggested by such indicators were generally seen as unilineal. Thus, a mixture of nondeciduous and semideciduous forest was interpreted as the penetration of the drier form, indicating a process of drying out, and a mixture of "forest" and "savanna" species on a forest margin suggested the penetration of savanna (e.g., Adam 1968).

The possibility of basing vegetation history on deduction from present observations—of process from form—contrasted with the impossibility of using historical sources at the time. Not only were few such sources available, but their use was generally rejected; Aubréville was explicit, suggesting that the subjectivity of the term *forest,* and the indetermination of locality prevented comparison of present data with written sources. Oral history was generally used only to illustrate processes already known "more scientifically"—"it confirms what we supposed" (Adam 1968)—rather than being explored in a systematic way. Opinion was more often solicited from white expatriates than from inhabitants.

Although these methods and deductive paths for determining vegetation history and the impact of land use were refined by early foresters for West Africa, they depended on specific theories and assumptions that were current in European and Indian forestry circles long prior to African colonization (cf. Grove 1997). This is exemplified by the theories concerning the climatic impact of farming so endlessly reelaborated in West African forestry documents: that deforestation leads to climatic desiccation, that today's climate is in part drier than that historically due to deforestation to date, and that further deforestation will lead to further desiccation. Thus, when Chevalier envisaged a desiccation scenario for Côte d'Ivoire, he cited the pessimistic assessment for Mauritius made by the eighteenth-century French natural philosopher Poivre (Chevalier 1909; cf. Grove 1995):

> Nature made every effort for the Isle de France [Mauritius]; man destroyed everything there. The magnificent forests which covered the soil in their movement once shook the passing clouds and made them dissolve into a fecund rain. The lands which are still uncultivated have not ceased to experience the same favours of nature but the plains which were the first to be cleared and which

were so by fire and without any woodland being saved ... are today of a surprising aridity and consequently less fertile; even the rivers considerably diminished are insufficient throughout the year to irrigate their thirsty sides; the sky in refusing them rain, abundant elsewhere, seems to be avenging the outrages made to nature and to reason. (Poivre, *Voyages d'un philosophe*, cited in Chevalier 1909:44; our translation)

Poivre had, in fact, probably got it the wrong way round. There was no forest because the climate was drier rather than the land being drier because there was no forest. Nevertheless, it was his analysis, globalized by the eighteenth century, which dominated from the first. The centrality of climate to forestry and forest policy was clear in the work of Moloney, the first governor of Lagos (Nigeria), in his *Sketch of the Forestry of West Africa*, where he analyzes rainfall statistics for Sierra Leone, tabulating their steady decline over the four years 1878–82:

> It is desirable that the attention of the community be drawn to the facts ... showing a remarkable and steady decrease in the amount of rainfall in this district during the last four years. ... The only cause that can be assigned for this decrease is the wholesale destruction of the woods and forests, which are at once the collectors and reservoirs of its water supply. This has occurred on other tropical regions, and when the cause was learned, by fatal experience through famine, the result of drought, then the forests were taken under Government protection and replanted, with the best results, but at great expense. ... I have added the rainfall statistics for 1883, 1884 and 1885 which point to an improvement in the direction of greater conservancy or more extended planting: perhaps of both. (1887:240–41)[2]

The tight network of scientists within which such ideas were applied in West Africa helps explain the rapid establishment of a pan–West African orthodoxy concerning forest history. The network that contributed to coherence in British West African forest analysis was strongly linked to India and Burma. For example, H.N. Thompson, the first trained forester in the Nigerian Forest Department, had earlier worked in Burma; he was employed to detail Ghana's forests in 1908 and to found Ghana's department, whose first employee was M. McLeod, one of Thompson's subordinates in Nigeria and earlier trained in India. As Ghana's Forest Department grew, the trained foresters it hired—Chipp, Gent, Moor—had all served earlier in India or Burma. Similarly tight networks characterized French West African forestry circles. Between 1900 and 1913, Chevalier made extensive tours of Guinea, Côte d'Ivoire, Togo, and Benin, and the uniformity in his reports became the basis for common policy in each of these countries, anyway linked within the AOF (Afrique Occidentale Française) . Chevalier's students authored important early works on forest botany, ecology, and history: F. Fleury (concerning Cameroon and Côte d'Ivoire), M. Hedin (Cameroon), and R. Portères (West Côte d'Ivoire) (Chevalier 1933:9). Strong links and mutual

---

[2] In much of this, he is quoting a Dr. Hart of the *West African Reporter*.

citations were common between French and British forestry services from
their earliest and were maintained throughout the colonial period. In short,
the world of forest policy and management in West Africa was small and
analyses easily became standardized. And, as internationalism in forestry
within West Africa grew, standardized analyses were consolidated.

Standardized deforestation analyses became institutionalized within
West Africa's developing forestry administrations and polices. Throughout
the colonial period, key forest scientists were also key administrators. The
need for forestry departments and conservation measures was justified by
these scientists largely on the basis of farmer-led deforestation and the threat
it posed to climate and future land productivity. Concepts central to early
forest ecology were concretized in forest policy and law, supporting—and
coming to be supported by—the forms that forest reservation took. Perhaps
the clearest example is in the relationship between the notion of "primary"
(climax) forest and tenure. In proposing the reservation of the "residual"
forests of the Togo mountains, for example, Aubréville explicitly linked the
idea of primary forest with a justification for state possession, as under
French colonial law the state had rights to "vacant" land:

> Jusqu'a ces dernières années les indigènes négligaient, en géneral, ces forêts des
> pentes. Ce sont des forets primaires dans lesquelles jamais l'indigène n'a exercé
> d'autre droit de jouissance que celui de quelques usages secondaires, tel que
> récolte de fruits et de menus produits. Elles sont donc indubitablement le carac-
> tère de forêts vacantes et sans maître. La composition de la flore permet de dire,
> avec précision, si une forêt est vierge ou d'origine secondaire. Dans ce second cas
> seulement, l'indigène peut prétendre à la rigeur avoir quelques droits d'occupa-
> tion assez mal défini. La forêt primaire appartient à l'Etat. (1937:100)

Equally, early ideas concerning both ongoing savannization and the relationship between forest and climate became embodied in the location of reserves and their management. The notion of ongoing southward savannization underlay the creation of "curtains" of reserves in both Francophone and
Anglophone countries to defend the forests against savannization. Reserves
were also created or planned around the headwaters of many rivers to protect hydrological relations deemed important for maintaining "normal"
river flow. In Guinea, a program was elaborated in the early 1930s to protect
the headwaters of the Niger River. In Ghana, several headwater reserves
were established, protecting the Afram, Fum, Fure, Klemu, Ochi, and
Pompo Rivers. Shelter-belt reserves were also established with the intention
of humidifying farmland downwind. The relationship between forest and
climate was inscribed at a more general level in the area of forest that planners aimed to reserve. It became "a rule of thumb" that in all parts of the
forest zone 20 percent of the area needed to be under forest (and reserved to
ensure this) to maintain climatic conditions—a rule derived from eighteenth-century German forestry (cf. Unwin 1920).

In Ghana, tree tenure law in the 1949 forest policy forged a strong distinction between natural and planted trees, with the state and its concessionaires asserting control over the former. In operation, this law defined all trees indigenous to the region as natural, and where they occurred in fields or around villages they were considered to be relics of original vegetation. The policy thus served to institutionalize the view that inhabitants were merely using, not actively managing, indigenous trees and to institutionalize the tendency to overlook landscape enrichment that might involve "hitching a ride" on ecological processes (cf. Richards 1987), without actually planting. Forest Department interest in controlling trees was mutually supportive of the scientific definition of them as relics of former natural vegetation.

The financing mechanisms for colonial forestry departments, while varying nationally, generally served to support prevailing analyses of forest cover change and its causes. West African forestry services derived revenues from the sale of permits and licenses for timber and wildlife exploitation, and from fines for breaking state laws. They were able to do this only by removing control over the management of resources such as trees, fire, and areas of forested land from inhabitants, deeming the latter inadequate resource custodians whose activities, destructive of forest, require repressive regulation. Revenues were thus ensured by a reading of the landscape as deforested. The economic structures within which forest services operated can thus be seen to have helped frame the production of knowledge about forestry problems and to produce localities accordingly.

Interpretations of forest history in social science

The analysis of forest cover change in terms of a relatively recent, rapid, and accelerating decline also accorded with colonial views of population history, the nature of African society, economy, ethnicity, and the character of migrants. In this sense, analyses within the disciplines of botany and forestry were mutually supportive of those within the emergent social sciences of the period, and together they served to produce particular images of social and demographic history, community, and ethnicity.

In many cases, the supposed existence of recent, intact forest cover was linked to the presentation of the forest zone as only recently inhabited significantly by agriculturalists. The image was of a zone that once housed only sparse hunter-gatherer or minimal root crop cultivator populations with a benign impact on forest cover awaiting the introduction of exotic cereal crops and iron technology as enabling conditions for population expansion, beginning gradually around 1500 and awaiting the twentieth century for its major impact. Slavery and internecine warfare were thought to have limited early population growth and, inversely, colonial Pax Brittanica and French "pacification" to have unleashed it.

Population histories were commonly framed in terms of the histories of particular peoples, as constituted according to colonial perceptions of "tribal boundaries". Environmental behavior was an important characterizing feature used in differentiating ethnic groups. In particular, Mande societies, whose origins were traced to the northern savannas, were identified as "savanna peoples" in contradistinction to "forest peoples", long-term inhabitants whose origins could not be traced outside of the forest zone and whose history was assumed to be forest benign. These differentiations articulated with forestry discourse in at least two ways. First, the southward migration of savanna peoples was linked to the progressive southward savannization of the forest zone. Thus, in Guinea the recession of the forest belt was seen to be accelerated by the migration from the north of Maninka savanna people with a particular proclivity for fire setting and forest clearance, as contrasted with the "native" Kissi and Toma, whose practices were more forest benign (Adam 1948). Colonial stereotypes concerning ethnicity could thus be linked to forest-related practices, and analysis of forest loss could serve, in a mutually supportive way, to reinforce ethnicity, for instance, by feeding into local discourses concerning ethnic identity. Second, southward immigration was seen not only to be expanding the population of the forest zone but also to be transforming its earlier economy by introducing new technologies (iron, cereals) and economic relations (trade).

Analyses that drew strong distinctions between "original inhabitants" (autochtones) and "migrants from the north" in turn became important in forest policy and administration. Forestry administrations sought to identify "rightful owners" in negotiating reserve boundaries and, in the case of British indirect rule, administering reserves. The idea of autochtone or landowner was very useful in this respect. The construction of "original inhabitants" implied, in turn, the construction of all subsequent arrivals as migrants and different in their orientation toward forested land. Forestry policy commonly strove to limit the perceived nefarious impact of migrants' practices. In Guinea in 1932, for instance, administrators arguing for stronger forest law pushed the case that, while in France even the rights of owners were restricted by state legislation, in Guinea even immigrant, non-landholding strangers were unconstrained from abusing forest resources as they liked. As Sharpe has argued for Cameroon, forestry services may have helped in this way to concretize local social structures and the autochtone-migrant distinction (1996).

Many of the tenets of these colonial analyses linking forest loss to social and demographic change remain alive and well in contemporary social science and development. Thus, with shifting cultivation usually seen as the key proximate cause, today's social science analyses of deforestation highlight issues such as immigration into forest areas, technological change, poverty, tenure insecurity, and population growth. The images of rural society thus created are well exemplified in a World Bank overview of deforestation in West and Central Africa, which argued that:

traditional farming and livestock husbandry practices, traditional dependency on wood for energy and for building material, traditional land tenure arrangements and traditional burdens on rural women worked well when population densities were low and population grew slowly. With the shock of extremely rapid population growth ... these practices could not evolve fast enough. Thus they became the major source of forest destruction and degradation of the rural environment. (Cleaver 1992:67)

Analyses informing today's forestry and conservation projects frequently invoke the idea of "communities" that once lived in benign harmony with extensive forest cover, succumbing to pressures from immigration and modernity, with analyses of immigration still perpetuating distinctions between "migrants" and "indigenous people". These distinctions have sometimes been reproduced in articulation with local discourses concerning ethnicity, as, for example, in Guinea, where colonial and now modern stereotypes between Kissi "forest people" and Malinke "savanna people" are reproduced by Kissia themselves in some social contexts, despite their irrelevance to day to day land use practice (Fairhead and Leach 1996).

We now go on to explore this mutual embedding of forestry and social science analysis in more detail in the country cases of Ghana and Côte d'Ivoire, showing how localities in each case have been produced through images of forest loss. But for each case we also show that counter-interpretations are possible, arguing, indeed, that the weight of historical evidence suggests the need to retheorize landscape history in fundamental ways. Linked to such retheorizing are implied shifts in policy and material control over resources, with strong implications for people's livelihoods.

Ghana

The estimates of past forest cover and subsequent deforestation that dominate forest conservation and policy literature suggest that Ghana's "forest zone" (ca. eight million ha) was more or less intact forest in the 1880s. Most assessments suggest a forest cover figure of eight to ten million ha around 1900, now reduced to less than two million ha.[3] These are, for example, the figures that appear in the British Overseas Development Administration Forest Inventory Project Seminar Proceedings, where Frimpong-Mensah asserts that at:

the turn of the century, Ghana had over 8,800,000 ha of forests. ... Only 4,200,000 ha of this remained by about 1950. The estimate for 1980 puts the forest area at about 1,900,000 ha. This means that Ghana has lost over 75% of its tropical forest within this century, due to inefficient agricultural practices (shifting cultivation) and over-exploitation. (1989:72)

---

[3] Estimations of present forest cover generally fall between 1.5 and 2 million ha. The variation hinges on variable estimations of the forest area outside the reserves.

In keeping with such figures, the World Bank (1988) has estimated that closed forest has been lost at an annual rate of 75,000 ha since the turn of this century. If, as most authors point out, deforestation has recently declined, this is simply because little forest remains and what remains is reserved.

Most modern sources concerning Ghana suggest that significant defor-estation began in the late nineteenth century. They imply that this region was hardly populated prior to this time, that it was not heavily farmed, and that what is today defined as the forest zone was indeed a zone of forest in the nineteenth century. Hall (1987), for example, suggests that the popula-tion of the closed forest zone increased from about 250,000 in 1850 to 750,000 in 1900 to 5 million today, implying a more or less exponential rise. But this demographic analysis is now long superseded. Indeed, recent work by his-torians suggests that the nineteenth century was a time when populations in the forest region generally fell rather than increased. And the historical and archaeological record suggests that processes of depopulation in Ghana's forest zone date back several centuries earlier, linked to Ghana's experience of the slave trade.

One can begin with the work of Wilks (1975, 1977), who estimates on the basis of the contemporary estimations of Bowdich and Freeman that the population in the early nineteenth century was 500,000 to 725,000. He sug-gests that during the nineteenth century metropolitan Asante's population declined to 250,000 to 375,000 (1975:87–93). While the early Asante popula-tion has been the subject of debate (cf. Johnson 1978, 1981; Wilks 1978), the state of the forest at the turn of the twentieth century, as we indicate below, would give Wilks's higher figures some support. Many early European visi-tors romanticized the "forest" vegetation of Ashante, but not all. When Huppenbauer visited Kumasi in 1881, for example, he noted that "The actual land of Asante is not forest, as Akyem for instance, but mostly culti-vated" (1881, cited in A. Wilks 1978–79:52).

The decline in Asante populations seems to have been a dominant trend throughout the nineteenth century, but it certainly accelerated from 1863 to 1911, years of military campaigns, civil war, and displacement. As the British governor commented in 1891:

> [A] part of Koranza, and also a part of Mampon, together with Dadiassi, Kokofu, and Inquanta, all powerful tribes, have crossed from Ashanti and sought refuge in the British Protectorate, and the countries they have left are being rapidly overrun by bush and forest, farming and trade operations having ceased in them. … Adansi also, is without population. The country is fast becoming forest. (cited in Wilks 1975:91)

It therefore seems that the population of Asante alone, let alone other areas of the forest zone, was around 350,000 in 1900 and that earlier in the nine-teenth century this figure would have been larger, not smaller. Parts of the "forest zone" around 1900 were therefore covered with forest regenerating on lands depopulated during the nineteenth century or before, while land settlement early in the twentieth century might partly have represented the return of populations to land they had vacated some years before.

Western Ghana also has a richly documented history and one of depopulation during the eighteenth century. Prior to the expansion of the Asante state in the eighteenth century, much of Ghana's forest region had already fallen under the control of the earlier Denkyira state, which Asante defeated in 1701. Wars between 1680 and 1715 left Denkyira depopulated and the Asante in control of the land between the Tano and Bia River. As Fuller described it, this tract of land became the game reserve of the Asante Kings, the New Forest of the conquerors ([1921] 1967). One might be tempted to suggest that despite this history of warfare the level of eighteenth-century Aowin populations had little lasting impact on the forests of this region. Yet the error of such deductions is clear from the description of Thompson, who during his 1908 survey of forests visited the area west of the Tano River, to the north and northeast of Enchi:

> I was rather disappointed in these forests as we were led to understand by the guides that they were extensive and practically virgin in character. This we found to be very far from the case, and the whole tract of country showed unmistakable signs of villages, having been once pretty well inhabited. Large tracts of forest were found to be of secondary origin, and signs of villages having once existed here were also not wanting. In fact, on our return to the village (Tomento, east bank of Tano) the chief admitted that a very long time ago the country had been inhabited by a people who had since moved westwards. (1910:46)

Similar historical evidence exists for other areas. Clearly, the full history of Ghana's pre-nineteenth-century population and economy is beyond the scope of this essay, but even on the basis of this evidence it would seem to us that populations in Ghana's forest region prior to the twentieth century were far higher than have been credited in the forestry literature. In many cases, these higher populations were perhaps more relocated than destroyed, so each instance of depopulation in one region implied another of repopulation elsewhere. But that "elsewhere" was often the Indies and Americas. The impact of the Atlantic slave trade on West Africa in general has been the subject of strong and heated debate. But, in sum, simulation of its demographic impact suggests an absolute African population decline, not just a decline relative to what there would have been otherwise (cf. Manning 1992; Lovejoy 1989). Certainly, those who consider the forest region to have been hardly inhabited prior to the cocoa boom overlook the evidence that it held considerable farming populations in the mid- to late nineteenth century and that populations around 1800 and before may well have been larger, not smaller. Of a forest zone of around 7 million ha, possibly 1.5 to 2 were farmed around 1880 and more earlier in the century. Furthermore, it is quite possible that the twentieth century is seeing the second clearance of Ghana's forest, the first having taken place from ca. 1000 to 1600 A.D. It remains possible that there is more forest in Ghana today than in the seventeenth century.

The extent and impact of early populations of the forest zone at the turn of the century, and the legacy of depopulation, can be gauged from the report on Ghana's forests made by Thompson in 1908. His findings, aimed at rationalizing timber exploitation, spotlighted the "problem" of shifting cultivation, but not as a recent phenomenon, and it was certainly not the raison d'être for his visit. Thompson's report certainly does not give the impression of entering a recent forest "frontier". Indeed, Thompson had to go looking for forests, writing of his choice of route that: "it was more important to discover what forests were left intact, and to explore wooded areas about which but little was known" (1910:6). This comment alone suggests the extent to which Ghana's forest zone was not high forest in 1908. In many southern regions of the forest zone, Thompson observed old forest only on ridges and hills or in patches "here and there". Such forest as there was, was generally either of small extent, old secondary forest on territorially disputed land, or on hill ranges. The only areas where there were extensive tracts of forest were southwest, west, and northwest of Kumasi. All the forest that could be reserved was reserved, and, as Thompson's report underscores, much land that was reserved carried only recent secondary forests. As he writes:

> When selecting forests for reservation, it will be found that comparatively few tracts are covered with so-called primeval or virgin forest; the majority of forests on the Gold Coast and in Asante consist of secondary irregular growth that has sprung up on areas previously cleared for farms by the natives. In places, such forests have, since they re-occupied the abandoned farms, been left untouched for such a long time that a sufficient interval has elapsed for the trees composing them to have grown into large trees of very nearly as good growth as the original ones that were felled. (1910:147)

The condition of Ghana's forests early in the twentieth century prior to their reservation was thus anything but pristine, yet it is against the image of pristine forest—not the state of forest when reserved—that today's forest condition is assessed (cf. Hawthorne and Musah 1995).

Historical evidence also questions the conviction in scientific and policy circles that the margins of Ghana's forest zone have been experiencing progressive savannization. Many authors have suggested that large areas of present savanna could support forest and did "originally" until forest was lost to "derived" savanna. On this basis, Sayer et al. (1992) suggest that Ghana's closed forest zone "originally" covered some 14.5 million ha, a view in keeping with most early foresters, who also assumed that the Guinea savanna zone would be largely forest covered were it not for the activities of farmers (e.g., Chipp 1922; Gent 1925). As Hall (1987) notes, belief in the derived savanna hypothesis during the 1920s and 1930s, and of the climatic impact of deforestation, was almost religious.

That vegetation in the transition zone of the center-west region reflects the ongoing degradation of forest to savanna has been accepted by several authors (Taylor 1952; Asare 1962). More recently, authors such as Swaine et al. (1976) have questioned this, suggesting that the vegetation appears to be more stable. They question the idea that savannas are the result of farming, pointing out that forest patches tend to be more heavily farmed than savannas. In other words, farming can enhance the presence of secondary thicket in savanna rather than degrading existing thicket to savanna (cf. also Markham and Babbedge 1979).

Other evidence supports the possibility that the region has been changing from a predominantly savanna landscape to a predominantly forested one in historical times. Indeed, certain foresters prior to independence appear to have held such views. The chief conservator of forests, Foggie, noted the advance of the forests in this region, writing that "In the north-west, the savannah at one time extended much further south. The forest reserves north and west of Sunyani are rapidly changing from savanna woodland back to closed forest" (1953:132). While Foggie attributed this forest advance to depopulation following warfare in the early 1800s, an earlier colonial forester, Vigne, who had noticed cases of closed forest encroaching on "savanna forest", attributed it to climatic change, considering that the Gold Coast was at the time experiencing a wet cycle, as indicated by a rise in the water level of Lake Bosumtwi. As he put it:

> In tension zones, relatively small climatic changes may have important influences, and it is difficult to account for the large extent of "savanna forest" in the area I studied by assuming a larger population in the past; I consider it is due partly to drier climatic conditions in the past, specially as measured in rainfall and humidity over the fairly short dry season. (1937:93–94)

That the forest was advancing in this region was hypothesized by one of its earliest "explorers", Freeman (1898:164–65), and, as we detail elsewhere, this can be supported from several historical accounts. This is not the place for a location by location critique of assessments of savannization arguments (see Fairhead and Leach 1998). Suffice it to say that the evidence for forest-savanna dynamics in Ghana certainly does not point to massive and recent savannization, and in many instances it points to the opposite.

The possibility that much of Ghana's forest zone (especially in the dry, semi-deciduous, and southern marginal zones) was savanna in historical times is tantalizingly suggested by several further data sources. First, the archaeologist Posnansky suggests that savanna inliers as found near Kumasi may well be relics of a past drier climate pre-9000 B.C., perhaps maintained by settlement, while other climate history data suggest that there have also been more recent dry phases (three to four thousand years ago and two to five hundred years ago) from which forest may be recovering (e.g., Talbot 1981; Nicholson 1979). Second, evidence from termite mounds suggests that areas under forest in the eastern region were in an earlier period under more open vegetation. Charter noted that scattered throughout Ghana's forest zone are the remains of old termitaria (*Macrotermes sp.*) built, he suggested, when the forest was more open either for climatic or biotic reasons (1946; see also Jones 1956 for Nigeria). Third, and intriguingly, there is also oral testimony concerning a savanna past nearer to Kumasi, at the time of the rise of the sixteenth- and seventeenth-century Denkyira state. Ivorian Akan mythology concerning their historic leader, Ano Ansema, suggests that: "He appeared among the Denkyira at Apibweso" and "Before, at Apibweso, there were no trees, there was nothing, only short grass" (Perrot 1982:40). Fourth, Hawthorne argues on the basis of the contrast between high species diversity in the wet evergreen zone of the extreme southwest, and the genetic paucity of the moist semi-deciduous forest, that the latter may be "scar tissue, a recently assembled group of mainly widespread, well-dispersed species, covering up after some immense disruption of this area and barely infiltrated by rarer species which could occur there". He suggests that "Perhaps widespread farming, elephant damage, or fire and drought (e.g., 1500 A.D., 3000 B.P. or 8000 B.P.) has been responsible" (1996:138).

The vegetation history of Ghana's transition zone is clearly extremely complicated. It is certain, however, that a broad brush analysis showing relentless one-way savannization does no justice to this complexity. It proves incorrect to assume that large tracts of savanna are "derived" and that savannas in what is today classified within the closed forest zone, or on its margins, were once forest. Indeed, such views obscure demonstrable instances of forest (or forest fallow) advance over savannas in certain regions, whether due to purposeful enrichment (e.g., in establishing palm forests or cocoa groves in savannas), climatic rehumidification, depopulation, or forest reservation. And they obscure ways in which farmers' own practices have been contributing to the spread of woody vegetation and the density of valued species. Many state forest reserves in Ghana, founded and located on the basis of earlier scientific assumptions, continue to attract local resentment and resistance. Other policies orginating in the same ideas, including government control over supposedly "natural" trees, have inhibited farmers from realizing economic benefits from trees standing in their fields and made them unwilling to protect them or encourage more to grow (Amanor 1994; Afikorah-Danquah 1997). Reforms to tree tenure law in Ghana are currently under way, but these will need to be linked to a rethinking of ecological dynamics and the social dynamics linked to them if policies are to become fully supportive of farmers' landscape-enriching practices and livelihood concerns.

Côte d'Ivoire

Following the United Nations' Food and Agriculture Organization's (FAO) forest resource assessment for 1980 (FAO 1981), Côte d'Ivoire acquired the reputation of having the highest rate of deforestation in the tropics. Numer-

ous articles appeared between 1978 and 1984 that used the same data—
summarized in table 1—and drew the same conclusions (e.g., Monnier 1981;
Bertrand 1983; Arnaud and Sournia 1979; Myers 1980). These works have
continued to influence more recent statements concerning Ivorian forest
cover change during the present century (e.g., Parren and de Graaf 1995;
Sayer et al. 1992; Myers 1994). These works, like their predecessors, have
produced Ivorian localities in conformity with their statements, whether in
discussing the effects of shifting cultivation in savannizing forest, or in their
discussion of the one-way conversion of high forest to other land uses.

Table 1. *Dense humid forest cover change in Côte d'Ivoire*

Date	Area (millions ha)
ca. 1900	14.5
end of 1955	11.8
end of 1965	9.0
end of 1973	6.2
end of 1980	4.0
end of 1990[a]	2.7
Total loss since 1900	11.8

[a] Figure from Sayer et al. (1992).
Source: FAO 1981.

From the figures given in table 1, it appears that the bulk of forest loss has
occurred during the twentieth century. Most analysts link this to a common
set of causes: the introduction of cash crops (cocoa and coffee); immigration
into the forest zone and population increase there, assisted by logging roads,
land clearance, and the timber industry; and development projects assisting
agricultural development at the expense of forests. Localities seen to succumb to these processes are thus produced in a way that conforms with the
estimates of forest cover change. Several authors suggest a slightly deeper
history, which Lanly summarized: "Until the 17th century the area of forest
was not modified by the inhabitants who lived in close symbiosis with the
natural environment. From the 17th century, and more particularly since the
start of the colonial period, the introduction of food crops, export crops and
population increase due to immigration" were responsible for a major reduction of forest area (1969:46).

The data for forest area in 1955 and 1965 were produced following a
major evaluation by the Centre Technique Forestier Tropical (CTFT) of
national timber potential (CTFT 1966; Lanly 1969). The data for 1980 were
based on the 1966 analysis but updated in the light of agricultural land use
statistics. FAO 1981 supplies no evidence for the figure for 1900, it being the
presumed area of the "forest zone", considered then to be more or less
"intact" by the authors. Yet, we shall argue, neither the figure for 1900 nor
those produced for 1955, 1965, and 1980 can be accepted. All subsequent

analyses therefore have serious shortcomings in their analysis of past vegetation cover.

The 1966 study used the complete air photographic cover for Côte d'Ivoire in 1954–56, and maps based on it, to calculate the area of the forest zone and the percentage of dense forest. The forest cover for 1966 was then calculated by comparing the proportion of land under forest in a sample of comparable 1956 and 1966 photos. It was concluded that the Ivorian forest cover had declined from 75 to 54 percent of a zone of 13.1 million ha (Lanly 1969). The study did not cover the extreme southwest, which contained a higher proportion of forest than the rest of the country, but when this had been taken into account it was found that forest cover had declined from 11.8 million ha in 1955 to 9.0 in 1966.

There are, however, several reasons to doubt this figure. First, it seems hard to match the assertion that 75 percent of Côte d'Ivoire's forest zone was forest in 1955 with the observations made in the 1950s by both Aubréville and Mangenot, colonial foresters who knew Côte d'Ivoire rather well. Mangenot suggested in 1955 that, excepting the southwest (ca. 2 million ha of forest), "the reserves of the forest department represent almost the only intact specimens permitting today the study of the dense forest" (1955:6–7). Forest reserves (then at 4.5 million ha) and forest in the southwest together represented only 6.5 million ha. Aubréville estimated the area of Ivorian forest in 1950 at only 7 million ha (1956).

These authors were, it seems, distinguishing forest (logged or not) from other land types. Mangenot was explicit about the illusions that the incautious observer might hold about Ivorian forest cover. "Seen from the summit of a hill", he wrote,

> the landscape appears as a sea of trees. ... But when one ... travels over it following the tracks, one sees that over vast areas this actually corresponds with a corpse: the forest has been destroyed, with only a few large trees surviving, in whose shade are palm, coffee, cocoa, and cola plantains, and fields of manioc and yam. Each village is therefore at the centre of a zone not *dewooded*—large trees exist everywhere, and species cultivated are small trees, bushes and giant forbes, but *deforested*. High forest has been replaced by a mosaic of plantations, fields and bush fallows of small secondary woods. (1955:4).

That the CTFT-derived study exaggerated forest cover in 1955 is also suggested by several other sources. A second analysis of Ivorian forest cover was conducted almost simultaneously in 1967 (Guillaumet and Adjanohoun 1971), from whose maps an estimate of forest cover could be made (although the authors themselves do not make such an area estimation). FAO 1981 argues that the data from this study confirm the assessment in Lanly 1969, as from its maps it appears that a similar forest cover of ca. 8.8 million ha can be inferred. Yet the forest cover map in Guillaumet and Adjanohoun was actually drawn up from the 1954–56 air photographs. Thus, rather than supporting the figure in Lanly 1969 of forest cover for 1966, they undermine its figure for 1955 by suggesting that in 1955 Côte d'Ivoire had ca. 8.8 million ha of forest.

A further reason to question FAO estimates of Ivorian deforestation derives from early colonial assessments of forest cover. While certain global forest analysts such as Breschin (1902) and Zon and Sparhawk (1923) assumed that the majority of the Ivorian forest zone was forest—works that have misled certain modern authors—other early Ivorian specialists were not as naive. Meniaud was explicit. Under the heading "Statistical errors concerning the surface of 'Grand Forêt' the empty spaces in the interior of the extreme limits, and the reasons for these spaces", he suggested that: "The areas given generally in statistics as being occupied by the high forest are calculated according to the extreme limits [that is] 11 million hectares for the high forest of Côte d'Ivoire". By taking into account farmed and savanna areas in the forest zone, the author calculates that "the primary forest, or that which is exploited only by the export timber industry" totaled only 8 million ha (1933:539).

Consideration of the relationships between population and land use should also force us to question assertions of such high forest cover in 1955. West of the Bandama River, rice is a major crop, and in these regions Chevalier (1909) calculated that population densities of only about seven people per km^2 were sufficient to cultivate the entire territory under rotational bush fallowing with preferred fallow lengths of ca. twelve to fifteen years. In 1955, three regions west of the Bandama much exceeded such population densities, while two regions in the extreme southwest had smaller populations, in keeping with descriptions of these areas, which were the last forest reserves. The east also had significant populations. Some areas would have been more, and others less, inhabited, leading to a vegetational mosaic of forests and fallows of different ages, merging with kola, palm, plantain, and other plantations. But it would be completely incorrect to consider this region to have been merely forest.

Furthermore, it is unlikely that forest cover in 1900 was much more than that in 1955; quite possibly, it was less. Large parts of the forest existing in the mid-twentieth century may have been regrowth on land earlier under cultivation, given evidence that many parts of the forest region were depopulated during the colonial wars between French forces and the Baoule, Dan, Bete, Guro, and Dida peoples between 1900 and 1912 (e.g., Weiskel 1980:208–9).The demographic shocks of the early colonial period were compounded by illnesses such as the 1918–19 flu epidemic, labor shortages engendered by conscription of soldiers for the 1914–18 war, and forced labor for porterage and road building. Areas thus abandoned would have succumbed to forest regrowth. While many observers noted huge forest loss near roadsides before independence, this is largely because the colonial regime had forced the Ivorian population to move to roadside villages and abandon earlier settlements.

Each of these historical sources might be questioned, whether on the grounds of subjectivity, site specificity, or ambiguity of definitions. But together they certainly challenge present-day assertions that Côte d'Ivoire had 14.5 million ha of forest in 1900 and seriously question the figure of 11.8 million ha for 1955.

Questions also emerge concerning the area of the "forest zone". In 1909, Chevalier asserted that the area was around 12 million ha, while others provided figures of between 11 and 13 million. All modern analysts, however, suggest that the forest zone is significantly larger, at around 15.7 million ha. This discrepancy can be explained partly by early authors' tendency to consider a strip of land 20 km inland from the coast (ca. 1 million ha) as lying outside the forest zone, as it lacked forest due to its lagoons, littoral savannas, or farmlands. More importantly, ca. 2 million ha on the northern margins of the forest zone that carry savanna or a forest-savanna mosaic are today considered to lie within the forest zone (as they "ought to carry forest"). Yet in 1912 they were classified as outside it. Early descriptions of the landscape show clearly that these areas did not carry extensive forest at the turn of the century. To say that they have been deforested since 1900, as many modern authors assert, is therefore erroneous.

Moreover, there is evidence that the area of the zone dominated by forest (and forest fallow) vegetation in Côte d'Ivoire has been increasing in recent centuries, not contracting in relation to savannas. Several oral accounts suggest that areas now well within the forest zone were savanna in the recent historical past. In a taped interview in 1981, 85-years-old Eonan Messou suggested to Ekanza (1981) that the Moronou region, within the forest zone, was savanna around three hundred years ago. Ekanza rejected Messou's evidence as impossible. But such a vegetation history is entirely consistent with recent ecological findings in Côte d'Ivoire. In the Baoule savannas, villagers suggest that "where one cultivates, the forest advances", and research on forest dynamics at the forest-savanna boundary show just that (Spichiger and Blanc-Pamard 1973). And elders affirmed to Adjanohoun (1964) that savannas once found within the dense forest have today disappeared under forest. Rather than suffering savannization, "it is the forest which gains on the savanna, and this despite their action" (cf. also Spichiger and Lassailly 1981).

Each of these issues forces one to question orthodox figures concerning national deforestation since 1900 and the ways localities within Côte d'Ivoire have been imaged in conformity with these. On the basis of the critique outlined above, we forward tentatively the alternative forest cover scenario elaborated in table 2. This suggests that forest cover loss in Côte d'Ivoire during the present century may have been only half of what present authors consider it to have been.

Table 2. *Alternative forest cover change estimates for Côte d'Ivoire*

Date	Area (millions ha)[a]	Reconsideration (millions ha)
ca. 1900	14.5	ca. 7–8
end of 1955	11.8	ca. 7–8.8
end of 1965	9.0	ca. 6.3
end of 1973	6.2	ca. 5.5
end of 1980	4.0	–
end of 1990[b]	2.7	2.7
Total	11.8	ca. 5.3–7.3

[a] Figure from FAO 1981.
[b] Figure from Sayer et al. 1992.

Table 3. *Suggested revisions to deforestation estimates since 1900*

Country	Orthodox	Suggested
Sierra Leone (since 1820)	0.8–5.0	ca. 0
Liberia	4.0–4.5	1.3
Côte d'Ivoire	13.0	4.3–5.3
Ghana	7.0	3.9
Togo	0.0	0.0
Benin	0.7	0.0
Total	25.5–30.2	9.5–10.5

Note: For a fuller analysis generating the figures presented here, see Fairhead and Leach 1998.

Conclusions

We have tried to indicate how the colonial gaze on social and demographic history, and its extensions in more recent social science analysis, have articulated with scientific methods, theories, and their institutionalization within West African forest policy in an enduring way. This nexus has produced knowledge concerning the causes, rate, and extent of forest loss that has presented localities in very particular ways and been highly functional to would-be-stewards of natural resources. Yet the weight of evidence from other sources—including historical evidence in more recent studies and perspectives emanating from local farmers and their oral accounts—points strongly to counter-interpretations.

Indeed, drawing together evidence from the two cases presented in this essay and the other countries we have studied (Fairhead and Leach 1998), we calculate that deforestation during the twentieth century has been significantly exaggerated across a large part of West Africa. As table 3 summarizes, it may well be only about a third of the figures used by international organizations and climatologists such as Zheng and Eltahir (1997) cited at the beginning of this essay. Cassandra is using poor data for her predictions.

It also seems likely that 1900–1920 was a high point in forest cover in several countries (certainly in Ghana and perhaps also in Côte d'Ivoire) following the decline of earlier farming populations. Forest loss might therefore appear as even less were it possible to take an earlier baseline.

As each of the cases exemplifies in different ways, the history of West African forests is long, involving phases of peopling, management, depopulation, and repopulation. Taylor's concern with European forest history is certainly pertinent to African conditions.

> The idea of great areas of primeval woodland, whose clearance in Saxon, medieval and even later times which is such a feature of Professor Hoskins' work and is still repeated endlessly today, continues to mislead us. ... We shall never understand the history of the English landscape until we remove from our minds the concept of primeval woodland that our prehistoric ancestors had largely removed from the landscape by 1000 B.C. (Taylor 1988 in Hoskins 1955/1988:8).

Exaggerated claims of deforestation have misled ecologists. They obscure how far present forest ecology and composition may reflect less "nature and its degradation", than real histories of climatic fluctuations in interaction with past land management. In West Africa in particular, claims of one-way deforestation have completely obscured what seems to have been a large increase in the area of the forest zone in recent centuries. Exaggerated estimates of deforestation on this scale will also mislead regional and global climatic modeling.

Exaggerated estimates of deforestation have other, more nefarious consequences. They obscure appreciation of how farmers may have been enriching and managing their landscapes in sustainable ways. They obscure the historical experience of inhabitants and the origins of their claims to land. They obscure locality as it has been lived and is understood by those living it. And, most significantly, exaggerated rates of forest loss have often unjustly supported draconian environmental policies that further impoverish people in what is already a poor region.

References

Adam, J. G. 1948. Les reliques boisées et les essences des savanes dans la zone préforestière en Guinée Française . *Bulletin de la Société Botanique Française* 98:22–26.

———. 1968. Flore et végétation de la lisière de la forêt dense en Guinée. *Bulletin d'IFAN*. Série A, 30(3):920–52.

Adjanohoun, E. 1964. *Végétation des savanes et rochers découverts en Côte d'Ivoire Centrale*, Mémoire ORSTOM 7, Paris.

Afikorah-Danquah, S. 1997. Local resource management in the forest-savanna transition zone: The case of Wenchi District, Ghana. *IDS Bulletin* 28(4):36–46.

Amanor, K.S. 1994. *The new frontier: Farmer responses to land degradation*. London: UNRISD and Zed Books.

Arnaud, J-C., and G. Sournia. 1979. Les forêts de Côte d'Ivoire: Une richesse naturelle en voie de disparition. *Cahiers d'Outre Mer* 127:281–301.

Asare, E.O. 1962. A note on the vegetation of the transition zone of the Tain Basin in Ghana. *Ghana Journal of Science* 2:60–373.

Aubréville, A. 1937. Les forêts du Dahomey et du Togo. *Bull. du Com. d'Etud. Hist. et Scient. de l'A.O.F.*, 20(1–2):1–221.

——. 1938. *La forêt coloniale: Les forêts de l'Afrique Occidentale Française*. Annales d'Académie des Sciences Coloniales, no. 9. Paris: Société d'Editions Géographiques, Maritimes, et Coloniales.

——. 1949. *Climats, forêts, et désertification de l'Afrique tropicale*. Paris: Société d'Edition de Géographie Maritime, et Coloniale.

——. 1956. Tropical Africa. In *A world geography of forest resources*, ed. S. Haden-Guest, J.K. Wright, and E.M. Teclaff. New York: Ronald Press.

Bertrand, A. 1983. La déforestation en zone de forêt en Côte d'Ivoire'. *Bois et Forêts des Tropiques* 202:3–17.

Breschin, A. 1902. La forêt tropicale en Afrique, principalement dans les Colonies Françaises. *La Geographie* 5:431–50, 6:27–39.

Charter, C. 1946. *Annual report of the soil scientist*. Ghana: West African Cacao Research Institute.

Chevalier, A. 1900. Les zones et les provinces botaniques de l'Afrique Occidentale Française. *Comptes Rendues des Séances de l'Académie des Sciences* 130(18):1202–08.

——. 1909. *Les végétaux utiles de l'Afrique Tropicale Française: Premiere étude sur les bois de la Côte d'Ivoire*. Etudes Scientifique et Agronomiques 5.

——. 1910. Le pays des Hollis et les regions voisines. *La Geographie* 21: 427–33.

——. 1911. Essai d'une carte botanique forestière, et pastorale de l'A.O.F. *Comptes. Rendus de l'Académie des Sciences* 152(6):1614–17.

——. 1912. Carte botanique, forestière, et pastorale de l'Afrique Occidentale Française. *La Géographie* 26(4):4–26.

——. 1920. *Exploration botanique de l'A.O.F.* Paris.

——. 1933. Le territoire géobotanique de l'Afrique tropicale nord-occidentale et ses subdivisions. *Bulletin de la Société Botanique de France*, séance du 13 Janvier 1933.

Chipp, T. 1922. *The forest officer's handbook of the Gold Coast, Ashanti, and northern Territories*. London: Crown Agents.

Cleaver, K. 1992. Deforestation in the western and central African forest: The agricultural and demographic causes and some solutions. In *Conservation of West and Central African Rainforests*, ed. K. Cleaver et al. World Bank Environment Papers, no.1. Washington, DC: World Bank.

Clements, F.E. 1916. Plant succession: An analysis of the development of vegetation. *Carnegie Institute Washington Publications*, 242:1–512.

CMI (Church Missionary Intelligencer). 1870. Journal notes by the Rev. A. Menzies of an expedition to the Mende country, with a missionary of the American society, 1869'. *Church Missionary Intelligencer* new series, 6:84–96.

CTFT (Centre Technique Forestier Tropical). 1966. *Ressources forestières et marché du bois en Côte d'Ivoire*. Abidjan: CTFT/SODEFOR.

Ekanza, S-P. 1981. Le Moronou a l'époque de l'administrateur Marchand: Aspects physiques et economiques. *Annales d'Université. Abidjan, Serie I, Histoire*, 9:55–70.

Fairhead, J., and M. Leach. 1996. *Misreading the African landscape: Society and ecology in a forest-savanna mosaic*. Cambridge and New York: Cambridge University Press.

——. 1997. Deforestation in question: dialogue and dissonance in ecological, social, and historical knowledge of West Africa, cases from Liberia and Sierra Leone. *Paideuma* 43:193–225.

——. 1998. *Reframing deforestation: Global analyes and local realities: Studies in West Africa*. London: Routledge.

FAO (Food and Agriculture Organisation of the United Nations). 1981. Forest resources of tropical Africa, part I and II (Country Briefs). Tropical Forest Resources Assessment Project, Rome.

Foggie, A. 1953. Forestry problems in the closed forest zone of Ghana. *Journal of the West African Science Association* 3:141–47.

Freeman, R.A. 1898. *Travels and life in Ashanti and Jaman*. London: Archibald Constable.

Frimpong-Mensah, K. 1989. Requirement of the timber industry. In *Ghana Forest Inventory Proceedings*. Accra: Ghana Forest Department/ODA.

Fuller, F. [1921] 1967. *A vanished dynasty: Ashanti.* 2d edition. London: Cass.

Gent, J.R.P. 1925. Why protect our forests? *Journal of the Gold Coast Agricultural and Commercial Society* 4(1):46–51.

Gornitz, V., and NASA. 1985. A survey of anthropogenic vegetation changes in West Africa during the last century: Climatic implications. *Climatic Change* 7:285–325.

Grove, R.H. 1995. *Green imperialism: Colonial expansion, tropical island edens and the origins of environmentalism, 1600–1860.* Cambridge: Cambridge University Press.

———. 1997. *Ecology, climate and empire: Colonialism and global environmental history, 1400–1940.* Cambridge: White Horse Press.

Guillaumet, J. L., and E. Adjanohoun. 1971. La végétation de la Côte d'Ivoire: Le milieu naturel de la Côte d'Ivoire. *Memoirs ORSTOM* 50:156–263.

Hall, J.B. 1987. Conservation of forest in Ghana. *Universitas* (University of Ghana at Legon) 8:33–42.

Hawthorne, W.D. 1996. Holes and the sums of parts in Ghanaian forest: Regeneration, scale, and sustainable use. *Proceedings of the Royal Society of Edinburgh* 104B:75–176.

Hawthorne, W.D., and A. J. Musah. 1995. *Forest protection in Ghana with particular reference to vegetation and plant species.* Gland, Switzerland, and Cambridge, UK: IUCN in collaboration with ODA and the Forest Department, Republic of Ghana.

Hoskins, W.G. 1955/1988. *The making of the English landscape.* London: Guild Publishing. 1988 edition with introduction by C. Taylor.

Johnson, M. 1978. The population of Asante, 1817–1921: A reconsideration. *Asantesem* 8:22–28.

———. 1981. Elephants for want of towns. In *African historical demography,* ed. C. Fyfe and D. McMaster. Vol. 2. Proceedings of a Seminar at the Centre of African Studies, Edinburgh, April 1981.

Jones, E.W. 1956. The plateau forest of the Okomu forest reserve. *The Journal of Ecology* 53/54.

Lanly, J.P. 1969. La regression de la forêt dense in Côte d'Ivoire. *BFT* 127:45–59.

Lovejoy, P. 1989. The impact of the Atlantic slave trade on Africa: A review of the literature. *Journal of African History* 30:365–94.

Mangenot, G. 1955. Etude sur les forêts des plaines et plateaux de la Côte d'Ivoire. *Etudes Eburnéennes* 4:5–61

Manning, P. 1992. The slave trade: The formal demography of a global system. In *The Atlantic slave trade: Effects on economies, societies, and peoples in Africa, the Americas, and Europe,* ed. J.E. Inikori and S.L. Engerman. Durham and London: Duke University Press.

Markham, R.H., and A.J. Babbedge. 1979. Soil and vegetation catenas on the forest-savanna boundary in Ghana. *Biotropica* 11(3):224–34.

Meniaud, J. 1933. L'arbre et le forêt en Afrique noire. *Academie des Sciences Coloniales: Comptes Rendus Mensuels des Séances de l'Académie des Séances Coloniales: Communications,* vol. 14.

Moloney, A. 1887. *Sketch of the forestry of West Africa, with particular reference to its present principal commercial products.* London: Sampson Low.

Monnier, Y. 1981. *La poussière et la cendre: Paysages, dynamique des formations vegetatales, et strategies des societes en Afrique de l'Ouest.* Paris: Agence de Cooperation Culturelle et Technique.

Myers, N. 1980. *Conversion of tropical moist forests.* Washington DC: National Academy of Sciences.

———. 1994. Tropical deforestation: rates and patterns. In *The causes of tropical deforestation: The economic and statistical analysis of factors giving rise to the loss of tropical forests,* ed. K. Brown and D. W. Pearce. London: UCL Press.

Nicholson, S.E. 1979. The methodology of historical climate reconstruction and its application to Africa. *Journal of African History* 20(1):31–49.

Parren, M.P.E., and N.R. de Graaf. 1995. *The quest for natural forest management in Ghana, Côte d'Ivoire, and Liberia.* Wageningen: Wageningen Agricultural University.

Pearce, F. 1997. Lost forests leave West Africa dry. *New Scientist,* 18 January 1997.

Perrot, C.H. 1982. *Les Anyi-Ndenye.* Abidjan and Paris: CEDA/Sorbonne.

Posnansky, M. 1982. Archaeological and linguistic reconstruction in Ghana. In *The archaeological and linguistic reconstruction of African history,* ed. C. Ehret and M. Posnansky. Berkeley: University of California Press.

Richards, P. 1987. On the south side of the Garden of Eden: Creativity and innovation in sub-Saharan Africa. Department of Anthropology, University College. London. Manuscript.

Rodney, W. 1970. *A history of the upper Guinea coast, 1545–1800.* New York: Monthly Review Press.

Sayer, J., C.S. Harcourt, and N.M. Collins. 1992. *Conservation atlas of tropical forests: Africa.* Cambridge: World Conservation Monitoring Centre and IUCN.

Sharpe, B. 1996. First the forest … settlement history, conservation and visions of the future in SW Cameroon. Paper presented at the conference Contested Terrain: West African Forestry Relations, Landscapes, and Processes, University of Birmingham, April.

Spichiger, R., and C. Blanc-Pamard. 1973. Recherches sur le contact forêt-savane en Côte d'Ivoire: Etude du recru forestier sur des parcelles cultivées en lisière d'un îlôt forestier dans le sud du pays baoulé. *Candollea* 28:21–37.

Spichiger, R., and V. Lassailly. 1981. Recherches sur le contact foret-savane en Côte d'Ivoire: Note sur l'evolution de la vegetation dans la region de Beoumi (Côte d'Ivoire centrale). *Candollea* 36:145–53.

Swaine, M.D., J.B. Hall, and J.M. Lock. 1976. The forest-savanna boundary in west-central Ghana. *Ghana Journal of Science* 16(1):35–52.

Talbot, M.R. 1981. Holocene changes in tropical wind intensity and rainfall: Evidence from southeast Ghana. *Quaternary Research* 16:201–20.

Taylor, C.J. 1952. *The vegetation zones of the Gold Coast.* Accra: Government Printing Department.

Thompson, H. 1910. *Gold Coast: Report on forests.* Colonial Reports, Miscellaneous, no. 66. London: HMSO.

Unwin, A. H. 1909. *Report on the forests and forestry problems in Sierra Leone.* London: Waterlow and Sons.

———. 1920. *West African forests and forestry.* London: Fisher Unwin.

Vigne, C. 1937. Letter to the editor of the *Empire Forestry Journal. Empire Forestry Journal* 16:93–94.

Voorhoeve, A.G. 1979. *Liberian high forest trees.* Reports, no. 652. Wageningen: Centre for Agricultural Publishing.

Weiskel, T. C. 1980. *French colonial rule and the Baoule peoples, 1889–1911.* Oxford: Clarendon.

Wilks, A.A. 1978–79. Huppenbauer's account of Kumasi in 1881. *Asantesem* 8:50–52; 9:58–62 and 10:59–63.

Wilks, I. 1975. *Asante in the nineteenth century: The structure and evolution of a political order.* Cambridge: Cambridge University Press.

———. 1977. Land, labour, capital and the forest kingdom of Asante: A model of early change. In *The Evolution of Social Systems,* ed. J. Friedman and M. J. Rowlands. London: Duckworth.

———. 1978. The population of Asante, 1817–1921: A rejoinder. *Asantesem* 8:28–35.

Wilson, J. L. 1836. Extracts from the journal of Mr. Wilson. *Missionary Herald,* May, 193–97, June, 242–48, July, 387.

World Bank. 1988. Ghana Forest Resource Management Project. Working Papers, nos. 1–6. Working paper.

Zheng, X., and E.A.B. Eltahir. 1997. The response to deforestation and desertification in a model of West African monsoons. *Geophysical Research Letters* 24(2):155–58.

Zon, R., and W.N. Sparhawk. 1923. *Forest resources of the world.* New York: McGraw-Hill.

The Other Side of "Nature": Expanding Tourism, Changing Landscapes, and Problems of Privacy in Urban Zanzibar

Kjersti Larsen

The term *nature*, in contrast to *culture*, usually refers to a dimension existing beyond human production. Nature is usually perceived as being in opposition to culture, as outside human constructions and interpretations. Perceptions of nature are, however, also mediated. Nature, like culture, is produced, interpreted, and defined (see also Ingold 1990; and Strathern 1992) and includes, for instance, understandings of the body. Within Western discourses, the body, in contrast to the mind, is usually seen as part of nature. Hence, the body or bodies have until recently been perceived as unmediated entities. More recent literature challenges this conceptualization of the body and places it within the domain of culture (Csordas 1990; Sharma 1996; Bell 1992; Ingold 1990; Scheper-Hugh and Lock 1987; Lakoff 1987; Johnson 1987). In situations in which there is an exchange of perspectives across cultural boundaries, people tend to perceive the other in terms of a nature-culture equation, often placing the other within the category of nature. However, the kinds of environment, behavior, habits, and tastes that are taken as indicators of natureness or cultureness clearly vary according to socioeconomic and cultural differences.

   In this essay I want to draw upon ethnographic material from Zanzibar Town and explore how contradictions in notions of nature, space and the body are brought forward in a situation of rapidly expanding tourism. Recent tourism[1] and encounters with other peoples and past and present policies have changed the urban environment and affected the ways in which women and men relate to each other within their habitats. A central question

This essay is based on field material collected in Zanzibar from 1984 to the present. Furthermore, I wish to thank Vigdis Broch-Due, James Fairhead, Ingjerd Hoëm, Simone Abram, and Grete Benjaminsen for valuable comments on this essay.

[1] Large-volume international tourism is a phenomenon of the last fifty years. Furthermore, mass tourism to developing countries has developed on a large scale in the last two decades (Brown et al. 1997). According to the World Tourism Organisation (1991), the number of tourist arrivals has, at a global level, risen from slightly over 25 million in the 1950s to 443 million in 1990. Thus, generally speaking, mass tourism, a recent global phenomenon that encourages a flow of people between national and territorial borders, is the fastest growing industry in the world today. This industry has impacts on the life and livelihood of people in local communities, whether rural or urban.

to ask is to what extent notions of body and habitat are intertwined. The question relates to a perspective that argues that geographical location becomes place only when it is associated with meanings based on human experiences and understanding. It concerns "the sense of place" (see Hirsch 1995) and perceptions of the physical world implicated in the interpretations and definitions of cultural values, meanings, and identities. Thus, in many cultural constructions of people and place certain kinds of bodies are typically perceived to belong "naturally" to certain kinds of habitats or landscapes. I will use the term *landscape* in order to stress that definitions of environments or natures are always articulations of certain productions of a "nature-culture" contrast and thus are perceived through particular cultural values, meanings, and identities (see Blunt and Rose 1994; and Pratt 1992). Arguing that perceptions of bodies and social and imaginary landscapes are interrelated, I will also discuss what I call embodiment of landscapes. Embodiment of landscapes refers, in this context, to the fact that bodies are perceived as grounded in certain habitats, and that people's perceptions incorporate sensory and sensual characteristics of the place—its smells, colors, tastes, and sounds. This physical and perceptual configuration of matter is carried by the notion of "bodyscape", coined by Broch-Due (1993).

In the following, I shall explore how a particular landscape, including its cultural values, meanings, and identities has been produced over time. In order to exemplify my discussion, I will, first and foremost, examine the present encounters taking place within a framework of expanding tourism. Within this framework, the phenomenon of spirit possession provides a setting where cultural and social change are identified and negotiated. Furthermore, the female initiation ritual *unyago* will be used to illustrate the importance of privacy in recent discourses on Zanzibariness. Thus, it seems that social memories and experiences from earlier and recent encounters are tied together and negotiated—although in different ways—through both the phenomenon of spirit possession and the female initiation ritual called *unyago*.

Tourism is about people traveling in order to expand their experiential, imaginary, and ideological landscapes. Yet it is also about the effects this form of traveling has on the landscapes of the communities receiving the travelers. Given the exchange of perspectives that is involved in these multifaceted encounters, tourism can be analyzed as an interaction of landscapes, a line of analysis that will explore in the course of this essay. The paradoxical point I hope will become clear is that, while tourists construct Zanzibar and Zanzibaris in terms of "nature" from their perspective on the natural and cultural Zanzibaris tend to place tourists on the side of nature as well. Moreover, in a Zanzibari setting linkages between bodies and landscapes become particularly illuminated by the phenomenon of spirit possession, in which a human body may be inhabited by a series of spirits, each of which is believed to originate in a particular landscape and ethnicity, often distant from the contemporary shores of Zanzibar. Thus, through local forms of spiritual experience persons may for periods of time belong to different imaginary landscapes, dramatizing the sense of these foreign places in their performances. While the dramas of spirit possession tend to replicate such multidimensional landscapes, other ritual events work more radically on these embodied ideas. As I shall elaborate on in some detail, the female initiation rituals, *unyago*, have the potential of transforming the social landscape and by extension local perceptions of bodies. Most significantly, both spirit possession and the female initiation ritual are phenomena that locally relate to discourses on Arabness and Africanness. Despite the great importance of Indian influence on local aesthetics and lifestyles (Pearson 1998), it has been the delineation of Arabness and Africanness that has dominated formations of changing Zanzibari social landscapes. What role the contemporary influence of Western tourism will play in future reconfiguration remains to be seen, but the contours of it will be appear throughout this essay.

Contextualizing present discourses on tourism and changing landscapes

Zanzibar, a semiautonomous polity in the United Republic of Tanzania, consists of two major islands, Unguja and Pemba, and is inhabited by approximately 750,000 people. Almost the entire population is Muslim. Recently, in the wake of shifting economic and political structures, the islands have attracted increasing numbers of tourist entrepreneurs and tourists, mainly Europeans and South Africans. Besides beaches, palm trees, and other attractions a tourist may look for, Zanzibar is associated with the romance of exotic urbanism. The capital, Zanzibar Town, is situated on the island Unguja, and is divided into Mji Mkongwe, or Stone Town, and Ng'ambo, or "the other side (of the Stone Town)". While Stone Town consists of three- and four-storied Arab-style buildings,[2] Ng'ambo is characterized by one-story houses built from either mud, stone and lime, or cement, with thatch or corrugated iron roofs, and the more recently built three- or four-story apartment buildings called the Trains (*traini*). The dual character of Swahili urbanism is rooted in the historical-material context of merchant capitalism and plantation slavery, and the inherited inequalities between the two portions of the city should not be underestimated (see Myers 1994). It is, however, the part of Zanzibar Town called Stone Town that is considered a first-class tourist attraction. Stone Town was in 1988 declared a conservation area and was , according to Meffert (1995:111) considered: "comparable to the Victoria Falls or Ngorongoro Crater".[3] Interestingly the tourist-driven con-

---

[2] Until the socialist revolution in January 1964, Stone Town was the seat of the sultan of Zanzibar, who was of Omani origin.

[3] The declaration of Zanzibar Stone Town as a conservation area in 1988 was not followed by a legal interpretation of the term *conservation area*. Thus, the declaration is challenged as "being null and void, and contradictory to the laws of Zanzibar" (Meffert 1995:109).

servation of the Stone Town has emphasized the Arabness of Zanzibar and its "thousand and one nights" images, which means that only a particular Zanzibar is conserved. *Zinjibar* or *Zanguebar* means "the land of the blacks", and according to Abdul Sheriff it refers to "a far-off island paradise where aromatic spices and ivory, princes and slaves were intertwined as fantasies partly based on historical reality" (1995:1). To many Zanzibaris, this image is at best ambiguous, as it relates to both the former period of slavery and the violent 1964 socialistic revolution—historical events that to most Zanzibaris echo differentiation, power relations, and discrimination based on ethnic identities.

Within the present context of tourism, Stone Town is, however, perceived as a symbol of exoticism, former grandeur, and affluence, while Ng'ambo is perceived in terms of triviality and poverty. This image of Zanzibar seems to be very much appreciated by tourists, who carry with them a westernized poverty discourse through which they label places, peoples, and identities as poor or prosperous, exotic or trivial (Broch-Due 1995). In order to confront these notions, it is important to note that the same images are not necessarily evoked among local people. Considering the dual character of Zanzibar Town, Stone Town had, at the time of the 1964 revolution, sewerage, drainage, street lighting, electricity, and indoor plumbing, while with a few exceptions Ng'ambo had none of these. Furthermore, land and housing values were higher in Stone Town than they were in Ng'ambo (Myers 1994:a). This being the situation, the regime that came to power following the 1964 revolution codified the differences between Stone Town and Ng'ambo and decided that Ng'ambo was to be Westernized. This project was part of an effort initiated by the postrevolution regime to provide the entire population with Western-style apartments (ibid.).

Through this project, the new regime wanted to eradicate what was defined as the colonial legacy, whether Arab or British. The so-called colonial legacy was claimed to have "used the Ng'ambo–Stone Town divide to categorise and create a readable, racial pattern of local power relationships in the landscape" (Myers 1994a:454). The new apartments that were built were out of character with preexisiting local customs and ways of life. The inhabitants have, however, over the years managed to use the space created by the state (Larsen 1990) to "remake the landscape through a series of adaptations and transformations"(Myers 1994a:442). Ironically, these blocks of apartments, which were built in order to Westernize this side of Zanzibar Town, are by today's tourists and expatriates found aesthetically undesirable (442). This attitude also illustrates the fact that recent forms such as alternative[4]

---

[4] So-called alternative tourism highlights a diversity of cultures and environments, which are, however, seen as smaller scale, and, according to Brown et al. (1997:316), "with more local opportunities, less economic leakage, and fewer undesirable impacts". The tourists undertaking this type of trip are, as pointed out by Dearden and Harron (1994), usually interested in particular attractions such as certain animals, mountains, cultural sites, or indigenous peoples as well as peoples whose ways of life are defined as different from the tourist's own.

and cultural tourism often include a particular perception of time the "sentiments of the bygone" take on spatial dimensions (Dearden and Harron, 1994; Brown et al. 1997). The past is turned into a foreign country (Lowenthal 1985), environment or culture encompassing an idea of the natural, the authentic, and the erotic.

However, as people are forced to confront with increasing frequency the interests of tourism, their perceptions of landscapes, organizations of space, and ways of life are also affected by this industry. The more recent liberalization of the economy, the privatization policy, and the introduction of political reforms have turned Zanzibar Town into a trade and tourist center. Although many countries regard tourism as having the potential to improve their economies, mass tourism does have its disadvantages. Impacts of tourism on the environment and local social, economic, and cultural aspects of life are in many cases undesirable. This being a more general situation, it is also important to bear in mind that Zanzibar has for centuries been a cosmopolitan society in terms of commerce, immigration, and tourism (in the broad sense of that term) (Sheriff 1987, 1995). Zanzibaris refer to places of origin beyond Zanzibar such as Arabia, mainland Tanzania, the Comoro Islands, India, and Iran, and with reference to these various places of origin they claim membership in different tribes (*makabila*). Another line of differentiation relates to economic status, and, as plantation economy and commerce have played a significant role in the island economy with regard to the development of an urban center, there exists a socioeconomic differentiation, which has linked ethnic origin with economic status. Hence, from a Zanzibari point of view the world as well as their society is multifaceted with regard to "ways of life", or what they call habits (*tabia*), culture (*mila*), and customs or traditions (*utamaduni*). Seen from this angle, Zanzibar can best be described as a "contact-zone" (Pratt 1992) shaped through the interaction of various social and imaginary landscapes. Let me now briefly recollect some past events that have impacted changing Zanzibari landscapes.

Some historical moments

Around 1840, the Indian Ocean trade expanded and the then sultan of Oman, Seyyid Said bin Sultan, transferred his capital from Muscat to Zanzibar Town and established the Zanzibar Sultanate. The slave trade, which initially was on a small scale, increased markedly during the nineteenth century for both export and domestic labor, on the newly established plantations in Zanzibar and other places along the coast (Sheriff 1995). The economic policy of the Omani Sultanate was based on a plantation economy, shipping facilities located at Zanzibar port, and a monopoly over the overland trade that reached the coast (Sheriff 1987, 1995; Myers 1995). The plantation and mercantile economy of the nineteenth century altered the Zanzibari landscape and created a complex urban center. While Zanzibar Town at the beginning of the nineteenth century was a small, influential trading center, it was by midcentury probably the largest town on the coast (Myers 1995). Simultaneously, Zanzibar Town became the seat of political and economic power in East Africa (Sheriff 1987). During this period, the typical dual nature of Swahili urbanism became part of the social landscape. The division between the wealthy and poor became spatially constituted; wealthy and politically influential people settled and built Stone Town, while those who were outside formal political spheres, less wealthy, and poor settled, built, and lived in Ng'ambo.

During the period of slavery, which lasted at least until 1907, people were divided into freeborn and slaves and, in addition, into Arabs, Asians, and Africans—a line of discrimination that increased during the period of British colonialism. These various sociopolitical and economic changes again altered the Zanzibari landscape. When slavery was abolished, the former slaves became workers on the clove plantations, servants or manual workers, peasants, or fishermen together with immigrants from the mainland. Gradually, the free as well as the slave population became part of a cultural universe in which Islam represented ideals and values inherent in an encompassing lifestyle. Thus, again Zanzibaris were confronted with changing economic, political, and ideological currents that impacted on their social landscape and their positions within it. Yet, although distinctions were never clear-cut, there were in the period from about 1840 to 1964 four racial groups that were officially recognized: Arabs, Indians, Africans, and Europeans. Within the political structure, the Arabs were the privileged group, especially Arabs from Oman, while Africans were in this context perceived as less civilized or "cultured".

An event that dramatically altered this landscape was the revolution of 1964. This revolution can be described as a struggle between propertied and nonpropertied classes, largely defined at the time on racial grounds. Many land and property owners, mainly of Arab descent, were killed or fled the island. This event also had dramatic effects on the lives of people of Comoro and Asian descent (Saleh 1995). Again, the social landscape was remade and new prestige and power relationships were inscribed in the Zanzibari landscape. After the revolution, a socialist government of mainly African descent ruled Zanzibar: Arabness became disqualifying, Africanness was qualifying, and the more public aspects of Islamic activity and rituals were discouraged. Zanzibar was meant to become a socialist state where all people would be defined as Afro-Shirazi. This term was constructed in order to rewrite existing identities conceptualized in terms of *makabila* (tribes) and to create a nation of Afro-Shirazi people, that is, people of mixed African and Persian origin. In this imaginary landscape, recollections of Arab origins and Arabness were officially discarded.

However, even today, Zanzibaris perceive themselves in terms of tribes (*makabila*). A person's tribe indicates her or his kin's place of origin outside Zanzibar such as Arabia, India, the Comoro Islands, or various places on the

mainland. Yet, as mentioned above, since the 1964 revolution the official policy has been to downplay discourses on tribal or ethnic differentiation.
Hence, themes concerned with differences with regard to tribes and places
of origin have continued through what could be called muted discourses, for
instance, through the phenomenon of spirit possession and in relation to the
female initiation ritual *unyago*. Below I will briefly illustrate how the phenomenon of spirit possession plays a significant role in discourses on identity and difference and thus, participates in creating a landscape that actually could, in many cases, best be described as a contact zone (see Pratt
1992).

The presence of spirits belonging to different tribes

"Others" have always been present within a Zanzibari landscape. Thus, although mass tourism in its present form is new, meeting other people, other
peoples, other ways of life, and hence other landscapes is not. Among the
foreigners present in Zanzibar are the spirits. They carry with them other
imaginary landscapes and rumors of a multifaceted Zanzibari social landscape and are at the same time agents who evoke social memory. In Zanzibar, the spirits called *masheitani* or *majini* form part of an Islamic cosmology.
Spirits, like human beings, are said to be created and sent to earth by God
(Larsen 1995). They are understood to have the same psychological and
emotional constitution as human beings, but they have no bodies in the
human world. In this situation, spirits are said to have the ability to inhabit
human bodies for limited periods of time. Only in this way can spirits mate-
rialize in the human world. The spirits, like human beings, are said to be
gendered, belong to different tribes and hold different religions (ibid.).
Moreover, the various tribes of spirits and individual spirits within tribes
are classified according to local notions of being poor and prosperous as
well as more or less civilized (*ustaraabu*) (Larsen 1998). According to their
tribe or place of origin, spirits have different habits, tastes, languages, and so
on and prefer, for instance, different kinds of food, music, colors, and
movement. The different tribes of spirits form part of different imaginary
landscapes.

When a woman or a man becomes embodied by a spirit, the spirit is
understood to take control over and use the person's body, and for a period
of time the spirit transforms that body according to her or his own social
landscape. In this instance, the body actually becomes the body of the particular spirit. Spirits acting and interacting in the human world through
human bodies present and represent various landscapes. Being embodied by
spirits belonging to different tribes, women and men may actually come to
embody several kinds of imaginary landscapes. Thus, the phenomenon of
spirit possession can be seen as a process through which women and men
can become familiar with different ways of being in the world. This process
is made possible both by accommodating the habits, demands, and wishes

of the spirits periodically inhabiting their own bodies and by observing and interacting with the spirits of family members and friends.

As I have discussed elsewhere (Larsen 1995, 1998) the spirits, called *masheitani ya rubamba* and *masheitani ya chang'ombe*, are seen as being of African origin. These tribes belong in what are perceived as African landscapes and have habits, tastes, and languages that are defined as less civilized and thus, from a Zanzibari point of view, "closer to nature". Let me give an example of the ways of *masheitani ya rubamba*.

Masheitani ya rubamba, are said to be pagan spirits from Pemba. They are spirits of African origin and are claimed to hold knowledge of healing and sorcery. Their Africanness is thus represented by origin and the kind of knowledge they hold but also by the spirits' personal names. They have names such as Muzi wa Sanda Wajinni wa Shariff wa Mkatamalini, which to the ears of townspeople sound like African names. Moreover, spirits belonging to this tribe are said to speak *kipemba* (the Pemba dialect of Swahili), which by townspeople is considered a less cultured language. During rituals performed on their behalf, the spirits move to the rhythm of two sticks beaten against each other. This is considered another expression of their Africanness. Moreover, *masheitani ya rubamba* prefer to wear the kind of black cloth that was commonly worn by slaves (*kaniki*) and they prefer food such as dried octopus, sugarcane that has not been peeled or cut into small pieces, raw eggs, cassava, and coffee. *Masheitani ya rubamba* represent and present non-Arabic features, Africanness, and hence what is perceived as an African landscape. They are seen as less civilized than those who display what is considered Arabic conduct. The ways of *masheitani ya rubamba* are considered non-Islamic, non-Arabic, and thus partly uncivilized, a part of "nature". In contrast, the habits, tastes, and languages carried by spirits that are understood to be part of Arabian, Malagasy, or European landscapes, that is, *masheitani ya ruhani, masheitani ya kibuki*, and *masheitani ya kizungu*, are perceived as civilized and "part of culture". Let me give an example of the ways of *masheitani ya ruhani*, who are spirits associated with an Arabic way of life: they are imaginary landscapes of Arabness.

Masheitani ya ruhani are said to be Muslim spirits from Arabia, and in contrast to *masheitani ya rubamba* they speak Arabic and only broken Swahili. They have typical Muslim names such as Jinni Sheik Suleman bin Mohammad bin Said and originate from Muscat and Jiddah. Spirits of this tribe are said to prefer white clothes, red or white turbans, and golden rings with green, red, or turquoise stones. Their habits and customs are considered refined and civilized. At the beginning of their ritual, verses from the Koran are recited and slowly the recitation turns into the singing of various songs in Swahili in order to praise Allah and the Prophet (*kasida*). The singing gradually turns into *dikhir* (*dhikiri*), that is, mentioning the names of Allah accompanied by the repetition of *Allah hai*—God the living one—as women and men perform special bodily movements. In contrast to *masheitani ya rubamba*, the knowledge of these spirits is focused on Islam. *Masheitani ya ruhani* prefer food such as *halua* (a kind of Turkish delight), small sweet bananas, boiled eggs, dates, sugarcane cut in small pieces, jasmine flowers, *halud* (alcohol-free perfume), rosewater, and coffee. Their food is, in contrast to the food preferred by *masheitani ya rubamba*, considered cultured. *Masheitani ya ruhani* reflect practices and attributes that people commonly associate with Arabness and being a Muslim, and thus the imaginary landscape inhabited and presented by this tribe of spirits is considered to be one of "culture".

As these examples illustrate, spirits, like human beings, are perceived as grounded in certain habitats. Furthermore, the examples show perceptions of various habitats or landscapes, sensory and sensual characteristics of the place—its smells, colors, tastes, movements, and sounds. Apart from being one kind of foreigner evoking rumors of foreign landscapes, the presence of various tribes of spirits also allows for recollections of earlier encounters and historical periods (see Stoller 1995; Boddy 1989; Kramer 1993; Lowenthal 1985; Connerton 1989; and Taussig 1993) within an ever-changing Zanzibari landscape. Furthermore, over time the tribes of spirits that are present in Zanzibar seem to vary and the ways of spirits belonging to these different tribes may change. Let me give an example.

Some years back, European spirits (*masheitani ya kizungu*) used to be quite common in Zanzibar and rituals were performed on their behalf. During the rituals, the spirits used to dance the waltz dressed in navy uniforms. These European spirits seem almost to have disappeared from Zanzibar today. European spirits still possess people, but their habits and tastes seem to have changed: they speak English, smoke cigarettes, and drink soft drinks and eat cakes.

What this brief example illustrates is that dominating discourses within Zanzibar carry their own stereotypes and "othering" devices in which the spirits play a significant role. The spirits evoke memories about various others; positions, values, and identities inherent in a Zanzibari social landscape over time; and simultaneously, to make possible discourses on cultural values, meanings and identities perceived as part of a particular landscape. Interestingly, in exploring their way of categorizing the spirits, such as the spirits of African, Arabic and European origin, Zanzibaris seem to hold an image of Africa as representing the ultimate other—an image that may not be very different from those of today's tourists. Most tourists, however, seem to include Zanzibar and Zanzibaris in this image. Zanzibaris associate European spirits with English, a language of dominance, a colonial past; they were dancing waltzes and wearing uniforms while today's European spirits are usually participating in what is seen as more "modern" activities such as smoking cigarettes and eating cakes with soft drinks. Thus, the European spirits seem to be quite acceptable in contrast to how the tourists in general are described by Zanzibaris. This view may reveal the tension that results from Europe being associated with both power and material wealth and immoral living. Thus, on both sides Zanzibaris and tourists alike perceive each other in terms of their own preconceptions (see Hoëm 1996), where the ultimate other is represented in an imaginary African landscape.

Local discussions of tourism and culture

How do Zanzibari women and men relate to changes in the social landscape following in the wake of expanding tourism? Below I shall explore contradictions brought to the fore by tourism, especially tensions revolving around the desire for economic growth and what is perceived as development and enduring cultural values, meanings, and identities. The contradictions can be found among Zanzibaris as well as between Zanzibaris and tourists, or, rather, visitors.

Concerning the division of Zanzibar Town into Stone Town and Ng'ambo, it is important to note that to most Zanzibaris today the strict differentiation does not exist. Actually, the oldest part of Ng'ambo is perceived as "downtown" (see Myers 1993, 1995). Moreover, Zanzibari women and men seem to prefer Ng'ambo as living quarters to Stone Town due to what they see as the relative privacy of Ng'ambo. Both women and men claim that with the increasing number of tourists in Stone Town there is more privacy in Ng'ambo. In Ng'ambo, they argue, it is still possible for both women and men to spend the afternoon and evening on the *baraza* (a bench just outside the house) without feeling that they are exposing themselves indecently. According to the dominant moral code in Zanzibar, women should not spend leisure time in public places where unfamiliar men may see them. Mzee Mohamed, a man in his forties working within tourism, told me with resignation one day after he had met a young male tourist in his neighborhood in Ng'ambo: "When tourists keep to the Stone Town and the hotels it is OK, but when they start to invade even our more private areas I really get worried". People also stress the better quality of the houses in Ng'ambo as well as the cleanliness (see Myers 1993; and Boissevain 1996). Thus, it seems that the Zanzibaris and tourists have quite different conceptualizations of this urban environment, and, more importantly, many Zanzibari women and men feel that tourism is affecting their lives beyond its economic aspects.

How do women and men in Zanzibar Town conceptualize and adjust to changes within their urban environment—changes caused by a rapidly expanding tourism, which is represented by the government as a prime generator of economic activities and benefits.

As mentioned above, the government of Zanzibar has recently focused on tourism as a main source of foreign exchange and thus of economic growth. Through economic and political liberalization, both local and foreign private companies or individuals are encouraged to invest in various kinds of tourist enterprises. As the crux of tourism is leisure and consumption, tourists consume services. Hence, tourism demands not only the provision of attractions, hotels, and restaurants but also construction and transportation workers, guides, consultants, musicians, dancers, craftpersons, instructors, gardeners, travel agents, and a host of others in both the formal and informal sectors. In other words, the types of employment afforded by tourism vary widely (see Benjaminsen and Wallevik 1998). The extent to which women and men are prepared to acquire the skills required in order to take up various positions varies as well as the cultural acceptability of the employment of women and men in different aspects of tourism.

Unemployment is defined nowadays as a problem faced by women and men in Zanzibar Town. Women and men of all age groups continuously discuss the problem of finding wage work at decent salaries. Most people seem to believe that their salaries are insufficient to meet what they perceive as their most basic needs. Many men cannot fulfill their expected roles as providers. In this situation, women—wives, mothers, sisters, and daughters—are obliged to operate within the informal sector. They start small businesses where they work as tailors or make various kinds of food and pastry, which they either sell from home or to cafes or restaurants, although ideally they should be provided for (Larsen 1990). Within this context, tourism often means greater occupational choices for women and men, the choices of course being differentiated by gender because local norms and values form a sexual division of labor. In the wake of expanding tourism, there are new economic opportunities, which gradually change aspects of gender relations and systems of redistribution within and beyond households. However, existing values and dominant notions of gender identity provoke resistance toward these changes in their landscapes. This resistance is reflected in the fact that women in particular are reluctant to work in hotels, bars, or as dancers performing in "cultural shows". This kind of work seems to conflict too much with notions of privacy, respect, and respectability. In these contexts, women would be exposed not only to unfamiliar men but to alcohol and dress codes that are considered indecent.

There is a strong tendency for women working within these sectors in Zanzibar to come from the mainland, while Zanzibari women are more likely to work in travel agencies, transport offices, shops, and or even as cleaners in hotels. However, when Zanzibari women do work in hotels, they work in those that are owned locally. In these hotels, most of the employees are relatives or men known to the employed women's families (see Benjaminsen and Wallevik 1998). Despite strict local evaluations of what constitutes suitable work and what kinds of work are suitable for women and men, tourism does mean increased employment, providing opportunities in situations in which most people have difficulty making ends meet. Yet, despite the fact that Zanzibari women and men living in Zanzibar Town often claim that tourism brings development and economic growth to the island, they express dissatisfaction with tourists "invading" and "transforming" what they see as their landscape. This dissatisfaction concerns in particular morality and lifestyle.

Resistance to undesired social and cultural change or undesired challenges of significant values finds its expression in Islamic movements. Ironically, although the main aim of these movements is to prevent certain kinds of social and cultural change, they evoke interpretations of central values that will eventually also alter their social and imaginary landscapes—although differently. Parkin (1995) discusses an Islamic fundamentalist movement that occurred in Zanzibar between December 1992 and January 1993 in which particularly marked expressions of Islamic fundamentalism took place. Parkin describes processions of young radical men and assemblies addressed by young acknowledged sheikhs as well as demonstrations demanding that, for instance, tourism should not be allowed in Zanzibar. The behavior of visitors and locals when meeting each other and their expectations concerning, for instance, "the culture" of the other is complex, as both hosts and guests may fall back on stereotypical views of each other. Nevertheless, these meetings may also be constructive.

Fuglesang (1994) discusses how women in Lamu, as a result, among other things of tourism, have begun to move into what has usually been conceptualized as male-oriented leisure spheres. Thus, the presence of tourists may also encourage new practices and activities with respect to both leisure activities and employment possibilities. However, in Swahili societies such as Lamu and Zanzibar increased visibility and mobility among women also releases ambivalent emotions among elders and younger men. As a way of handling competing claims and what is seen as moral degradation, there has therefore been a revival in Zanzibar Town of the importance of strict Islamic practices and symbols when Zanzibari culture is to be defined. On the other hand, Western ways of life are also associated with economic possibilities, economic growth, education, and power. As such, Western ways are fascinating for many young women and men who feel the injustice of not being able to take part in the affluent consumerism and what may appear to be the enjoyment of life and leisure.

Tensions between social and cultural consequences of economic growth and so-called development, the desire for things to remain unchanged, and nostalgic notions of authentic landscapes can be found within tourism. Actually, tourism highlights the Western paradox: the idea of never-ending economic growth understood in terms of development and a quest for the authentic landscape encompassing everything perceived as exotic and erotic (see Selwyn 1996). This quest may even allow people, in the name of tourism, to ignore and transgress the other's significant boundaries, for instance, those of privacy.

The urban environment and the importance of privacy

In Zanzibar Town, tourists and their way of being present in the urban environment highlight until now more or less taken for granted norms and values that define what it means to be a Zanzibari woman or man. Among these norms and values are privacy and concealment. Perceptions of the physical world are implicated in the interpretations of meaningful social identities, which are now challenged by the impact of tourists and tourism in the urban environment. At this point in history, it seems that, on the one hand, transformations of the physical world confuse the interpretation of social identities in Zanzibar and that this confusion is in particular directed toward gender relations, gender identities, and hence morality. Yet, on the other hand, from a Zanzibari perspective the tourists appear uncivilized, their behavior being seen as primitive rather than refined. The tourists' way of being in the world may appear to be "closer to nature", at least within a Zanzibari imaginary landscape. Moreover, most tourists who place Zanzibar and Zanzibaris within their particular image do not discover the elaborate Zanzibari aesthetics. And it is interesting to note that when many Zanzibari women and men go abroad as tourists they behave and dress in ways similar to those of tourists in Zanzibar and hence in ways they themselves would hardly promote within a Zanzibari setting. As tourists, they move in and accommodate their perceptions of an-other landscape. The tourist scene as such seems to present a setting in which people may explore ways of being and behaving that in everyday settings are unthinkable. Let me elaborate on this point. Zanzibar Town is a Muslim, sex-segregated society in which women and men are expected to maintain distance. This distance is made possible through the material environment, clothing, and other bodily practices as well as the division of labor by sex (for further elaboration, see Larsen 1990). Religious beliefs have a major influence on domestic and public architecture, the use and production of social space, the creation of identities, and thus on the formation of a landscape. Harvey proposes that we may see the "symbolic ordering of space as providing a framework for experience through which we learn who and what we are in society" (1990:214).

In Zanzibar, privacy is a significant dimension within the social landscape. Both women and men stress its importance: the privacy of others must be respected; and invasion, such as direct visual corridors into the private domains of others, is prohibited (see Larsen 1990; and Myers 1993). Zanzibari women and men strictly distinguish between public and private space, while in public spheres they stress the importance of distance between women and men, veiling of especially the female body, and the absence of affectionate jests between women and men. Thus, they behave according to notions of privacy and respectability. Although there are in Zanzibar Town many kinds of lifestyles and Zanzibaris also live in so-called Westernized ways, the behavior of many tourists may appear as inversions of a Zanzibari moral code; through their practices, from a Zanzibari perspective, they transform public into private space. Thus, public space becomes ambiguous for Zanzibaris. Female tourists especially are said to have no sense of decency and are blamed for ruining the culture by spreading promiscuity and fornication. Typically, poor, angry, alienated, religiously conservative, unemployed, and uneducated young men are, in their desire to protest, swept up by the more fundamentalist Islamic movements (Myers 1993). Moreover, discussions and statements from this discourse filter into the wider population and are repeated and discussed in the sitting rooms and on the *barazas* in the various neighborhoods.

Occupying public spheres, that is, the streets, parks, cafes, and restaurants in Zanzibar Town, tourists appear, according to Zanzibari standards, naked, unclean (by wearing slightly dirty clothes that have not been ironed), and sexually revealing. They do what Zanzibaris only do in private: women and men openly express affection toward each other, for instance, by holding hands or kissing in the streets; moreover, they may consume alcohol. From this perspective, they "pollute" public space. However, the problem is not so much what is seen to be their lack of moral standards but the fact that they transform the environment so that Zanzibari women and men feel uncomfortable there. One of the shop owners called on me one morning at the market. There was a group of tourists in which a young woman was wearing a short dress. The elderly shop owner asked me if I would be so kind to tell this woman that she had to dress properly because her presence in that dress made it difficult for Zanzibaris to feel comfortable and hence be present at the marketplace. Hence, for people inhabiting urban landscapes, the presence of mass tourism may be experienced as interventions in their social landscapes.

The following example illustrates the importance of concealment and privacy and the uneasiness created when that which is seen as belonging to private space is suddenly revealed in public. This example shows how eagerness to present Zanzibari "culture" to guests and tourists can actually be a critical endeavor. In this context, which was a conference organized by a secular women's organization, the focus was not on Islam and Islamic practices and symbols but rather on the female initiation ritual *unyago*, which was presented as "tradition" and valuable piece of exhibition for guests and tourists. More recently, several larger hotels both in Zanzibar Town and on the East Coast of Unguja have advertised "cultural shows" that present what are called "traditional Zanzibari dances".

Dances from the female initiation ritual as "cultural show"

Unyago is the name of the female initiation ritual performed in Zanzibar (Larsen 1990, 1991). Historically, *unyago* is said to have originated from the mainland and to have been incorporated into a Zanzibari culture through female slaves (Strobel 1979). *Unyago* is not universally performed but plays an important role in the formation of female gender identity and thus in discourses on morality more generally. The ritual should be surrounded by secrecy; the knowledge transferred through the ritual is available only to those initiated, and, moreover, as the ritual concerns sexuality and female-male relationships, it is associated with everything private (-*ya ndani*). Hence, the ritual should be concealed from the public. Moreover, although some Zanzibaris would claim that *unyago* is old-fashioned, and thus traditional in opposition to that which is modern, those who do participate and are initiated do not place *unyago* knowledge within the sphere of tradition (see Larsen 1991). Moreover, the knowledge transferred in this ritual context is in general, understood to be important and powerful. This attitude is also found with regard to the knowledge and abilities of the above-discussed spirits of African origin. This is so despite the fact that the ways of these spirits are seen by most Zanzibaris as less civilized and refined than those of other tribes of spirits as well as humans.

In January 1992, the Tanzania Media Women's Association (TAMWA), which is a women's organization consisting of well-educated, professional women, arranged a seminar on "Research methodology and empowerment", in Zanzibar Town. The seminar was attended by visitors from Africa and overseas. As a part of the "cultural entertainment", or rather as a "cultural show", one of the two *unyago* groups in Zanzibar Town was invited to perform dances in public for the first time. Perhaps surprisingly, the group agreed. The dancing was performed in the compound of one of the tourist hotels. Camera-toting tourists, including men, a journalist, and a photographer from one of Tanzania's main daily newspapers, the *Daily News*, therefore also witnessed this occasion. The next day a picture of the dancers was published on the newspaper's frontpage. It portrayed two women dressed in ordinary dresses with a *khanga*[5] covering their skirts and one *khanga* rolled together and tied around their hips (*kibebwe*). They were kneeling buttocks to buttocks with their hands on the ground in front of them. The ritual leader, *nyakanga*, played the drum called *dumbak*. The photo had the heading "*Unyago* as Zanzibari Women Know It". The caption read "Although experts say that initiation and puberty ceremonies do not form a major part of daily life, as was the case to Africans in the past, in some areas the presence of the traditional instructions of moral codes and practices known as '*jando na unyago*'[6] in Swahili are still there. One such place is Zanzibar where a 75-year old woman leads an unyago-ngoma-group and continues with the traditional rituals by performing at such ceremonies. ..." (*Daily News*, 22 January 1992). Through this photograph, female sexuality had in a way suddenly turned into public news.

---

[5] *Khanga* refers to a two lengths of colorful cotton, which women wrap around the lower and upper parts of the body, including the head.

[6] The term *jando* refers to male initiation rituals.

The photograph led to a dispute in the newspaper during the following days as well as among women in Zanzibar Town, who gave the impression of being shocked by it. Women of different ages, whether initiated or not, said that they found the picture indecent, although they also seemed to enjoy the event in the way most of us usually enjoy scandals of which we are not explicitly a part. The ritual leader was angry but also said that she felt shame (*aibu*) because of the picture. However, she claimed that the newspaper had it all wrong. She had never agreed to perform a *unyago* dance, only *msondo*, which is a dance also performed at weddings and as such not secret. She described the publication of the picture as a shameful affair.

Apparently, as far as I heard, the event was not commented on or discussed by men. However, those men I explicitly asked about the photograph said that the women who had performed *unyago* dances in public were without shyness/shame (*hawana haya*). By this, they insinuated that the women who had performed *unyago* dances "in front of people" (*mbele ya watu*) had exposed themselves to the world as sexually active and by so doing indicated that women in general do not embody moral responsibility. The reactions were also reflected in the subsequent newspaper articles and the course of events following the publication of the picture. The newspaper debate continued thus: at first, the editors made a public apology. They wrote that the photograph of the *unyago* dance was not meant to offend women but to portray the rich African culture of *unyago*. They referred to the fact that women had lodged complaints against the photograph and the article. The newspaper did not comment on these complaints, although it did account for these in the newspaper. It had received reactions from women such as "Why is *unyago* being exposed to the public eyes? Not only obscene, but it is also an insult to us [women]", "*Unyago* is some kind of traditional instructions of moral codes and practices for young African women and it is for women's eyes and ears only", and "Being the people's newspaper, I didn't expect the *Daily News* to publish such indecent photos" (*Daily News*, 26 January 1992).

As it turned out, the woman's group, TAMWA, responsible for staging the performance was not happy with the photograph either. It claimed that it was an insult to women and had adverse repercussions. The chairperson, explaining why the group had performed in public, said that the women from the *unyago* group were there only for the purpose of a "cultural show". She also stressed that *unyago* usually was performed in closed compounds for educational purposes (for young women). A press statement issued on behalf of the seminar stated that the photograph "undermined women's quest for respect and equality", thus blaming the *Daily News* for the article. But this was not the end of the story. The Zanzibar government condemned TAMWA for allowing the *unyago* to be performed publicly and photographed. A statement issued by the Ministry of Information, Culture, Tourism, and Youth and signed by the deputy minister read: "TAMWA should be in a better position to know and appreciate the importance of preserving the country's culture instead of misleading it". The statement stressed that *unyago* was a dance performed in secrecy in the presence of women only and only for those women whose presence is necessary. The statement continued: "When the dance is performed in public, the whole meaning and intent of the dance becomes irrelevant and is even insulting to women. The ministry therefore condemns the humiliation and scandalising of women in public and causing them to be photographed and published in the newspaper".

In response to this, a delegation from TAMWA traveled from Dar-es-Salaam to Zanzibar to apologize to the government and the people of Zanzibar for staging the *unyago* dances in public. Then, finally, the ritual leader whose group had performed was interviewed in the newspaper. She explained that she and two associates had been invited by TAMWA to perform what she described as a type of *unyago* dance called *msondo*, which is not exclusive to women alone. She said that no one with a clear mind would perform the secret *unyago* in public. She blamed a group of women at the seminar for overdoing the dance by engaging in the types of dances that are supposed to be secret. "The photographed group is not mine", she charged, blaming TAMWA and the chairman of the Zanzibar Art Council for not intervening when these women went too far. (Later it was said that the women who were photographed actually were members of TAMWA in Zanzibar.) She complained that the fuss over the issue had made her an object of ridicule and that most women, young and old, blamed her for tampering with the island's traditional values.

What was meant as a "cultural show" presenting and maybe celebrating Zanzibari women's "culture" actually resulted in a huge public debate on the immorality of exposing Zanzibari women to an audience beyond their control. The debate engaged and put claims on people in different positions. At stake were not only the honor of Zanzibari women but the importance of having the power to uphold certain boundaries between public and private space, boundaries produced by practices defined and controlled locally. Moreover, staging *unyago* as a "cultural show" illustrates contradictions within Western constructions of nature. The cultural show may be attractive to tourism because it is understood to portray, or also to express, the authentic, but by staging the ritual that which can be defined by Zanzibaris as an authentic dimension of their imaginary landscape could become stripped of its force.

What the above example also illustrates is how people can suddenly find themselves in situations in which some of their activities and knowledges are exhibited for the sake of tourism. Thus, by presenting an image in line with the desires of a voyeur, a local landscape is marketed to the tourist industry in a certain way and made attractive as a phenomenon to be observed, sampled, and even experienced in some depth. The particular event described above highlights how women and what is seen as women's traditional knowledge might be presented in order to reveal the strength and power of women's traditions in the face of modernity. Simultaneously, this presentation confronts a Zanzibari social landscape based in certain notions of what is to be exposed and what should remain concealed. Furthermore, it is possible that when the women agreed to perform in the first place this might actually have been understood as an attempt to move parts of the *unyago* dance from a private to a more public sphere, if only a *female* public sphere, in which case they did not know that there would be men, journalists, and photographers in the audience.

The kind of Zanzibar that attracts tourists is an image based on notions of Arabness: strict codes of etiquette, silence, and formal aesthetics. Yet the ritual displayed draws on images not only of a "natural" femininity associated with sexuality and strength but on notions of Africanness: ritual dances, drums, noise, and elaborate body movements. Thus, a "foreign country" represented through this kind of cultural tourism encompasses and vacillates between images of a cultivated past and the natural. Exoticism is ascribed to others. Also, from a Zanzibari perspective, others are seen and categorized according to a certain production of a nature–culture dichotomy. As mentioned, spirits existing in a Zanzibari universe are categorized according to tastes, customs, and habits perceived by Zanzibari women and men as being more or less beautiful, refined, and civilized. Spirits of mainland origin rank low within this system of categorization, yet their knowledge and abilities are admired, respected, and even feared (Larsen 1998). The female initiation ritual, *unyago*, is not classified by most Zanzibaris as a mainland ritual, although it is well known that it was introduced to Zanzibar by women brought there from the mainland as slaves. Rather, it is said today that the rituals have been transformed according to Zanzibari values and traditions and have become part of what constitutes a Zanzibari social and imaginary landscape. But the content of the ritual has to be concealed because it concerns knowledge and skills that confront Zanzibari notions of how people should present themselves to others and the kind of behavior, knowledge, and skills that should be revealed and openly presented within a civilized society.

The paradox brought to the forefront by this *unyago* event was that what was defined as tradition is actually part of a present landscape; it is part of a particular modernity. The knowledge that is seen to provide women with strength has to be concealed from the outer world; it is private, even secret knowledge. Thus, many Zanzibaris hold that this should not be exhibited in order to present a Zanzibari landscape in terms of visions, dreams, and desires that are defined by those not taking part in it—whether they are conference participants or tourists. Although Zanzibar for centuries has been a meeting place for people of different social and imaginary landscapes, a contact zone, this does not imply that they neglect their own desires and visions concerning the landscape of which they are a part.

By way of conclusion

Tourists may come to Zanzibar to experience nature and authenticity—be that "natural" beaches and forests, a natural Arabic landscape, natural people, or even women performing rituals in what is perceived to be an authentic landscape. Yet in the wake of their visit Zanzibaris see that central values and identities within what they define as their social and imaginary landscape—and thus, culture (*mila*)—are threatened. Hence, there is in tourism, at least in Zanzibar, an inherent tension between economic growth and increased employment possibilities and a desire to uphold a certain kind of landscape in which significant values such as privacy remain observed. As I have discussed, the distinction separating privacy from public space is central in the organization of everyday life and for upholding a sense of identity. This sense of identity relates to, for instance, the bench outside the front door where women and men can have social interaction through neighborliness in morally correct ways (Allen 1979; Donley-Reid 1982; Myers, 1994b). The distinction between the private and the public is according to Zanzibari women and men threatened by social and cultural changes in the wake of tourism and thus the presence of tourists.

Tourism is, beyond doubt, an agent of change (Butler 1996; De Kadt 1992; Pearce 1989; Mathieson and Wall 1982). The problem of tourism as a long-term source of foreign exchange is, however, that the sustainability of tourism is, as stated by Brown et al. (1997:316), "directly tied to maintaining the integrity of that attraction and mediating the interaction between the tourists and the attraction over time, so that interest is maintained". To be able to mediate a form of interaction that recognizes the social and imaginary landscapes constituted by people inhabiting places or attractions seems to be difficult. As tourists, they move in, and accommodate to their perceptions, an-other landscape. This being said, it is important to keep in mind that tourism is only one force for change. Often the changes resulting from tourism are not direct but indirect and filtered through many other forces of agency such as, for instance, global economics, movies, and television. However, competing claims and multiple livelihoods highlighted by the presence of tourism and tourists are what in their complexity make up what outsiders may see as the flexibility and multivocality of Zanzibariness and what insiders see as different positions within a moral landscape produced by local practices.

Through a focus on Arabness, certain kinds of claims on history and ways of life associated with the upper classes are contrasted with Africanness. Moreover, within the Swahili universe, what foreigners often perceive in terms of Arabness and Africanness may also correspond to ideas of the public and the private among Zanzibaris. Interesting in this context is the fact that the presence of tourism and the Zanzibari imaginary landscape produced in order to attract and satisfy certain dreams, visions, and desires evoke lines of distinction beyond that of Africanness and Arabness. In the wake of tourism, there is also a dramatic delineation between the Muslim and the Western way of life, between that which is defined as religious and morally good and that which is seen as a secular and morally bad. In this context, there is a tendency to embrace more fundamentalist interpretations of Islam.

To conclude, I will recollect an encounter I had some years ago as a tourist in Lamu. As I sat in a restaurant, a young girl selling jasmine flowers approached me. I bought some, and we began to talk about life in Lamu and Islamic values. She looked at me and said with a firm voice that because of their lack of respect (*heshima*) and shyness (*haya*) as well as their ignorance of Islamic values tourists could enjoy this life but would suffer in hell because of their ignorance. In contrast, she and other local Muslims living poor yet decent lives would enjoy paradise in the next life (see Waldren 1996). This brief example is meant to illustrate how tourists are conceptualized according to moral codes and values that are actually a negation of what is seen as Islamic. Observing tourists as the amoral other, the girl expressed general understandings, enhancing moral discourses that legitimize interpretations of Islamic values that are to be found in more fundamentalist movements. In this context, ongoing changes in productive and reproductive processes within Zanzibar, the different values, moral codes, and ways of life represented by tourists and Islamic fundamentalist movements, may also invoke new social and imaginary landscapes.

Dominant representations of the life of tourism and tourists as well as society as represented by Islamic fundamentalist movements are related to certain desires, visions, and dreams concerning the organization of society and one's own position and possibilities within it. Tourists may "experience" Zanzibar in terms of their own desires, visions, and dreams, which make them create new spaces for themselves. And in this case expanding economies and social and imaginary landscapes are not creating an ubiquitous form of homogenized life, as is often suggested within the globalization debate. Rather, what is striking is the way in which the dialectic of interaction may create sharper distinctions between Arab-African or Arab-Western oppositions.

References

Allen, James de Vere. 1979. The Swahili house: Cultural and ritual concepts underlying its plan and structure. *Art and Archaeology Papers*, special issue: 1–32.

Bell, Catherine. 1992. *Ritual theory, ritual practice*. Oxford: Oxford University Press.

Benjaminsen, Grete, and Hege B. Wallevik. 1998. Tourism in Zanzibar: A fool's paradise? M.Sc. thesis, Agricultural University of Norway.

Blunt, Allison and Gillian Rose, eds. 1994. *Writing women and space: Colonial and postcolonial geographies*. London: Guilford.

Boddy, Janice. 1989. *Wombs and alien spirits: Women, men, and the Zar Cult in northern Sudan*. Madison: University of Wisconsin Press.

Boissevain, Jeremy. 1996. *Coping with tourists: European reactions to mass tourism*. Providence, RI: Rerghahn Books.

Broch-Due, Vigdis. 1993. "Making meaning out of matter: Perceptions of sex, gender and bodies. In *Carved flesh cast selves: Gendered symbols and social practices*, ed. Vigdis Broch-Due, Ingrid Rudie, and Tone Bleie. Oxford: Berg.

———. 1995. Poverty and prosperity in Africa: Local and global perspectives. Occasional Papers Series, no. 1, Poverty & Prosperity. Uppsala: Nordiska Afrikainstitutet.

Brown, Kate, R.K. Turner, H. Hameed, and I. Bateman. 1997. Environmental carrying capacity and tourism development in the Maldives and Napal. *Environmental Conservation* 24(4):316–25.

Butler, Richard. W. 1996. The role of tourism in cultural transformation in developing countries. In *Tourism and culture: Global civilisation in change*, ed. Wiendu Nuryanti. Ulaksumur, Yogyakarta: Gadjah Mada University Press.

Connerton, Paul. 1989. *How societies remember*. Cambridge: Cambridge University Press.

Csordas, Thomas J. ed. 1990. *Embodiment and experience: The existential ground of culture and self*. Cambridge: Cambridge University Press.

De Kadt, E. 1992. Making the alternative sustainable: Lessons from development for tourism. In *Tourism alternatives*, ed. Valene. L. Smith and William. R. Eadington. Philadelphia: University of Pennsylvania Press.

Donley-Reid, L. 1982. House power: Swahili space and symbolic markers. In *Symbolic and structural archaeology*, ed. Ian Hodder. Cambridge: Cambridge University Press.

Drearden, O., and S. Harron. 1994. Alternative tourism and adaptive change. *Annals of Tourism Research* 21(1):81–102.

Fuglesang, Minou. 1994. *Veils and videos: Female youth culture on the Kenyan coast*. Stockholm Studies in Social Anthropology, 32.

Harvey, David. 1990. *The condition of post modernity: An enquiry into the origin of cultural change*. Oxford: Blackwell.

Hirsch, Eric. 1995. Landscape: Between place and space. In *The anthropology of landscape*, ed. Eric Hirsch and Michael O'Hanlon. Oxford: Clarendon Press.

Hoëm, Ingjerd. 1996. The scientific endeavor and the natives. In *Visions of empires, voyages, botany and representations of nature*, ed. David Miller and Hans Reill. Cambridge: Cambridge University Press.

Ingold, Tim. 1990. An anthropologist looks at biology. *Man* 25(2):208–29.

Johnson, Mark. 1987. *The body in the mind. The bodily basis of meaning, imagination and reason*. Chicago: University of Chicago Press.

Kramer, Fritz. 1993. *The red fez: Art and spirit possession in Africa*. London: Verso.

Lakoff, George. 1987. *Women, fire, and dangerous things: What categories reveal about the mind*. Chicago: University of Chicago Press.

Larsen, Kjersti. 1990. *Unyago—fra jente til kvinne: Utforming av kvinnelig kjønnsidentitet i lys av overgangsritualer, religiøsitet of modernisering*. Occasional Papers in Social Anthropology, no. 22. Oslo: University of Oslo.

———. 1991. Kvinnelige overgangsritualer i Zanzibar Town: En diskusjon om kjønnsidentitet og endring. In *Gender, culture and power in developing countries*, ed. KristiAnne Stølen. Vol. 1. Oslo: Centre for Development and the Environment (SUM), University of Oslo.

———. 1993. Kunnskap, kjønnsidentitet og endring: Om ulike former for kunnskap i Zanzibar Town. *Norsk Antropologisk Tidsskrift* 1:

———. 1995. *Where humans and spirits meet: Incorporating difference and experiencing otherness in Zanzibar Town*. Ph.D. thesis, University of Oslo.

———. 1998. Spirit possession as historical narrative: The production of identity and locality in Zanzibar Town. In *Locality and belonging*, ed. Nadia Lovell. London: Routledge.

Lowenthal, David. 1985. *The past is a foreign country*. Cambridge: Cambridge University Press.

Mathieson, Alister, and Geoffery Wall. 1982. *Tourism: Economic, social, and physical impacts.* New York: Longman.

Meffert, Eric, F. 1995. Will Zanzibar Stone Town survive? In The history and conservation of Zanzibar Stone Town, ed. Abdul Sheriff. London: James Currey.

Myers, Andrew G. 1993. Reconstructing Ng'ambo: Town planning on the other side of Zanzibar. Ph. D. diss., University of California at Los Angeles.

———. 1994a. Making the socialist city of Zanzibar. *Geographical Review* 4:451–64.

———. 1994b. Eurocentrism and African urbanisation: The case of Zanzibar's other side. *Antipode* 26(3):195–215.

———. 1995. The early history of the "other side" of Zanzibar Town. In *The history and conservation of Zanzibar Stone Town,* ed. Abdul Sheriff. London: James Currey.

Pratt, Mary L. 1992. *Imperial eyes: Travel writing and transculturation.* London: Routledge.

Parkin, David. 1995. Blank banners. Islamic consciousness in Zanzibar. In *Questions of consciousness,* ed. Anthony Cohen and Nigel Rapport. London: Routledge.

Pearce, Douglas G. 1989. *Tourist development.* Harlow: Longman.

Pearson, Michael N. 1998. *Port cities and intruders: The Swahili coast, India, and Portugal in the early modern era.* Baltimore: Johns Hopkins University Press.

Saleh, Mohammed A. 1995. La communauté Zanzibari d'origine comorienne premiers jalons d'une reserche en cours. In *Islam et sociétés au Sud du Sahara,* ed. Jean-Louis Triaud. Paris: Editions de la Maison des Sciences de l'Homme.

Scheper-Hugh, Nancy and Margareth Lock. 1987. The mindful body: A prolegomenon to future work in medical anthropology. *Medical Anthropology Quarterly* 1:6–41.

Selwyn, Tom. 1996. *The tourist image: Myths and myth making in tourism.* Chichester: Wiley.

Sharma, Ursula. 1996. Bringing the body back into the (social) action: Techniques of the body and (cultural) imagination. *Social Anthropology* 4(3):251–63.

Sheriff, Abdul. 1987. Slaves, spices and ivory in Zanzibar. London: James Currey.

———. 1995. An outline history of Zanzibar Stone Town. In *The history and conservation of Zanzibar Stone Town,* ed. Abdul Sheriff. London: James Currey.

Stoller, Paul. 1995. *Embodying colonial memories: Spirit possession, power, and the Hauka in West Africa.* London: Routledge.

Strathern, Marylin. 1992. *After nature: English kinship in the late twentieth century.* Cambridge: Cambridge University Press.

Strobel, Margareth. 1979. *Muslim women in Mombasa, 1890–1975.* New Haven: Yale University Press.

Taussig, Michael. 1993. *Mimesis and alterity: A particular history of the senses.* London: Routledge.

Waldren, Jacqueline. 1996. *Insiders and outsiders: Paradise and reality in Mallorca.* Providence, RI: Berghahn Books.

World Tourism Organisation 1991, Madrid.

Primitive Ideas: Protected Area Buffer Zones and the Politics of Land in Africa

Roderick P. Neumann

Introduction

In this essay I critically evaluate the conceptualization and implementation of participatory, integrated conservation and development programs in Africa. My focus is directed specifically on the interventions of international nongovernmental organizations (NGOs) into rural land use and access in communities bordering protected areas. These interventions are planned and implemented by conservation organizations such as the World Conservation Union (IUCN) and the World Wide Fund for Nature (WWF), headquartered in Europe and North America and operating on a global scale. The conservation programs of these organizations are in turn increasingly funded by bi- and multi-lateral donors like the World Bank, the European Community, and various national agencies from the First World. The geographic extent of protected areas alone makes an examination of international interventions crucial. Nine African countries, including Namibia, Tanzania, the Central African Republic, and Botswana, have 9 percent or more of their land under strict protection in national parks and game reserves.[1] Tanzania's total of nearly 130,000 square kilometers exceeds the combined territories of Holland, Slovakia, and Switzerland.

It is not merely the size of the land area in question, however, that makes an analysis of conservation interventions important. Efforts by conservation NGOs to include the lands surrounding protected areas as buffer zones under the jurisdiction of the state have major implications for the politics of land. In the cases of the international conservation interventions under examination, land politics can be viewed as operating on two geographical scales. The first is global and raises questions about the relations of power between rural communities in Africa and international conservation NGOs and about how power relations between local communities and the state are affected by global environmental agendas. In their conceptualization, global

This chapter was first published as an article in 1997 in *Development and Change* 28(3):559–82 and is reproduced here with the kind permission of Blackwell Publishers, Oxford.

I wish to thank Sara Berry, Caroline Cartier, Gail Hollander, Jody Solow, Richard Schroeder, and an anonymous reviewer for the journal *Development and Change* for their helpful criticisms of earlier drafts of this essay.

[1] Percentages compiled from data in IUCN 1991.

conservation strategies tend to gloss over the magnitude of the political change involved (Redclift 1984) and invest international conservation groups and allied states with increased authority to monitor and surveil rural communities (Luke 1994). Recent studies indicate that programs attempting to integrate conservation with development serve to extend state power into remote and formerly neglected rural areas (Hitchcock 1995; Lance 1995; Hill 1996). The second scale on which land politics is affected is at the intracommunity level. Many of the programs and projects under review here emphasize land registration and tenure reform in general as key to stimulating the adoption of more resource-conserving land use in buffer zones. Research indicates that land conflict in rural Africa has often been heightened by land tenure reform and registration efforts (e.g., Bassett and Crummey 1993; URT 1992). Conservation interventions will therefore undoubtedly engage with and influence ongoing negotiations and struggles over landownership and access within communities.

One of my aims in this essay is to understand the conceptualization and political consequences of conservation and development programs through an investigation of the discursive practices of their principal advocates. First, the analysis will lead us to examine the design and purposes of interventions described in the publications of several of the key organizations involved. Thus, in the first two sections I demonstrate that new forms of intervention, politically, tend to represent a continuity with rather than a cleavage from past practices. Second, the focus on conservation discourse requires an exploration of development during the colonial period of Western ideas of non-Western peoples and their relationships with nature. The cultural influence of the "age of empire" (Hobsbawm 1987) continues to reverberate into the present (Said 1994) to structure our understanding of the causes of environmental degradation and proposals for environmental conservation. The third section, then, shows how—like the imperial European interventions that preceded them—conservation interventions are impelled by powerful ideas about the Other. Nature conservation in Africa is deeply embedded in ambivalent Western constructions of the Other and the places "they" inhabit.

This is not to say simply that Western interventions are guided by Western stereotypes. Rather, it is to recognize, as Said (1994) does, that the struggle over geography, while central to the historical relationship of the West and the Third World, is not only about military conquest or economic dominance, "but also about ideas, about forms, about images and imaginings" (7). Specifically, it is absolutely critical that we understand how images and stereotypes of the Other guide Western prescriptions for biodiversity conservation and show how these play out in very concrete struggles over land. Thus, I examine the process through which conservationists alternately invoke images of the "good native" (traditional, nature conserving) or the "bad native" (modernized, nature destroying) and, by doing so, define "legitimate" claims to land in protected area buffer zones. These images are part of what Torgovnick (1990) labeled "primitivist discourse". She suggests that "conceptions of the primitive ... drive the modern and the postmodern across a wide range of fields and levels of culture: anthropology, psychology, literature, and art" (21). To this list I would add environmental conservation and its international institutions.

I am not arguing that advocates of conservation interventions possess a "false" idea of rural Africans for which we could substitute a "correct" one. Instead, I wish to demonstrate how older ideas and images of the Other can persist in modified form to shape programs for biodiversity protection. The persistence of these images results in the formulation of projects that are based on misguided assumptions about land tenure systems in Africa and are inherently contradictory. The fourth and penultimate section, therefore, illuminates some of the assumptions and contradictions within the new integrated conservation and development proposals. I conclude by offering a set of modest suggestions for research and for reorienting conservation interventions toward more participatory and socially equitable approaches.

The "new" approach to conservation in Africa

Calls to include "local participation" and "community development" as part of a comprehensive strategy for biodiversity protection in Africa are now ubiquitous, with organizations ranging from the World Bank to grassroots human rights activists offering endorsements (see Cleaver 1993; KIPOC 1992; Lusigi 1992a; Oitesoi ole-Ngulay 1993; and World Bank 1993). In outlining its lending policies, the World Bank (1993:41) emphasized that it would seek to integrate "forest conservation projects with ... macroeconomic goals" and involve "local people in forestry and conservation management". Writing for the IUCN, Oldfield asserted that "new ideas are needed" in biodiversity conservation because local people "all too often see parks as government-imposed restrictions on their traditional rights" (1988:1). In short, the redistribution of the material benefits of conservation and the resolution of conflicts between conservationists and local communities are central elements in a purported "new approach" (see Baskin 1994; Fletcher 1990; and Ramberg 1992) to conservation in Africa.

The revamped conservation philosophy in Africa is manifested in the proliferation of integrated conservation-development projects (ICDPs, from Wells and Brandon 1993). ICDPs take various forms, but all embody the idea that conservation and development are mutually interdependent and must be linked in conservation planning (see Kiss 1990; McNeely and Miller 1984; and Miller 1984). An important rationalization for these initiatives is that "conservation policies will work only if local communities receive sufficient benefits to change their behavior from taking wildlife to conserving it" (Gibson and Marks 1995:944). In other words, "the basic notion of an exchange of access for material consideration is central to ICDPs" (Barrett and

Arcese 1995). "Benefits" to local communities include those directly related to wildlife management (wages, income, meat), social services and infrastructure (clinics, schools, roads), and political empowerment through institutional development and legal strengthening of local land tenure (Ghai 1992; Gibson and Marks 1995; Makombe 1993). Additionally, ICDPs are often linked with cultural survival efforts and thus seek to incorporate indigenous knowledge and practices in conservation management (e.g., Alcorn 1993; Colchester 1994; Nepal and Weber 1995; World Bank 1993:73–75). Indigenous peoples, so the argument goes, have been living sustainably in relatively undisturbed habitats for generations and can thus be active participants in implementing conservation policy (e.g., Cleaver 1993; IUCN et al. 1991; Oldfield 1988; Sayer 1991).

The main features of ICDPs are embodied in protected area buffer zones, a particular land use designation that is gaining increasing currency within conservation circles in Africa. Government and nongovernment conservation officials support buffer zones as an ideal means of promoting environmental protection while simultaneously improving socioeconomic conditions on reserve boundaries. Buffer zones are now included in virtually all protected area plans (Wells and Brandon 1993:159) and are viewed, along with other participatory ICDPs, as *the* key strategy for the future of biodiversity maintenance in Africa (Wells and Brandon 1992; Omo-Fadaka 1992; Cleaver 1993; Baskin 1994). The buffer zone idea is most directly traceable to the Man and the Biosphere Program (MAB) biosphere reserve model of the United Nations Educational, Scientific, and Cultural Organzation (UNESCO), first proposed in 1968 (Batisse 1982). There are now numerous published definitions of *buffer zones* (e.g., Bloch 1991; Mackinnon et al. 1986; Oldfield 1988; Sayer 1991). Generally, they are lands adjacent to parks and reserves where human activities are restricted to those that will maintain the ecological security of the protected area while providing benefits to local communities. Though ecological and biological concerns have typically driven conservationists' designs for buffer zones and related strategies, they are increasingly presented as a means of strengthening local land and resource claims (e.g., Makombe 1993; Mbano et al. 1995; Newmark 1993). The buffer zone idea originally entailed the legal demarcation of boundaries that would separate land uses in transitional stages, though sometimes authors use the term less discriminatingly (e.g., Mwalyosi 1991; Mbano et al. 1995).[2]

Much of the writing on buffer zones has been light on analysis and evaluation, tending to be more "philosophical and prescriptive" (Bloch 1993:4). At the foundation of this "philosophy" is the notion that conservation will not succeed unless local communities participate in management of and receive material benefits from protected areas. Participation in buffer zones

---

[2] Bloch (1993) suggests that there has been a shift toward "buffering strategies" rather than geographical zonation. Nevertheless, establishment of legally designated and bounded buffer zones continues to increase across the continent.

can best be accomplished by first securing local people's rights to land and resources (Bloch 1991:4, 1993:6). Writing in a World Bank technical paper, Cleaver argues that a "key to success in better forest management [in Africa] will be local people's participation. ... This is best done through their owner-ship of land and of resources on the land" (1993:94). An additional rationale for supporting tenure reform as part of conservation planning is "that pri-vate investment in environmental protection increases with secure tenure" (World Bank 1993:88). Thus, many new conservation proposals seek to inte-grate land surveying, titling, and registration efforts to improve land tenure security for buffer zone residents.

The issue of local land tenure in buffer zones is also seen to converge with cultural survival/indigenous rights efforts among conservationists and development experts (see Colchester 1994; and Sayer 1991). Writing for the IUCN, Oldfield suggests that where "tribal and indigenous peoples" have customary land and resource rights, "buffer zones should be established by vesting title to the lands with the local communities at the level of either the village or ethnic group" (1988:4). Similarly, Cleaver recommends that where "traditional authority still exists, group land titles or secure long-term user rights should be provided" (1993:98). Rather than individual titling, most proposals suggest group titling to communities, so that land in a community "can continue to be allocated according to customary practice" (100). In gen-eral, the policy rhetoric of institutions and organizations such as the IUCN and the World Bank presents indigenous land rights as complementing the goals of ICDPs (IUCN, quoted in Colchester 1994:30; World Bank 1993:73–5).

A kinder, gentler conservation?

Despite the sympathetic treatment of local land rights and the emphasis on benefit sharing by buffer zone proponents, land alienations and local im-poverishment continue seemingly apace. Many of the projects sound alarm-ingly similar to the fortress-style approach to protected areas that they sup-posedly replace. Forced relocations, curtailment of resource access, abuses of power by conservation authorities, and increased government surveillance are reported more often than are successful integrations of local people into conservation management (see Colchester 1994; Ghimire 1994; Gibson and Marks 1995; and Hitchcock 1995). Rather than representing a new approach, many buffer zone projects and other ICDPs resemble colonial conservation practices in their socioeconomic and political consequences. In actuality, many buffer zones constitute a geographical *expansion* of state authority be-yond the boundaries of protected areas and into rural communities. Given the already substantial proportion of land placed in protected areas across Africa, the potential for spatially extending the reach of the state is tremen-dous. A few examples will illustrate.

Beginning with the case of Madagascar, we can observe how proposals to integrate conservation with rural development in buffer zones involve new forms of state intervention and restrictions on land use. The Madagascar Environmental Action Plan, developed with the assistance of the World Bank, aims:

> to help farmers to sedentarize and to incite them to invest in the medium term in soil conservation, agroforestry and reforestation. ... To discourage shifting culti-vation and other forms of deforestation, via integrated development in the zones surrounding protected areas. (quoted in Bloch 1993:5)

The bank's and the Madagascar government's rationales are virtually identi-cal to ill-fated colonial efforts across Africa to convert shifting cultivators into "progressive farmers" (e.g., Moore and Vaughan 1994). Paradoxically, current conservation advocates, like their colonial predecessors, conceive of *tavy* (the local term for shifting cultivation in Madagascar) not as "indig-enous knowledge" in practice but as a "long-lived habit" (Andriamam-pianina 1985:84, cited in Ghimire 1994:211), which must be eliminated. In-creased monitoring of land use activities by the state is required to imple-ment conservation agendas in buffer zones. In the country's Mananara Bio-sphere project, the state has substantially increased the number of forest guards, engaged the support of local police, and placed forestry extension agents with surveillance duties in buffer zone communities (Ghimire 1994). In general, Madagascar's present conservation policies "stress the need to remove villagers from within protected areas [and] to create larger buffer zones" (212). At Montagne d'Ambre National Park, the government recently added a buffer zone, which has expanded the park authority's control over village lands and resources. In effect, park management has been "encroaching upon local forest and land resources" (220).

Turning to Tanzania, several buffer zone projects have been proposed or implemented, with similar ramifications for local land and resource control. For instance, a buffer zone project is under way at the Selous Game Reserve, already the largest protected area on the continent at 50,000 square kilo-meters. In the 1980s, the Selous Conservation Programme was implemented under the aegis of the German organization Deutsche Gesellschaft für Tech-nische Zusammenarbeit (GTZ) in an attempt to address some of the conflicts between reserve authorities and local communities. A 1988 study produced for GTZ recommended that a buffer zone be established along the perimeter of the game reserve (Lerise and Schuler 1988). The authors of the study recommended that within the buffer zone "Game Authorities should have the final say. It should not be considered as part of village land" (130). The government subsequently established a buffer zone encompassing 3,630 square kilometers of adjacent forest, grazing pasture, and settlement under the jurisdiction of the reserve authorities (Ghimire 1994). Similarly, a pro-posed buffer zone at Lake Manyara National Park, Tanzania, would be managed by park authorities, who would oversee land use (Mwalyosi 1991). In this case, restrictions on adjacent land uses are seen as essential to

"minimize conflicts across boundaries between the Park and adjacent villages" (176). As a final example, the Serengeti Regional Conservation strategy, on the boundaries of Serengeti National Park, was launched in 1985.
The strategy includes three types of buffer zones, including "mandatory"
ones (Mbano et al. 1995). In these areas, the ultimate resolution for land use
conflicts is "the removal of land uses that are incompatible with conservation" (613).

A final case comes from Cameroon. Korup National Park and its "support zone" encompass 4,500 square kilometers of tropical rain forest in
southwestern Cameroon (Lance 1995). The implementation of the Korup
project, though formulated as a participatory ICDP, has meant an increase in
the policing capacity of the state. Consequently, the buffer zone now has a
much higher concentration of law enforcement officials than in any other
nearby government lands (ibid.). Once again, when compliance with conservation objectives is not forthcoming eviction and relocation are the ultimate solution. As Colchester points out, "the same laws that made resettlement from [Korup National Park] necessary would also apply in the buffer
zones to which the populations were relocated, making their presence there
equally illegal" (1994:16).

The above cases serve to reveal the relations of power between First
World conservationists and rural African communities embodied within the
new approach to conservation. As long as a "tradition" of living "in harmony with nature" is maintained in a manner suitable to buffer zone planners, local communities may remain on the land. It is the prerogative of First
World conservationists (backed up by the power of the state), however, to
determine whether land uses are compatible with their interests or suitable
for the purposes of the buffer zones. A recent IUCN publication uses an example from Nigeria to describe how this works in practice. "The SZDP
[Support Zone Development Programme] and the Park Management Service
will thus work closely together to monitor village behaviour, and to admin-
ister appropriate 'rewards' and 'punishments'" (Sayer 1991:32). Later, the
author elaborates a set of general procedures for park authorities:

> The documentation and monitoring of traditional uses is an essential first step.
> Measures must be applied to ensure that harvests do not exceed sustainable
> levels. It is particularly important to regulate access to the resources to authorised
> individuals or communities. Distinctions have to be made between harvesting for
> subsistence use and commercial exploitation for distant markets. (67)

In essence, these buffer zone management guidelines call for the geographical expansion of park authority to monitor and regulate the daily lives of
local community members and to force compliance through systems of rewards and punishments.

In sum, though the documents of international conservation NGOs present ICDPs and buffer zones as participatory and locally empowering, the

power to propose, design, and enforce buffer zones lies far distant from rural African communities. The concept of participation is severely limited and frequently based on an assumption that local indigenous communities live in harmony with their environment. In many proposals that suggest a place for people in buffer zones, the image of the Other as closer to nature is central. This image is best exemplified in an IUCN publication:

> Traditional lifestyles of indigenous people have often evolved in harmony with the local environmental conditions. ... Retaining the traditional lifestyles of indigenous people in buffer zones, where this is possible and appropriate, will encourage the long-term conservation of tropical forest protected areas. Protecting the rights of local communities ensures that they remain as guardians of the land and prevents the incursion of immigrants with less understanding of the local environment. (Oldfield 1988:12)

"They" belong in buffer zones because they have coevolved with the environment and will serve as protectors against the incursions of "outsiders", who have lost that harmonious relationship with nature. Indigenous peoples thus bear a tremendous burden—to demonstrate to outsiders (i.e., Western conservationists) a conservative, even curative, relationship with nature while risking the loss of their land rights should they fail. As Stearman observes, there is a growing danger that indigenous peoples "must demonstrate their stewardship qualities in order to 'qualify' for land entitlements from their respective governments" (1994:5). Their lifestyles must allow them to do what immigrants and, significantly, Westerners cannot—produce and reproduce in an ecologically benign way. Conservationists' ideas for indigenous participation in buffer zones are structured by a long history of western notions of the non-Western "primitive".

Ambivalent primitivism and participatory ICDPs

In her book, *Gone Primitive*, Torgovnick (1990) details how central the idea of the primitive Other is in the formulation of Western identity. She traces the use of the word *primitive* as a reference to non-European peoples to the late eighteenth century, noting that while the meaning of primitive changed through the years it always implied original, pure, and simple cultures. This conception is not only historical and evolutionist, but has a spatial dimension as well, places where the primitive can still be found. "Otherwise we cannot get to it, cannot find the magical spot where differences dissolve and harmony and the rest prevail" (187). Our feelings about the present and ideas for the future, Torgovnick argues, are projected outward toward these primitive spaces and their inhabitants. In this way, the production of images of the non-Western Other and its place is integral to the formation of a contrasting identity of self (see also Lutz and Collins 1993).

To a great degree today, the Third World, and particularly Africa, is constituted as primitive in the West. Western journalists and policymakers evaluate contemporary social and political problems as evidence of a "new

Barbarism" in Africa (Richards 1995), conditions that are inexplicable in rational, civilized terms. The nineteenth-century metaphor of Africa as the Dark Continent persists in reconstituted form to structure current Western discussions of AIDS in Africa (Jaroz 1992). The discursive construction of Africa as a place of danger, darkness, and irrationality "legitimates the status quo and perpetuates unequal relations of power" (105). Moreover, images of Africa, or the Third World in general, as primitive continue to have important political ramifications. For Torgovnick, demonstrating how cultural production and the production of knowledge continue to be guided by ideas of what is or is not primitive is critical for understanding First World interventions into the Third. She reminds us that Western involvement in Vietnam and the Persian Gulf would have been "less possible without operative notions of how groups or societies deemed primitive become available to 'higher' cultures for conquest, exploitation, or extermination" (1990:13).

Today, ideas about the Other are produced and reproduced in our museums (Haraway 1984), popular magazines like *National Geographic* (Lutz and Collins 1993), and popular films (Gordon 1992). Lutz and Collins's (1993) study of *National Geographic* photographs shows how the magazine consistently portrays peoples of the Third World "as *exotic ... idealized ... naturalized* and taken out of all but a single historical narrative" (89, italics in the original). They found that in nearly one-third of the photos they analyzed non-Westerners are presented against a visual background that provides no social context at all. Often these photos are set against a purely natural background. In these representations, non-Western peoples appear socially undifferentiated and are thus unaffected by conflicts of interest between genders, generations, or classes. The controversial film, *The Gods Must Be Crazy*, provides another illustration. The film begins with a voice-over narration describing the Kalahari "Bushmen" as "pretty, dainty ... little people" whose society knows "no crime, no punishment" and who "live in complete isolation without knowledge of outside people". The filmmaker intended no irony despite the Bushmen's long historical involvement in regional and global commodity circuits and their horrific experience with the genocidal policies of colonial administrations (Gordon 1992). The film, whose Bushman protagonist has never seen a soda bottle until one falls from the sky and strikes him on the head, presents an image of Bushmen that, in the context of the Namibian liberation war, had powerful political symbolism. Any society unable to recognize a soda bottle can hardly be capable of ruling itself.

Nor would it seem are the Bushmen capable of managing their natural resources. Western-initiated conservation interventions in Africa are structured by the same representations and tainted by the same stereotypes of the primitive as are other fields of knowledge and areas of cultural production. These perceptions likewise have significant political ramifications, especially for the politics of land. Primitivist discourse in land use interventions in

Africa has a long history. During colonial rule, the "primitive methods" of "backward" African farmers were condemned for their "inefficient" and "destructive" agricultural practices and massive state interventions for soil conservation were called for (Beinart 1984; Feierman 1990; Moore and Vaughan 1994). Pastoralists were likewise targeted for "development" (e.g., Hodgson 1995a; Neumann 1995b), and African hunting everywhere was characterized by wildlife conservation advocates as cruel and wasteful slaughter (MacKenzie 1987; Neumann 1996). In all cases, correcting "destructive" African society-environment interactions required expanding state power in rural areas through land use restrictions, hunting bans, destocking, evictions, and land alienations.

Then as now, however, primitivist discourse was ambivalent. There was also a more positive primitive stereotype, particularly in the area of wildlife conservation, which did not require such drastic sanctions and dislocations. Running counter to the image of the environmentally destructive native was the idea of the native living "amicably amongst the game" (quoted in Neumann 1995b:160). From the early twentieth century into the 1980s, various proposals were put forward for including "wild" Bushmen in Namibian game reserves where they could continue to live in ecologically benign "primitive affluence" (Gordon 1992). A colonial administrator in 1930s Tanganyika (now Tanzania) explained the aesthetic appeal of including "primitives" in national parks:

> [T]he pig-tailed [Maasai] "moran" poised on one leg with a spear for a prop, standing sentinel over his father's cattle, is a picturesque sight, and it is fitting that this human anachronism should make his home in [Serengeti National Park] the same country as the rhinoceros and other survivors of a bygone age. (Sayers 1933:441)

Here, pastoralists are quite literally equated with the fauna as part of the overall spectacle of "wild" Africa, an analogy repeated by conservationists up until the decade of independence (see Neumann 1995b). In this way, as Haraway notes, "primitive" Africans were "consigned to the Age of Mammals, prior to the Age of Man. That was [their] only claim to protection, and of course the ultimate justification for domination" (1984:41).

Since decolonialization, the language has changed markedly and conservation advocates increasingly see that the future of biodiversity protection in Africa lies in some form of alliance and cooperation between parks and nearby communities. As the following quote from a World Bank publication indicates, this new conceptualization includes a strong emphasis on securing local land tenure:

> In many forest areas, traditional tenure mechanisms are still potentially operative, and governments can divest control of forest land and the trees on them to local communities, through allocation of group titles or secure user rights. This must be done with care, however, since many of these communities have broken

down under the pressure of migration, logging, and government land ownership. (Cleaver 1993:100)

Clearly, there has been a tremendous shift in the language of conservation since colonial rule, when park administrators considered pastoralists "as part of our fauna" (quoted in Neumann 1995b:160). Nevertheless, this passage and the previous quote from Sayers contain certain continuities. Both are embedded in and reflect the relations of power between the West and Africa. At the time of authorship, Sayers was assistant chief secretary for Tanganyika Territory and Cleaver was agricultural division chief in the Africa Technical Department of the World Bank. More to the point, and perhaps less obviously in the contemporary quote, both reflect historically deep beliefs about and images of non-Western peoples. The caveat that Cleaver presents to his own suggestion of local empowerment is revealing. Forest communities are either traditional or broken down, similar to the way that "Africans were imagined as either 'spoiled' or 'unspoiled'", by early-twentieth-century hunter-naturalists (Haraway 1984:50). Traditional communities, in the context of forest conservation, are entitled to respect for existing property claims. Communities that have "broken down" are another matter. Once Africans modernize, become tainted by civilization, they have no place in the pre-cultural African landscape of parks, reserves, and buffer zones.

Conservationists' ambivalence over indigenous peoples as destroyers or protectors of nature[3] is an extension into modern contemporary conservation thought of a conflicting, and historically deep, set of Western stereotypes of the Other. This "ambivalent primitivism" prevents the coherent conceptualization and implementation of protected area buffer zones. Ambivalent primitivism also reflects a reluctance among conservationists to confront the essential tension between local communities and protected areas—the question of who holds the power to control access to land and resources. As Alcorn correctly notes, conservationists act "as gatekeepers to a discussion table that does not have a place set for those whose homeland's future hangs in the balance" (1993:426). Gatekeeping is to a significant degree a process of labeling. When we categorize rural African communities' land management systems as customary, traditional, or broken down or label individuals as indigenous, landless, or recent immigrants, we are in essence exercising the power to assign land rights.

Typically in buffer zone projects, when local people have failed the sustainability test they must be removed (Ghimire 1994; Mbano et al. 1995:613; see also Colchester 1994). As I pointed out earlier, most of the proposals for formally demarcated buffer zones vest control over land use in the state, which has the power to relocate. The Lake Manyara buffer zone and wildlife corridor proposal cited above (Mwalyosi 1991) called for the relocation of

---

[3] See Redford and Stearman 1993 and Alcorn 1993 for a brief introduction to the debate conducted in *Conservation Biology*.

existing settlements that interfered with conservation goals. Reporting on the situation in the support zone of Cross River National Park, Nigeria, J. Sayer predicts that traditional land uses "will destroy the biological integrity of the park" and calls for "radical changes in land use" (1991:30). In other regions, such judgements on the sustainability of local land uses has provided "a convenient excuse to divest [people] of their homelands" (Stearman 1994:4). Commonly, nonsustainable land use is explained by the "incursion of immigrants" or the breakdown of traditional society (Oldfield 1988; Cleaver 1993). In other cases, the very premise of indigenous management has been challenged. Some conservation biologists have vigorously contested the idea that "there can be harmony between wildlife and pastoral exploitation" (Prins 1992:117). Ultimately, the stereotypes and assumptions, both positive and negative, essentialize local people in such a way as to obscure the politics of land within buffer zones.

Questionable assumptions and contradictory designs

Who are the "local people"?

The conceptualization and design of new conservation interventions—because they are infused with essentialist notions of the primitive—are founded upon unsubstantiated assumptions and fraught with contradiction. To begin, the idea of "local people" is rarely rigorously examined, either in design or implementation. In most ICDPs, including buffer zones, local people are generally treated as a homogeneous entity, with little attention paid to gender, class, or ethnic differentiation. Almost no socioeconomic research has been conducted prior to implementation of projects (see Wells and Brandon 1992:13), and there is scant documentation of project impacts on different segments of society (Bloch 1993; cf. Stocking and Perkin 1992). When socioeconomic research is conducted, it is typically based on survey questionnaires (e.g., Newmark 1993), with little attention devoted to qualitative differences within communities. None of the various buffer zone proposals reviewed here recognize that rural communities are often politically fractured and socially differentiated in complex ways. Rather, people are categorized as indigenous, nontraditional, subsistence farmers, pastoralists, or any number of other labels that mask age, gender, class, and ethnic differences within communities.

Thus, while the notion of securing "local" land rights as a basis for participation is a significant advance in conservation thought, buffer zone plans rarely make explicit which rights are secured for whom. For example, contested land and resource claims between men and women can be a major source of intracommunity conflict. Across Africa, gender is one of the key factors in determining ownership of and access to land and resources (e.g., Carney and Watts 1990; Rocheleau 1988; Schroeder 1993; Sefa Dei 1994). Lines of differential access and ownership between men and women may be drawn depending upon the type of activity, type of resource, the species, the location, or the intended use of the resource. Interventions for conservation and development may favor one group over another and exacerbate inter-gender conflicts (Carney and Watts 1990; Schroeder 1993). Schroeder, for example, has documented a case in The Gambia in which the promotion of tree planting for environmental stabilization had the effect of usurping women's access rights in favor of men's (1995).

Wealth differentiation is another process that may result in competing claims among community members. Pronounced socioeconomic stratification within communities can lead to the formation of class interests that may conflict on the question of land and resource tenure. Land conflicts in Tanzania, for instance, often revolve around the improper transfer of property from poorer peasant farmers to well-connected local elites or wealthy outsiders (URT 1992). Where resource exploitation for market sales is promoted in ICDPs, profits may flow to the wealthy, who have the capital, knowledge, and status to mobilize labor and transport products to market (see Dove 1993). As Colchester (1994:34) notes, local elites will rarely willingly make way for "local people's" participation, but rather manipulate projects to advance their own political power. In Zambia, the Administrative Management Design for Game Management Areas (ADMADE) program identified "chiefs" as "traditional rulers". Subsequently: "Chiefs used these initiatives to secure more power for themselves rather than to facilitate local participation for wildlife" (Gibson and Marks 1995:947). In effect, where patron-client relations are strong, ICDPs can serve to perpetuate or reinforce those relations without substantially improving the livelihoods of the "local people" or promoting conservation.

Closely related to the question of who the local people are is the role of local institutions—particularly the rapidly proliferating "indigenous" NGOs—in defining and negotiating land tenure for rural communities. Locally based NGOs have acquired powerful cachet among international donors and development NGOs. While NGOs are undoubtedly now key players in rural politics in Africa (Clark 1991; Nyang'oro 1992; Wellard and Copestake 1993), questions remain about their capacity to promote democratic participation or offer alternative models for conservation and development (Bonner 1993; Ellis 1994; Hanlon 1991; Vivian 1994). Neither can we assume that local NGOs, which often play a central role in the new ICDPs, represent local peoples, interests for the purposes of regulating land use and access.[4] As Hodgson points out, in Tanzania a handful of well-educated men have positioned themselves as *the* representatives of "Maasai" interests to outside donors by virtue of their leadership of indigenous NGOs (1995b). These NGOs, some of which are involved in ICDPs in Tanzania (Neumann

---

[4] According to a study by Thomas-Slayter (1994), the emergence of grassroots organizations has led in some cases to increased equity and democratization and in others to increased social stratification.

1995a), marginalize the roles of women and usurp male elders' authority in project decision making (Hodgson 1995b). Too often, NGOs demonstrate the same inefficiencies and lack of attention to local needs and aspirations characteristic of state-run projects (Vivian 1994; Korten 1994). Relating Murombedzi's (1992) study of an ICDP in Zimbabwe, Derman points out that "the creation of a local NGO funded by national and external donors serves as a new layer of bureaucracy between the communal land residents … and their rights to manage their own resource base rather than as an effective vehicle for articulating community interests" (1995:207).

What is "customary land tenure"?

Another underlying assumption in many ICDPs is that there exists a locally agreed upon body of customary law regarding land access and ownership, which can be documented and preserved through legal registration. Commonly, projects seek to incorporate the protection of "indigenous" or "customary" land and resource rights within their objectives. For instance, several buffer zone projects or proposals in Tanzania have a land-titling component that overlaps with local (particularly Maasai) efforts to secure customary land rights (AWF 1989; KIPOC 1992; Makombe 1993:24; Mbano et al. 1995; Neumann 1995a; Newmark 1993). These proposals are based upon the supposition that titling of land leads to greater security in property rights and greater security will create the conditions for conservation (Oldfield 1988; Cleaver 1993). Evidence from land-titling and land tenure reform in general does not always support the first part of the hypothesis, and some findings support its converse—that land titling may threaten the security of many customary rights holders (URT 1992; Roth 1993; Vivian 1994). In part, this is because land registration programs are unlikely to address the complexity and flexibility of existing land and resource tenure. Research in the social relations of common property systems in Africa reveals a seemingly endless variety and complexity of rights, obligations, and rules, many of them ad hoc (Bassett and Crummey 1993). Rights to a particular area of land may have multiple claims upon it, both group and individual, and can include rights to water, fuel, grazing, and cultivation plots, which in turn may vary according to season, species, or intended usage (Fortmann and Bruce 1988; Peters 1987, 1994; Wilson 1989; Neumann 1992a; Campbell 1993).

Under such systems of land and resource tenure, questions of whose rights and which rights will be privatized become critical (Shipton and Goheen 1992). Peters (1994:177) has found that titling can be a way to legally protect groups or individuals *against* customary claims. In the case of Tanzania, where village land titling has been under way since the late 1980s, land conflicts are increasing rather than decreasing (Coldham 1995; Van Donge 1993; URT 1992). In Kenya, land reform has eliminated established access rights to trees for households that had relatively weak cultivation rights within their communities (Dewees 1995). In sum, customary land claims are not always readily identifiable nor consensually determined—the relative economic and political power of competing interest groups and individuals often determines which claims become documented in law.

The overarching problem with the conceptualization of the relationship between security of tenure and conservation goals in ICDPs is a critical lack of historical understanding of customary land law in Africa. Throughout much of British-ruled Africa, colonial administrators of the early twentieth century tolerated and often actively promoted the retention of "customary land law" in areas of African settlement and occupation (Berry 1992; Colson 1971; Ranger 1993). British colonial authorities across Africa researched, recorded, and legally recognized customary land law as part of the process of implementing indirect rule (Berry 1992; Colson 1971). Since African societies were often engaged in internal struggles over the power to control land, colonial authorities' efforts to record "traditional" property relations resulted in conflicting testimony (Berry 1992; Ranger 1993). One consequence of this history is conflict over the meaning of tradition and the power to define customary land use and control (Shipton and Goheen 1992). Consequently, colonial land policy generated "unresolved debates over the interpretation of tradition" (Berry 1992:336) and local struggles over the power to assign meanings of land (Peters 1987; Shipton and Goheen 1992). Hence, the notion of traditional land tenure is largely a result of colonial governance rather than an ancient feature of African property relations (Bassett 1993; Colson 1971; Berry 1992; Ranger 1993).

This is not to say that claims are simply invented out of thin air. They are not. Customary claims, in varying degrees, are derived from social practice. Locally derived and understood meanings attached to land and resources carry with them sets of obligations, responsibilities, and rights that apply differentially according to social position. The issue of "invented tradition" (Ranger 1983) in land tenure arises over questions of the power to designate categories and narrate history, which ultimately have the effect of assigning rights, responsibilities, and obligations (Shipton and Goheen 1992). In other words, it is essential that we understand whose version of history is being narrated and who has the power to make their version the legally sanctioned one. In attempting to secure property rights for local communities, ICDPs are in danger of igniting similar internal power struggles and generating the same sorts of conflicting claims as did colonial interventions into African property relations.

Contradictions in conception and implementation

The ambivalent conceptualizations of local people and local land uses produce contradictions in the conceptualization and implementation of buffer zone projects. Perhaps the most problematic contradiction in buffer zone proposals concerns the relationship between "traditional" land use practices and project goals for "community development". Much of the rationale for encouraging local participation is based on the idea of retaining "the traditional lifestyles of indigenous people in buffer zones" (Oldfield 1988:12). Yet many projects, and sometimes the same projects, promote increased wage labor, greater market integration, and the "modernization" of land use practices as the way to spread the "benefits" of conservation to local communities and relieve pressures on protected areas. In Guinea, for example, conservationists' buffer zone proposals encourage "agricultural intensification and the provision of off-farm employment" to reduce pressures on protected areas (Fairhead and Leach 1994:506). Often, economic "benefits" include training and employing local people as tour guides, game scouts, and protected area guards (e.g., Gibson and Marks 1995). In other words, the objective is to convert segments of local communities—whether hunter-gatherers, pastoralists, or peasants—into wage laborers in a "modern" cash economy.

A second contradiction arises between the conceptualization of buffer zones and their implementation. While the ICDP approach assumes that land use restrictions will be compatible with local economic development needs and be based upon local participation in management and decision making, this remains to be demonstrated (Bloch 1993; Fairhead and Leach 1994; Ghimire 1994; Barrett and Arcese 1995). In many ICDPs, the effects on local participation or the devolution of decision making is far from evident. In virtually all of the proposals for formally demarcated buffer zones, control over land use (and thus over people) rests with conservation agencies or related ministries. For example, an IUCN publication suggests that Zaire represents the ideal model, wherein the state is legally empowered to regulate land use in buffer zones as part of protected area legislation (Oldfield 1988:3). This particular legislation gave park authorities jurisdiction over "human activities *within 50 kilometers of the boundary* of gazetted protected areas" (Oldfield 1988:3, emphasis added). Recently, a senior UNESCO scientist writing about core areas with buffer zones suggested that these units "be managed as a single entity, with marked and patrolled boundaries and entry only through manned gates and access roads" (Lusigi 1992b:35). In other words, entire communities would be enclosed within a quasi-militarized boundary with land use activities closely monitored by central government authorities. Rather than improving the security of tenure of buffer zone residents, projects often extend state authority over settlement and land use well beyond protected area boundaries, thereby heightening the insecurity of local land tenure.

Conclusion

The persistence of primitivist discourse in the "new" conservation approach in Africa diminishes the possibilities for creative, socially just, and viable solutions to threats to biodiversity. Ideas of the primitive structure the implicit (and erroneous) assumptions of socially undifferentiated local communities whose land uses and access rights are ancient and internally un-contested. In the first instance, the persistent image of traditional society or indigenous peoples existing in harmony with nature precludes any analysis of social differentiation and agrarian change or an understanding of rural communities' linkages to a larger political economy. These are "the people without history" (Wolf 1982), who exist in some static premodern equilib-rium. This conceptualization ignores the historical forces that link under-development and environmental degradation in Africa. Second, the mean-ing of *traditional* and the identification of traditional beliefs and practices related to landownership and access are never straight forward. The process of identifying customary property rights is to a great degree a political one because it involves questions of the power to narrate history, to define tra-dition, and in the process to make claims to land and resources. Third, and perhaps most importantly for the successful integration of conservation and development, the negative image of the backward and destructive African peasant or pastoralist justifies the continuation of the historical pattern of expanding state control over land and increasingly restrictive interventions. Many of the buffer zone proposals represent a tremendous territorial expan-sion of state power and sometimes outright land alienation in the name of conservation. The land area that might fall under the jurisdiction of state conservation authorities is potentially enormous. Buffer zones typically entail a belt two to ten kilometers wide around protected area boundaries and up to fifty kilometers in the most extreme case of Zaire. Relocations and evictions, euphemistically referred to as the "removal of incompatible land uses", are key buffer zone management strategies, even as securing local land tenure is touted as an important benefit to adjacent communities.

In general, buffer zone proposals suffer from a failure to recognize, let alone analyze, unequal relations of power and how they relate to land and resource access and, ultimately, to the efficacy of conservation policies. This is a dangerous oversight, as these proposals remain subject to the same sorts of politically charged questions—how is access controlled, to what degree is the institution of control seen as legitimate by the community, how is the range of uses determined, and who has authority for monitoring compli-ance—as did the colonial interventions that preceded them. ICDP support-ers talk of empowering, but as with advocates of incorporating indigenous knowledge into development plans they "seldom emphasize that significant shifts in existing power relationships are crucial to development" (Agrawal 1995:416). Increasingly in contemporary cases, local groups, often through the formation of NGOs, are demanding autonomous control of land and re-sources, which they view as a customary property right that has been usurped by the state. In this context, "it is often socio-political claims, not land pressure *per se*, which motivate encroachments into the reserve" (Fairhead and Leach 1994:507). Local demands can be politically radical, and most conservationists and state authorities are reluctant to go so far as to grant sole control of forests and wildlife habitat to villages or other local political entities. Local participation and local benefit sharing, however, are not the same as local power to control use and access, which, in the end, is what many communities seek.

It is thus quite likely that many of the proposals and projects reviewed here will result in increased conflicts over land and resources, both within communities and between local communities and the state. Given the historical antagonisms between local communities and protected area administrations, it would be reasonable to expect the same sorts of conflicts and resistance tactics that have existed since the colonial period (see Neumann 1992b). Residents adjacent to protected areas are well aware of the historical continuities of conservation policies and their effects on local livelihoods. For example, Hitchcock recently reported that some Tyua in Botswana and Zimbabwe have observed that the state "was being replaced by international institutions which were pursuing the same kinds of policies of control and dispossession" (1995:193). These policies in buffer zone projects have been resisted through violent confrontations in Madagascar (Ghimire 1994), Uganda, and Cameroon, where people have angrily protested their dislocation from buffer zones (Colchester 1994).

Recognizing the persistence of notions of the primitive in buffer zone proposals offers an opening for reconceptualizing relations between conservation advocates and rural communities in Africa. The opportunity lies in breaking down the constructed boundary between modern and traditional, civilized and primitive, us and them. By abandoning these undifferentiated categories, we can see local indigenous societies as subject to some of the same troubling politics of class, ethnicity, and gender as confront us. Avoiding the temptation to either romanticize or demonize rural peoples in Africa, perhaps we can build a dialogue that is truly mutual and initiate institutions and policies that actually empower people to control their lives and improve the conditions under which they live.

First, we need to recognize that past and present conservation policies are complicit in creating the climate of land tenure insecurity within which many rural African communities operate. The establishment of virtually every national park in sub-Saharan Africa required either the outright removal of rural communities or, at the very least, the curtailment of access to land and resources. As a result, buffer zones extend the authority of the park to monitor and restrict the land and resource uses of populations already displaced by protected areas. Policies need to be reconceptualized as mechanisms for power sharing between local communities and state and international institutions rather than as opportunities for extending state control. Research needs to be directed toward identifying and developing institutional mechanisms for controlling access and use of lands and resources that are seen as legitimate by affected communities and that have a detectable effect on conservation goals.

Second, research and policy need to be directed toward identifying the lines of fracture in rural communities and how segments of a community are differentially and even adversely affected by conservation proposals. Specifically, we need to recognize that local communities are not homogeneous entities whose members share a common set of interests regarding land and resource rights and that conservation interventions, almost by definition, will produce winners and losers in struggles over access. Local politics in rural Africa often revolves around the competing land claims of men versus women, or the poor versus more well to do peasants, within villages or even within households. Most importantly, we need to problematize the notion of traditional or customary land tenure as the product of years of intracommunity struggle over rights, not a set of ancient laws frozen in time.

Finally, we need to understand how the development interventions in buffer zones relate to conservation. Many of the projects reviewed are designed not to improve livelihoods but merely to defuse local opposition. This is a very short-sighted and short-lived "solution" which simply "buys" the support of (some segments of) local communities rather than integrating conservation with development. Whether the "benefits" of conservation are reaching the people most directly involved in activities that threaten protected areas or, if they are, whether they have any marked effect on their land and resource decisions remains an open question. Research focused on the politics of land is needed to demonstrate the link between conservation and the improvement of local livelihoods. Moving in these directions, I believe, will lead us closer to a truly new approach to biodiversity conservation.

References

Agrawal, Arun. 1995. Dismantling the divide between indigenous and scientific knowledge. *Development and Change* 26:413–39.

Alcorn, J. 1993. Indigenous peoples and conservation. *Conservation Biology* 7(2):424–26.

Andriamampianina, J. 1985. Traditional land-use and nature conservation in Madagascar. In *Culture and conservation: The human dimension in environmental planning*, ed. J. McNeely and D. Pitt. Kent: Croom Helm.

AWF. 1989. Protected areas, neighbors as partners: African Wildlife Foundation community conservation projects. Paper presented at the workshop Wildlife Resource Management with Local Participation, Harare.

Barrett, Christopher B., and Peter Arcese. 1995. Are integrated conservation-development projects (ICDPs) sustainable? On the conservation of large mammals in sub-Saharan Africa. *World Development* 23(7):1073–84.

Baskin, Yvonne. 1994. There's a new wildlife policy in Kenya: Use it or lose it. *Science* 265:733–4.

Bassett, T.J. 1988. The political ecology of peasant-herder conflicts in the northern Ivory Coast. *Annals of the Association of American Geographers* 78(3):453–72.

———. 1993. Introduction: the land question and agricultural transformation in sub-Saharan Africa. In *Land in African agrarian systems*, ed. T.J. Bassett and Donald E. Crummey. Madison: University of Wisconsin Press.

Bassett, T.J., and Donald E. Crummey, eds. 1993. *Land in African agrarian systems*. Madison: University of Wisconsin Press.

Batisse, Michel. 1982. The biosphere reserve: A tool for environmental conservation and management. *Environmental Conservation* 9(2):101–11.

Beinart, William. 1984. Soil erosion, conservationism, and ideas about development: A southern African exploration, 1900–1960. *Journal of Southern African Studies* 11(1):52–83.

Berry, S. 1992. Hegemony on a shoestring: Indirect rule and access to agricultural land. *Africa* 62(3):327–55.

Blaut, J.M. 1993. *The colonizer's model of the world: Geographical diffusionism and Eurocentric history.* New York: Guilford.

Bloch, P. 1991. Buffer zone management: Access rules and economic realities. Paper prepared for LTC workshop Tenure and the Management of Natural Resources in Sub-Saharan Africa, Madison, Land Tenure Center.

———. 1993. *Buffer zones, buffering strategies, resource tenure, and human-natural resource interaction in the peripheral zones of protected areas in sub-Saharan Africa.* Madison: Land Tenure Center.

Bonner, Raymond. 1993. *At the hand of man: Peril and hope for Africa's wildlife.* New York: Knopf.

Byers, Bruce. 1994. Armed forces and the conservation of biological diversity. In *Green security or militarized environment,* ed. Jyrki Kakonen. Brookfield: Dartmouth.

Campbell, David. 1993. Land as ours, land as mine: Economic, political, and ecological marginalization in Kajiado District. In *Being Maasai: Ethnicity and identity in East Africa,* ed. T. Spear, and R. Waller. London: James Currey.

Carney, Judith, and Michael Watts. 1990. Manufacturing dissent: Work, gender, and the politics of meaning in peasant society. *Africa* 60(2):207–41.

Clark, John. 1991. *Democratizing development: The role of voluntary organizations.* London: Earthscan.

Cleaver, Kevin. 1993. *A strategy to develop agriculture in sub-Saharan Africa and a focus for the World Bank.* World Bank Technical Papers, no. 203. Washington, DC: World Bank.

Colchester, Marcus. 1994. *Salvaging nature: Indigenous peoples, protected areas, and biodiversity conservation.* UNRISD Discussion Paper. Geneva: United Nations Research Institute for Social Development.

Coldham, Simon. 1995. Land tenure reform in Tanzania: Legal problems and perspectives. *Journal of Modern African Studies* 33(2):227–42.

Colson, Elizabeth. 1971. The impact of the colonial period on the definition of land rights. In *Colonialism in Africa, 1870–1960,* ed. Victor Turner. Cambridge: Cambridge University Press.

Derman, B. 1995. Environmental NGOs, dispossession, and the state: The ideology and praxis of African nature and development. *Human Ecology* 23(2):199–215.

Dewees, Peter. 1995. Trees and farm boundaries: Farm forestry, land tenure, and reform in Kenya. *Africa* 65(2):215–35.

Dove, Michael. 1993. A revisionist view of tropical deforestation and development. *Environmental Conservation* 20(1):17–24.

Ellis, Stephen. 1994. Of elephants and men: Politics and nature conservation in South Africa. *Journal of Southern African Studies* 20(1):53–69.

Fairhead, James, and Melissa Leach. 1994. Contested forests: Modern conservation and historical land use in Guinea's Ziama Reserve. *African Affairs* 93:481–512.

Feierman, Steven. 1990. *Peasant intellectuals: Anthropology and history in Tanzania.* Madison: University of Wisconsin Press.

Fletcher, Susan. 1990. Parks, protected areas, and local populations: New international issues and imperatives. *Landscape and Urban Planning* 19:197–201.

Fortmann, Louise, and John Bruce, eds. 1988. *Whose trees? Proprietary dimensions of forestry.* Boulder: Westview.

Ghai, Dharam. 1992. *Conservation, livelihood, and democracy: Social dynamics of environmental changes in Africa.* UNRISD Discussion Papers, no. 33. Geneva: United Nations Research Institute for Social Development.

Ghimire, Krishna. 1994. Parks and people: Livelihood issues in national park management in Thailand and Madagascar. *Development and Change* 25:195–229.

Gibson, C.C., and S.A. Marks. 1995. Transforming rural hunters into conservationists: An assessment of community-based wildlife management programs in Africa. *World Development* 23(6):941–57.

Gordon, Robert. 1992. *The Bushman myth: The making of a Namibian underclass.* Boulder: Westview.

Hanlon, Joseph. 1991. *Mozambique: Who calls the shots?* Indianapolis: Indiana University Press.

Haraway, Donna. 1984. Teddy bear patriarchy: Taxidermy in the Garden of Eden, New York City, 1908–1936. *Social Text* 4(2):20–64.

Hill, Kevin. 1996. Zimbabwe's wildlife utilization programs: Grassroots democracy or an extension of state power. *African Studies Review* 39(1):103–21.

Hitchcock, R.K. 1995. Centralization, resource depletion, and coercive conservation among the Tyua of the northeastern Kalahari. *Human Ecology* 23(2):169–98.

Hobsbawm, Eric. 1987. *The age of empire, 1875–1914.* New York: Pantheon.

Hodgson, D. 1995a. The politics of gender, ethnicity and "development": Images, interventions, and the reconfiguration of Maasai identities in Tanzania, 1916–1993. Ph.D. diss., Department of Anthropology, University of Michigan.

——. 1995b. Critical interventions: The politics of studying "indigenous" development. Paper presented at the annual meetings of the American Anthropological Association, Washington, DC.

IUCN. 1991. *Protected areas of the world: A review of national systems.* Vol. 3: *Afrotropical.* Gland, Switzerland: IUCN.

IUCN, United Nations Environment Program, and World Wide Fund for Nature. 1991. *Caring for the earth: A strategy for sustainable living.* Gland, Switzerland: IUCN.

Jaroz, L. 1992. Constructing the Dark Continent: Metaphor as geographic representation of Africa. *Geografiska Annaler* 74 B(2):105–15.

KIPOC. 1992. *The foundation program: Program profile and rationale.* Principal Document no. 4. Loliondo, Tanzania: Korongoro Integrated Peoples Oriented to Conservation.

Kiss, A. 1990. Living with wildlife: Wildlife resource management with local participation in Africa. World Bank Technical Papers, no. 130.

Korten, Frances. 1994. Questioning the call for environmental loans: A critical examination of forestry lending in the Philippines. *World Development* 22 (7):971–81.

Lance, T. 1995. Conservation politics and resource control in Cameroon: The case of Korup National Park and its support zone. Paper presented at the African Studies Association Annual Meeting, 4 November, Orlando, Florida.

Lerise, F., and U. Schuler. 1988. Conflicts between wildlife and people: Village development planning for three settlements bordering the Selous Game Reserve. Consultants' Report for GTZ.

Luke, Timothy W. 1994. Watching at the limits of growth. *Capitalism, Nature, and Socialism* 5(2):43–63.

Lusigi, W., ed. 1992a. *Managing protected areas in Africa.* Paris: UNESCO.

——. 1992b. New approaches to wildlife conservation. In *Managing protected areas in Africa,* ed. W. Lusigi. Paris: UNESCO.

Lutz, Catherine, and Jane Collins. 1993. *Reading National Geographic.* Chicago: University of Chicago Press.

MacKenzie, J. 1987. Chivalry, social Darwinism and ritualized killing: The hunting ethos in Central Africa up to 1914. In *Conservation in Africa: People, policies and practice,* ed. D. Anderson and R. Grove. Cambridge: Cambridge University Press.

Mackinnon, J., K. Mackinnon, G. Child, and J. Thorsell. 1986. *Managing protected areas in the tropics.* Gland, Switzerland: IUCN.

Makombe, K., ed. 1993. *Sharing the land: Wildlife, people, and development in Africa.* IUCN/ROSA Environmental Series, no. 1. Harare, Zimbabwe: IUCN.

Mbano, B.N., R.C. Malpas, M.K. Maige, P.A. Symonds, and D.M. Thompson. 1995. The Serengeti regional conservation strategy. In *Serengeti II: Dynamics, management, and conservation of an ecosystem,* ed. A.R. Sinclair and P. Arcese. Chicago: University of Chicago Press.

McNeely, J. and K. Miller, eds. 1984. *National parks, conservation, and development.* Washington, DC: Smithsonian Institution Press.

Miller, K. 1984. Regional planning for rural development. In *Sustaining tomorrow: A strategy for world conservation and development,* ed. F.R. Thibodeau and H.H. Field. Hanover and London: University Press of New England.

Moore, Donald. 1993. Contesting terrain in Zimbabwe's eastern highlands: Political ecology, ethnography, and peasant resource struggles. *Economic Geography* 69:380–401.

Moore, Henrietta, and Megan Vaughan. 1994. *Cutting down trees: Gender, nutrition, and agricultural change in the Northern Province of Zambia, 1890–1990.* London: James Currey.

Murombedzi, J. 1992. Decentralization or recentralization? Implementing CAMPFIRE in the Omay communal lands. CASS Working Papers, no. 2. Harare: Center for Applied Social Science, University of Zimbabwe.

Mwalyosi, R.B. 1991. Ecological evaluation for wildlife corridors and buffer zones for Lake Manyara National Park, Tanzania, and its immediate environment. *Biological Conservation* 57:171–86.

Nepal, S.K., and K.E. Weber. 1995. Managing resources and resolving conflicts: National parks and local people. *International Journal of Sustainable Development and World Ecology* 2:11–25.

Neumann, R.P. 1992a. Political ecology of wildlife conservation in the Mt. Meru area of northeastern Tanzania. *Land Degradation and Rehabilitation* 3:85–98.

——. 1992b. *The social origins of natural resource conflict in Arusha National Park, Tanzania.* Ph.D. diss., University of California, Berkeley.

——. 1995a. Local challenges to global agendas: Conservation, economic liberalization, and the pastoralists rights movement in Tanzania. *Antipode* 27(4):363–82.

——. 1995b. Ways of seeing Africa: Colonial recasting of African society and landscape in Serengeti National Park. *Ecumene* 2(2):149–69.

——. 1996. Dukes, earls, and ersatz Edens: Aristocratic nature preservationists in colonial Africa. *Society and Space* 14:79–98.

Newmark, William. 1993. The role and design of wildlife corridors with examples from Tanzania. *Ambio* 22:500–504.

Nyang'oro, Julius. 1992. *The receding role of the state and the emerging role of NGOs in African development.* Nairobi: All Africa Conference of Churches Research and Development Consulting Service. Manuscript.

Oitesoi ole-Ngulay, Saruni. 1993. Inyuat e-Maa/Maa pastoralists development organization: Aims and possibilities. Paper presented at the IWGIA-CDR conference The Question of Indigenous Peoples in Africa, Greve, Denmark, 1–3 June.

Oldfield, Sara. 1988. *Buffer zone management in tropical moist forests: Case studies and guidelines.* Gland, Switzerland: IUCN.

Omo-Fadaka, Jimoh. 1992. The role of protected areas in the sustainable development of surrounding regions. In *Managing protected areas in Africa,* ed. W. Lusigi. Paris: UNESCO.

Peluso, N.L. 1993. Coercing conservation? The politics of state resource control. *Global Environmental Change* 3(2):199–217.

Peters, P. 1987. Embedded systems and rooted models. In *The question of the commons: The culture and ecology of communal resources,* ed. B. McKay and J. Acheson. Tucson: University of Arizona Press.

——. 1994. *Dividing the commons: Politics, policy, and culture in Botswana.* Charlottesville: University of Virginia Press.

Prins, Herbert. 1992. The pastoral road to extinction: Competition between wildlife and traditional pastoralism in East Africa. *Environmental Conservation* 19(2):117–23.

Ramberg, Lars. 1992. Wildlife conservation or utilization: New approaches in Africa. *Ambio* 21:438–39.

Ranger, T. 1983. The invention of tradition in colonial African. In *The invention of tradition,* ed. E. Hobsbawm and T. Ranger. Cambridge: Cambridge University Press.

——. 1993. The communal areas of Zimbabwe. In *Land in African agrarian systems,* ed. Thomas J. Bassett and Donald E. Crummey. Madison: University of Wisconsin Press.

Redclift, Michael. 1984. *Development and the environmental crisis: Red or green alternatives?* London: Methuen.

Redford, K.H. and A.M. Stearman. 1993. Forest-dwelling native Amazonians and the conservation of biodiversity. *Conservation Biology* 7(2):248–55.

Richards, P. 1995. Fighting for the rainforest. Paper presented at the annual meeting of the African Studies Association, 3–6 November, Orlando, Florida.

Rocheleau, Dianne. 1988. *Women, trees, and tenure: Implications for agroforestry*. In *Whose trees? Proprietary dimensions of forestry*, ed. Louise Fortmann and John Bruce. Boulder: Westview.

Roth, Michael. 1993. Somalia land policies and tenure impacts. In *Land in African agrarian systems*, ed. Thomas J. Basset and Donald E. Crummey. Madison: University of Wisconsin Press.

Said, Edward. 1994. *Culture and imperialism*. New York: Knopf.

Sayer, J. 1991. *Rainforest buffer zones: Guidelines for protected area managers*. Cambridge: IUCN Forest Conservation Programme.

Sayers, G.F. 1933. A visit to the Serengeti Game Reserve. *Country Life* 28 (October):440–41.

Schoepf, Brooke G. 1984. Man and biosphere in Zaire. In *The politics of agriculture in tropical Africa*, ed. J. Barker. Beverly Hills: Sage.

Schroeder, R. 1993. Shady practice: Gender and the political ecology of resource stabilization in Gambian garden/orchards. *Economic Geography* 69(4):349–65.

———. 1995. Contradictions along the commodity road to environmental stabilization: Foresting Gambian gardens. *Antipode* 27(4):325–42.

Sefa Dei, G.J. 1994. The women of a Ghanaian village: A study of social change. *African Studies Review* 37(2):121–45.

Shipton, Parker, and Mitzi Goheen. 1992. Introduction: Understanding African land-holding—power, wealth, and meaning. *Africa* 62(3):307–25.

Stearman, A.M. 1994. Revisiting the myth of the ecologically noble savage in Amazonia: Implications for indigenous land rights, *Bulletin of the Culture and Agriculture Group* 49:2–6.

Stocking, M. and S. Perkin. 1992. Conservation-with-development: An application of the concept in the Usambara Mountains, Tanzania. *Transactions of the Institute of British Geographers* 17:337–49.

TANAPA. 1994. *National policies for national parks in Tanzania*. Arusha, Tanzania: Tanzania National Parks.

Thomas-Slayter, Barbara. 1994. Structural change, power politics, and community organizations in Africa: Challenging the patterns, puzzles, and paradoxes. *World Development* 22(10):1479–90.

Torgovnick, Marianna. 1990. *Gone primitive: Savage intellects, modern lives*. Chicago: University of Chicago Press.

URT. 1992. *Report of the Presidential Committee of Inquiry into land matters*. Vol. 1. Dar Es Salaam: United Republic of Tanzania.

Van Donge, Jan. 1993. The arbitrary state in the Uluguru Mountains: Legal areas and land disputes in Tanzania. *Journal of Modern African Studies* 31(3):431–48.

Vivian, Jessica. 1994. NGOs and sustainable development in Zimbabwe: No magic bullets. *Development and Change* 25:167–93.

Wellard, Kate, and James Copestake. 1993. *Non-governmental organizations and the state in Africa: Rethinking roles in sustainable agricultural development*. London: Routledge.

Wells, M., and K. Brandon. 1992. *People and parks: Linking protected area management with local communities*. Washington, DC: World Bank, WWF and USAID.

———. 1993. The principles and practice of buffer zones and local participation in biodiversity conservation. *Ambio* 22(2–3):157–62.

Wilson, K.B. 1989. Trees in fields in southern Zimbabwe. *Journal of Southern African Studies* 15(2):367–83.

Wolf, E. 1982. *Europe and the people without history*. Berkeley: University of California Press.

World Bank. 1993. *The World Bank and the environment: Fiscal 1993*. Washington, DC: World Bank.

Eroded Consensus: Donors and the Dilemmas of Degradation in Kondoa, Central Tanzania

Wilhelm Östberg

Problems of land degradation have a long history in the central parts of Kondoa District, Tanzania. Accelerated soil erosion can be dated as far back in time as 900 years (Eriksson 1998). Apart from man-induced erosion, geological processes during much longer time periods have also contributed to the stripping of the Kondoa Irangi Hills.

Symptoms of land degradation are evident throughout the Kondoa Irangi Hills. Large parts of the slopes are dissected by deep gullies. Hillsides have been cleared of natural vegetation and are intensively cultivated. During the rains, this leads to excessive surface runoff. In places, sheet erosion has removed two to three meters of topsoil, and badlands are common (Christiansson 1988; Payton and Shishira 1994). But people still have to make their livings in these degraded environments. The Tanzanian government supports them through projects intended to rehabilitate depleted lands in a number of areas and also to promote sustainable forms of land use. Some of these initiatives have attracted foreign support.

The HADO project (*Hifadhi Ardhi Do*doma, or Land Rehabilitation in Dodoma Region) was launched as the direct result of a visit by President Julius Nyerere to Kondoa District in 1969. Alarmed by the severe land degradation in the central parts of the district, the president directed the Forestry Department to take urgent action to remedy the situation (Kawa 1993:103; Kikula and Mung'ong'o 1993:4). This subsequently led to the establishment of the HADO project in 1973. It is active in three districts of

Research was carried out under the umbrella of the research program "Man Land Interrelations in Semiarid Central Tanzania", jointly run by the Institute of Resource Assessment, University of Dar es Salaam, and the Environment and Development Studies Unit, School of Geography, Stockholm University. I am indebted to the program's co-ordinators, Professors Carl Christiansson and Idris S. Kikula, for practical support, but in particular for stimulating discussions and comments. Comments by Dr. Ingvar Backéus, Dr. Mats Eriksson, and Richard Kangalawe, as well as from one of the editors of this volume, Dr. Richard Schroeder, were likewise very helpful. Research clearance from the Tanzanian Commission for Science and Technology is gratefully acknowledged, as is financial support provided by the Swedish Council for Planning and Coordination of Research (FRN) and by Sida. Mrs. Margaret Cornell read and corrected my English with admirable exactness and sensibility. Forester Joseph Mduma joined me for most of the time in the field and contributed not only his wide knowledge of environmental rehabilitation work but also his exceptional ability to make research issues accessible to people with no previous experience of research work. I am most grateful. The usual disclaimers apply.

Dodoma Region. The discussion in this essay is focused on the situation in the central parts of Kondoa District, the so-called Kondoa Eroded Area (KEA), where the main activities are taking place. Typical of the time, the project's initial approach to the erosion problem was technical, involving the construction of contour bunds and check dams, the production and distribution of seedlings, and tree planting. In their analysis of the causes of land degradation in Kondoa, the project's management emphasized adverse natural conditions in combination with faulty agricultural practices (Mbegu and Mlenge 1983). In this, they joined a long tradition in Tanzania, and in Africa, of advocating state intervention to correct what are taken to be destructive local ways of farming.[1] However, other factors, such as the effects of misdirected external interventions and the wider sociopolitical context generally, as well as processes of commercialization, have also been highlighted as important causes of the severe land degradation (e.g., Banyikwa and Kikula 1981; Kesby 1982; Östberg 1986; Christiansson et al. 1992; Kawa 1993; Barr 1994; Mung'ong'o 1995; Dejene et al. 1997). These studies of the Kondoa situation reflect an ongoing reorientation in research on natural resource management. The first major move away from technical treatises was toward the political ecology of land degradation (Blaikie 1985; Richards 1985) and the linkages between poverty and environmental stress (Chambers 1983; Redclift 1984). More recently the diversity, dynamics, and flexibility of small-scale farming in Africa (e.g. Berry 1993; Netting 1993; Niemeijer 1996) have come into focus along with a deeper understanding of semiarid areas as inherently unstable but also more resilient than has usually been assumed (Behnke et al. 1993; Scoones 1995; Dahlberg 1996).

Twenty-five years ago land rehabilitation was largely understood to mean tree planting and landscape engineering. Therefore it was logical that the Forestry Department should be made responsible for the new project. As rehabilitation and conservation gradually evolved into a broader program of land husbandry, the foresters working for the HADO project widened their professional competence considerably both in work experience and service training.

The decisive moment for the HADO project came in 1979 when an area of 1,256 km[2], the so-called Kondoa Eroded Area (KEA),[2] was closed to grazing, thereby transforming the farming system and the economy of the people of the area (Östberg 1986; Mung'ong'o 1995). Local bylaws prohibit cutting of green-wood in the area and the clearing of forests for cultivation. A comprehensive rehabilitation program includes tree planting, reseeding of badlands, and support for on-farm conservation and stall feeding of improved livestock.

---

[1] This theme has been analyzed in, for instance, Feierman 1990; Moore and Vaughan 1994; Maddox et al. 1996; and Leach and Mearns 1996.

[2] The acronym KEA today also stands for Kondoa Enclosed Area. The area is not physically fenced, but its exact borders are delineated in the relevant bylaw.

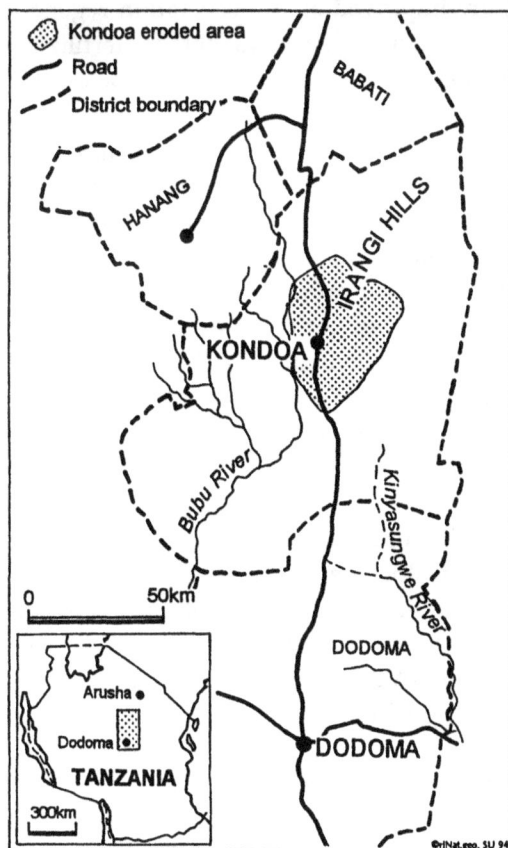

Kondoa District is situated along the main road between Dodoma and Arusha. The so-called Kondoa Eroded Area is roughly congruent with the Irangi Hills.

(Map: Dept. of Physical Geography, Stockholm University)

The decision to evict all livestock from the KEA was taken after the project had established experimentally that where contour bunds had been constructed good ground cover could be expected after two years if the area was protected from grazing (Mbegu and Mlenge 1983:41).[3] The project management reasoned, why not temporarily close the whole area to grazing in order to achieve a breakthrough in the rehabilitation effort? The idea won the support of the dominant political organization of the time, the ruling Chama Cha Mapinduzi (CCM) Party. It was carried out in an uncompromising manner, just like the compulsory villagization that had occurred some five years earlier (Östberg 1986:chap. 6). The project became responsible for seeing that the bylaws protecting the environment were complied with and for prosecuting offenders in court. The legacy of the HADO project is thus one of planning from the top down and of firm execution.

The scope of the HADO project and the approach to land rehabilitation adopted by its managers had few precedents in Africa. Initial results were

---

[3] Previous experience could also have been relied upon. The rotational grazing schemes of the 1930s proved that vegetation quickly recovers with careful pasture management (Banyikwa and Kikula 1981:12).

dramatic. Conditions for agriculture improved as the vegetation regenerated, despite the fact that manure is by and large not available for fertilizing
the fields.[4] Study groups came from near and far to learn about the experiment.

Several factors combined to give the project considerable stamina: the
visible changes the area underwent, the project's independent role as a national pilot project responsible directly to the ministry in Dar es Salaam instead of to the district and regional administrations,[5] and the resolute political backup from the CCM Party during the project's crucial early years.
However, with time these same factors turned out to have their drawbacks.
The project became trapped in its own apparent success. By the late 1980s
and early 1990s, the political conditions that had made compulsory destocking possible evaporated and a program that was once a major governmental
priority came under attack as a stagnant, hierarchical project with little innovative power (cf. Norén 1995:27).

As a multiparty system began to be discussed seriously in Tanzania in
the early 1990s, local opposition to destocking was able to express itself
much more openly (Mung'ong'o 1995:148). Changing ideologies in extension work suggested that the project should give up its policing role.[6] This
opinion also met with support within HADO itself. Some members of the
staff were convinced that stall feeding of livestock combined with grass
strips on contours in cultivated fields was such a profitable enterprise that
with time livestock owners would themselves come to defend the quarantine. A new policy was in the making. However, long before an effective
alternative way of keeping check on the number of livestock illegally entering the area had been worked out, the livestock issue had developed into a
matter that was very difficult to handle.

The principal outside donor to the HADO project, the Swedish International Development Agency (SIDA), was preparing to pull out as a large-
scale and long-term Dutch-financed integrated rural development project
established itself in Kondoa. The newcomers showed no interest in HADO;
quite the contrary. They had introduced themselves as representing a more
democratic and locally anchored approach than the supposedly topdown
HADO project. HADO had for years enjoyed a privileged position, which

---

[4] While it is not possible to monitor developments with the help of production statistics from
the district offices, a number of published reports document the fact that the farmers themselves found the situation improved and also that outside observers were of the same opinion.
See for instance, Östberg 1986; Kikula and Mung'ong'o 1993; Mung'ong'o 1995; Kangalawe
1996; and Liwenga and Kikula 1997.

[5] The project's standing in Dar es Salaam was reinforced when E.M. Mnzava became director of
forestry. He had been a member of the study team recommending that a special land rehabilitation project should be established at Kondoa.

[6] That this responsibility ought to be transferred to the district authorities was one of the recommendations of the 1995 external evaluation of the HADO project (Erikson et al. 1995:47). The
evaluators were of the opinion that no material support from SIDA was needed to implement
this recommendation (48).

some people in the district administration felt had also been associated with a certain aloofness.[7] Such reasons made it difficult to turn to the District Council for support when HADO's external funds were cut. Few councillors wanted to become associated with HADO's policy of restricting natural resource use. And in any case the council could not afford to take on the many staff members that HADO had on its payroll. Meanwhile, support from Dar es Salaam for HADO dwindled with the appointment of a new director of forestry, who turned out to have little interest in land management issues in smallholder areas. The possibilities of keeping the HADO project operative rapidly diminished.

On 1 March 1997, the leading Tanzanian newspaper, the *Daily News* reported that the "achievements by the country's pioneer conservation project [HADO] ... are likely to go down the drain. ... [L]arge stocks of cattle have started being grazed in the project areas, charcoal operators are in full swing, while felling of trees for brick making is now the order of the day" (Kitururu 1997). This was in glaring contrast to the opinions of a joint Swedish-Tanzanian (SIDA and the Ministry of Tourism, Natural Resources, and Environment) evaluation of the HADO project, which two years earlier had concluded that "there are good prospects for the service on soil conservation to continue without SIDA assistance in the future" (Erikson et al. 1995:45).

We shall examine the situation as it existed during the last few ears of the 1990s and discuss possible strategies for the immediate future. Natural resource management in the KEA is, as we shall see, a hot issue. But first let us look back briefly at the history of soil conservation in Kondoa, as this is of immediate relevance for today's situation.

Resistance to soil conservation

Kondoa has been the focus of land rehabilitation attempts since 1927 when clearing of forests began as a combined effort to combat sleeping sickness and provide new land for people living in denuded areas (Fosbrooke 1950–51; Banyikwa and Kikula 1981). Since that time, a wide variety of reclamation methods have been tried. In the early 1930s, ridge cultivation was introduced to reduce runoff on cultivated land. Contour bunds were constructed on slopes. Sisal and other agaves were planted along contours and check dams installed in gullies. Schemes of rotational grazing were organized together with compulsory reduction of livestock numbers. After the Second World War, the government again embarked on clearing the plains to combat *tsetse* and to provide new land for residents of areas considered to be beyond possible rehabilitation (Fosbrooke 1950–51:169ff.; Lambert 1957:58; Banyikwa and Kikula 1981:13; Kikula and Mung'ong'o 1993:23ff.,

---

[7] In a comment on HADO's institutional environment, Kawa (1993:216) goes so far as to allege that it had been characterized by tension and paranoia rather than cooperation.

33). Organized clearing continued into the mid-1960s (Kawa 1993:109), and in fact the process still continues as farmers themselves clear forest to obtain new farms (Mung'ong'o 1995: 85ff.; Östberg 1995:chap. 6).

Summing up the results, one may note that contour ridges on cultivated land have become a regular feature of farming in the area. Other innovations have been less persistent, notably the attempts to introduce quotas for the number of livestock allowed in a given area.

Few government programs have such direct and radical impacts on local communities as conservation policies, with their demand for labor inputs, their restrictions on the use of natural resources, and even their wholesale takeover of land by moving people out of areas declared in need of protection. The Kondoa case can in this respect be said to form part of the widespread practice of states expanding their control over rural areas by means of conservation schemes. Throughout, the Rangi people[8] have by and large remained suspicious of soil conservation initiated by the government. During the 1940s, the opposition to soil conservation became linked to a more general political resentment against colonial rule (Kesby 1982:72; Mung'ong'o 1995:45ff., 71).[9]

As the nationalist struggle intensified in the late 1950s it became impossible to enforce the communal conscription of labor and conservation work came to a halt (Berry and Townshend 1973:244; Mung'ong'o 1995:73). Meanwhile, a profitable market for finger millet had developed. Kondoa became a major supplier to the beer trade in Arusha and Kilimanjaro Regions (Kawa 1993:77). Forests were cleared to provide land for a rapidly expanding crop production, just as is happening today on the plains east of the Kondoa Eroded Area.

During the 1960s, critical attitudes toward soil conservation persisted. Bunds and terracing reminded people of a time when the government had directed them into compulsory communal work. Independence, by contrast, had brought hopes of access to new land and increased possibilities of making money. Land degradation thus continued, and in the mid-1970s it was further aggravated by the program of compulsory villagization (cf. Kikula 1997b). In the central parts of Kondoa District, land degradation had by then reached the alarming state that led to President Nyerere's intervention and the advent of the HADO project.

Throughout all these events, the residents of the Kondoa Hills for their part have approached the problems of soil erosion from a different angle from that of the administration. In this area of unpredictable rainfall and

---

[8] The Rangi are a Bantu-speaking people who form the majority population of Kondoa District (cf. Kesby 1982; Östberg 1986; Mung'ong'o 1995). The Kondoa Eroded Area includes their core area.

[9] In other parts of East Africa, political opposition also was galvanized by opposition to erosion control (see for instance, Young and Fosbrooke 1960; Rosberg and Nottingham 1966; Tignor 1976; Blaikie 1985; Throup 1988; Feierman 1990).

fragile soils of comparatively low fertility, many farmers are doubtful that investments in soil conservation can be recovered. Even worse, conservation structures may increase soil erosion. When there are torrential rains, these structures may give way and the concentrated runoff may play havoc with the crops growing in the fields. During dry periods, on the other hand, bunds are considered to be of limited value. Furthermore, conservation makes demands on family labor that many households are unable to meet.[10] Instead of investing in engineering-type soil conservation, many Rangi farmers prefer to expand their areas of operation. When pressure on the land increases and harvests dwindle, some villagers will leave to take up new land elsewhere while other members of the family remain behind. The Rangi as a group can be said to practice an expansionist, permanent agriculture (Östberg 1986:29). Their principal solution to problems of soil erosion thus is to move temporarily, or permanently, to areas with better conditions, while also aiming to retain a foot in the Rangi core areas.

Thus, the administration and large parts of the Rangi community have held fundamentally different opinions about how to cope with the fragile environment of the Kondoa Irangi Hills. While the extension service tried to safeguard each and every piece of land against the evils of soil erosion, the local communities aimed at ever wider spheres of operation. In the view of the conservationists, this simply meant moving the problems of land degradation to other areas. They argue that the Rangi way of coping with land degradation only aggravates the problem. The administration has therefore time and again favored restrictions. For many of those hit by the various regulations, conservation became synonymous with the execution of power by outsiders. Farmers felt that they were being deprived of both their civil freedoms and a decent chance to make a living. Small wonder that many villagers opposed the conservation schemes. To some, it became tantamount to a patriotic obligation to be hostile toward conservation. Land rehabilitation became politicized.

When the HADO project attempted to solve the soil erosion problem by means of afforestation, agricultural engineering, and destocking, it was reiterating solutions that had already been tried, with little success, during the colonial period. HADO had little reason to expect that its approach to land rehabilitation would be met with enthusiasm. Nevertheless, some remarkable results were recorded.

A transformed agricultural potential

The removal of livestock in 1979 soon resulted in a notable regeneration of vegetation within the enclosed area (Tosi et al. 1982:15). However, the vegetation remained in the early successional stages. The process was slow be-

---

[10] Feierman (1990:188ff.) has discussed in careful detail a comparable situation in the Usambara Mountains.

cause of the degraded soils, and the grasses generally have poor grazing value (Backéus et al. 1994, 1996). All the same, the new ground cover was important in protecting the soil from erosion, and it also had other impacts. The sediment carried in the streams appeared to have decreased. The watercourses became much narrower. Villagers noted that streams flew longer into the dry season than was formerly the case; in the foothill communities below the Irangi escarpment the mountain streams were reported to provide water for a longer period of the year and under more controlled conditions.[11] Rainfall was being infiltrated into the area on a totally different scale from that before the destocking. As sand fans stabilized, new areas become available for cultivation. Sugarcane, sweet potatoes, and vegetables were valuable money earners in such areas.[12] The enclosed area was feeding a larger population than ever before.[13] At the same time, vermin were said to attack crops more frequently than before and fears were voiced locally that *tsetse* may return as woody vegetation reestablished itself in the area (Mung'ong'o 1995:10ff.).

The change had not taken place without strong feelings being expressed. The expulsion of livestock was undertaken amid an atmosphere of hostility and opposition from large parts of the population of the area. However, a number of people are able to maintain more than one household—one based on arable agriculture in the destocked hills and another based on free grazing and a somewhat hazardous cultivation on the plains outside the KEA. For many, this has turned out to be a very successful combination, and these households are accumulating more and more wealth in their hands (Mung'ong'o 1995:85ff., 98ff.; Östberg 1986:69ff.; Östberg 1995:chap. 6).

Development in an area like Kondoa, with its precarious farming conditions, seems to be best supported by attempting to improve the general productivity of the area—which is what has happened since 1979 and which is currently being threatened as free-grazing livestock is again becoming a fact in the enclosed area. At an individual level, agricultural investments are difficult to recover reliably given the erratic, scarce, and occasionally violent rainfall.[14] An overall environmental improvement was for a number of years gradually making the situation somewhat less precarious. It is such small-scale improvements, together with increased reliability in food production,

---

[11] On the other hand, it was a common complaint in some parts of the Irangi Hills that a lot of water was being consumed by the abundant vegetation and therefore less available for other uses. To weigh the pros and cons is a complicated matter. However, something all agreed on was that the situation had stabilized and the problem with flash floods was much reduced.

[12] Cf. Östberg 1986:chap. 4. Mbegu (1996:151ff.) lists a number of additional benefits following from the rehabilitation of the Kondoa Eroded Area and provides examples of the economic value of these for individual farmers.

[13] The demographic situation has been analyzed by Madulu (1990, 1996).

[14] Available rainfall data for Kondoa have been analyzed by Ngana (1996) and Pinner (1996). Ngana (1992) has calculated the probability that annual rainfall in Kondoa is less than or equal to 450 mm is about 20 percent, making agriculture a somewhat hazardous undertaking.

that may improve the district's economic performance. Environmental stability is thus basic to all attempts at improvement. This is precisely what is being threatened by the collapse of the HADO project. A few examples will illustrate the situation.

A former forest plantation

At Gubali, some ten kilometers north of Kondoa town, the HADO project planted three hundred hectares with eucalyptus. This forest plantation has remained a constant source of conflict between the HADO project, which aims to rehabilitate and put to productive use a severely degraded area, and villagers unwilling to give up what they consider to be their land.

During the process of compulsory villagization in the mid-1970s, the people of Gubali had been removed from their area. This occurred without their consent and the process took four years to implement (Östberg 1986:61ff.). Opinion is widespread in the village that it was all a big mistake. A typical comment is that "Nyerere never intended that people should be moved away from areas along the roads", and, since Gubali is transected by the road connecting the major towns of Arusha and Dodoma "people living in the bush should have been moved here instead". However, the principal reason for putting an end to habitation and farming in Gubali was in fact that the area was degraded, and from the standpoint of the administration it was logical to ask the HADO project to solve the Gubali problem by afforesting the area.[15] In effect, the HADO project came to expropriate land that the residents of Gubali considered to be theirs. Conservation became a political matter, in this case as in so many others.

For years, forest fires were a constant menace to the Gubali plantation. Eventually, the HADO project stopped clearing fire lines and replanting. It did not take long before the first cultivation appeared. A small canteen was opened where the road from Haubi village connects with the main road. Today a small center of well-built brick houses has sprung up there. A newly built primary school is in its first year of operation. The unauthorized settlement at Gubali has become recognized as a subvillage under Kolo village.

The valley and parts of the surrounding pediments are now cultivated. Only a few isolated full-grown eucalyptus trees testify to there having recently been a forest plantation there. The HADO project never managed to harvest these trees. When people started to move back to Gubali, they quickly transformed the trees into sawn timber, building poles, and firewood for brickmaking.

---

[15] This was logical not only because HADO was part of the Forestry Department but also because at the time it had among its objectives the promotion of the *ujamaa* policy. *Ujamaa*, a Swahili term that translates as "family" but also as "socialism", was the central notion in Tanzanian development thinking during 1960s and 1970s.

Villagers at the unauthorised settlement at Gubali inside the enclosed area are constructing a school for their children. A village is emerging in an area which according to the HADO Project should be rested from permanent cultivation.   (Photograph by Wilhelm Östberg, 1997)

The fields at Gubali illustrate the different ways the settlers avail themselves of the renewed land resource. Some settlers retain the contour bunds and cultivate the land in between. Some make use of the tied ridges to plant fruit trees in these microcatchments. Others use the bunds for growing sweet potatoes, which means that the bunds will disappear with time. Most farmers who keep the bunds either remove the stabilizing vegetation that the HADO project had planted on them or cut it down because they fear that snakes and rodents will hide in it. It is common to see a cleared bund kept intact for some twenty to thirty meters and then demolished for another stretch, demonstrating the likely fate of the conservation structures in the area. As cultivation continues, the bunds will gradually disappear. The impression one gets is that farmers in this area appreciate the value of the bunds but not to the extent that they are prepared to keep them vegetated and even less to make an effort to keep them in good repair. With time, they will go, as indeed has happened to the similar structures built during the period of British rule.

In discussions with villagers on the changes in Gubali, they emphasize that development is now reaching the area. The years when the land was rested were, in many villagers' eyes, time lost. When they were removed from their village, they had been in the process of building a school. Now they are resuming the work they were forced to abandon more than twenty years ago. Looking back on the period when their area was deserted, they see little value in this. "Here was just bush. People traveling to Haubi were in danger of both wild animals and robbers along the road. Who could you call on for help during those years? There was no one living here". Now things are changing for the better as people move in.

In conversations with outsiders, the settlers are keen to emphasize that conditions in Gubali are excellent. Given the long history of environmental intervention in Kondoa, most villagers are anxious to forestall discussions about deforestation and land degradation. Strategic thinking aside, their comments also reflect characteristic traits in Rangi social life. When uncultivated areas become settled, fields and roads replace bush, and schools and mosques are built, life is improved in the eyes of most Rangi. Planted fruit trees are of more value than undifferentiated wilderness. It is better to have cooperating relatives as neighbors than monkeys invading the fields. Ordered society emerges almost instantly as the Rangi establish themselves in a new area (Östberg 1995:175ff.). The question of what level of exploitation the land can sustain is far less important than the fact that a community is coming into being and surpluses are being produced.

Gubali was given a new start when the formerly degraded lands were rested and trees were planted. This resource is now being taxed, and transformed, by men and livestock. Standing on the hills above Gubali, one can see herds of animals grazing in the dissected slopes toward Bolisa village. This is done quite openly and in a place that is about as much in the center of the Kondoa Enclosed Area as one can get. There was a time when a limited number of animals were brought into the more peripheral parts of the enclosed area for grazing during the night. Such precautions are no longer necessary. All over the KEA, livestock were actively grazing the conservation bunds, where vegetation abounds.[16] In many gullies, we saw goats finishing off the last patches of grass.

The badlands at Chakwe

People living in Chakwe, just south of Gubali, were already resettled in the 1940s as their area was judged to be beyond recovery (Banyikwa and Kikula 1981:12; Östberg 1986:89ff.). When the HADO project was launched, stone check dams were built in the gullies and contour bunds were constructed across the slopes. This area still bears incontestable evidence of how badly depleted it is. Nevertheless, some vegetation has established itself and wildlife has entered the area. In the mid-1980s, HADO had a small bulking plot here growing the nitrogen-fixing herb *Crotolaria ochroleuca*, but it never managed to harvest anything as the crop was eaten by wildlife. The healing

---

[16] I am reporting on the areas I last visited during 1997. Other fieldworkers of the MALISATA research program have made identical observations at Baura, Mafai, Mulua and Thawi. There is no doubt that in the late 1990s livestock were grazing everywhere in the enclosed area.

process is now coming to an end as livestock are freely grazed in the area. A few first fields have been taken up for cultivation, and trees have been cut for charcoal making. Where a few years ago we saw the prints of lion's paws on the ground, and where the botanists working in the area had to keep armed guard while making inventories of the changing vegetation, one now walks among goat droppings.

Kome Forest Reserve

Saw pits are in evidence. Selective logging has resulted in only aged individuals remaining from valuable tree species. The forest is diminishing as people clear it to expand their fields. New settlements are appearing within the reserved area.

A seed bank analysis carried out over three consecutive years showed that all endemic or near endemic taxa were absent; the uniqueness of the Kome Forest Reserve is about to be lost (Lyaruu et al. 1997). With the destruction of the reserve, the water regime in the streams serving the Haubi and Mafai villages will be affected, as will the villages at the foot of the escarpment. The same is true for the Ntomoko spring from which water is piped to more than a dozen villages out on the plains.

A related uncontrolled exploitation is occurring a few kilometers southward along the escarpment. There the forest of the steep upper slopes is being cleared for short-lived finger millet fields. The area is now dotted with clearings, which are very visible and certainly an invitation to farmers of the foothill communities to open up the much less steep pediment slopes below, which up to now have been closed to cultivation. The clearing along the escarpment at Mwisanga threatens both the Mwisanga Dam (cf. Yanda 1995) and the important Ntomoko water resource as well as the controlled provisioning of water to the foothill areas.

Encroachment on the Kome Forest Reserve has been going on for a number of years and is indeed indicative of a process that is occurring all over the area. Forests are being cleared for cultivation, timber, and firewood (cf. Mung'ong'o 1995:85ff.; Östberg 1995:chap. 6). The effects are being felt only gradually, but the signs are evident.[17]

Stall feeding of livestock in the KEA

Farmers who want to keep stalled improved cows in the KEA can obtain permits to do so as well as support from the HADO project. More recently, the Livestock Production Research Institute at Mpwapwa has introduced zero grazing of goats in the KEA.

---

[17] For instance, the road between Mnenia and Itololo is now badly damaged by gullies and deposited sand. Bridges have been washed away by runoff in several places. During the years when the conservation legislation was systematically enforced, this road remained in good repair. It seems reasonable to link the current deterioration to the recent felling of trees and grazing in the reserved forest above the escarpment.

Economically, stall feeding of large and smallstock is a good idea. In this area, improved cows produce an average of eight to ten liters of milk a day, while stalled zebu cows produce three to four liters, compared to about a liter from those kept under free grazing (Ogle 1991:21; Larsson 1993:32; Ulotu 1994:33; Ogle et al. 1996:130). Selling five to seven liters a day gives a monthly income of about 30,000 Tanzanian shillings (TSh).[18] To this is added the value of the milk consumed by the family plus the income from selling manure and calves. Eriksson (1993:20) reports on farmers obtaining higher incomes from their improved livestock than the extension staff advising them earn. It is difficult to think of an alternative way of earning a similar income in rural Kondoa.

There is little indication that the market for milk is becoming saturated.[19] Disease is the main risk involved, and currently veterinary support is not well developed in Kondoa. However, the situation is not totally out of hand; a comparative study of the economics of free grazing and stall feeding in Kondoa District documented that the health problems of livestock grazing freely were greater than those of the stall-fed animals (Ulotu, 1994:37).

Free-grazing animals in the area would in all likelihood put an end to the stall-feeding project. Few farmers seem prepared to carry feed and water to their animals if others are allowed to graze. And even if they were prepared to do so their workload would increase considerably, as there would be less grass available to cut. Even the fodder they cultivate would very likely be finished off by the grazing animals.

Policing the KEA

For some fifteen years, the HADO project attempted to keep a check on the number of animals entering the area. This obligation has now been taken over by the local administration, mainly by the village and ward executive officers (VEOs and WEOs), hopefully with support from the village councils. These local-level administrators find themselves trapped in a difficult situa-

---

[18] The milk price during the rainy season of 1997 varied between 150 and 200 TSh per liter. The difference in yield between wet and dry seasons for stalled animals is surprisingly low (Larsson 1993:33). Ulotu (1994:51) found that for keeping one cow the net income over running costs, including own labor costs and depreciation, was already far more than the official minimum monthly wage. He notes that the system not only produces good returns but it has the additional advantage that the greatest amount of labor is needed at slack periods when the opportunity cost of labor is low (53).

[19] Ogle et al. (1996:130) noted that the "demand for fresh milk is being satisfied, at least in one or two of the villages where the project concentrated its initial activities", but they did not analyze the situation further. The farmers we spoke to in the wet season of 1997 said that they could easily find customers for their surplus milk. Eriksson (1993:24) calculated that to produce even such a modest quantity as 0.1 liter of milk per person in Bolisa and Haubi, two pioneering villages in the stall feeding scheme, milk production needed to be increased twenty-three times. The number of stall-fed cows has not changed much since she made her calculation.

tion.[20] They are supposed to keep their respective areas free of grazing live-stock and to stop the illegal felling of trees, charcoal burning, and the cutting of green wood. Since many of these practices are now widespread, a visit to their area by a high-ranking official can result in local staffpersons being fired on the spot for not fulfilling their duties. However, attempts to influence the situation are beset with difficulties.

If officials approach a farmer known to be keeping livestock in the enclosed area, they are immediately rebuked: "Why do you pester me about this when Kondoa town is full of livestock belonging to *wakubwa* (big people)?" Whatever the truth of such allegations, they have been important in guiding many people's decisions to bring livestock back into the enclosed area. And, attitudes aside, every time we visited Kondoa town during 1997 livestock were observed grazing outside the D.C.'s office, the HADO office, the police station, and so on. Since this goes on undisturbed, farmers, of course, ask why they should abide by the law while influential people do not do so. A diligent VEO in fact has no case to argue.

To carry out his duties successfully, a VEO needs to be on good terms with the village chairman and council. There is every likelihood that people in these positions keep livestock. The VEO thus has to confront precisely those people with whom he needs to cooperate to carry out his many other obligations.[21] Often he is also related to people who keep livestock, who will, of course, ask why he makes difficulties for his own people. "Don't be so sure about that government you serve so diligently. All of a sudden something can happen and you will be out of work. Where will you turn? You will come to us and then it's no bad thing if we have a few cows so that we can help you out."

Some of the poor households in the enclosed area cooperate with live-stock owners on the plains. They keep livestock in the Irangi Hills in return for milk, manure, draft power, and perhaps the odd heifer. Rich and poor become allied in taxing the resources of the enclosed area.

Erosion of consensus: competing viewpoints on the future of Kondoa

The opinion that HADO should no longer be involved in seeing that the by-laws governing natural resource management are complied with has stronger support in the donor community than among environmental scientists. The donors ask themselves whether it is reasonable to expect politicians who are dependent on the popular vote to be in the forefront of enforcing unpopular decisions like grazing restrictions. The primary concern for

---

[20] This section gives details of difficulties encountered during the current mode of operation. This is not to say that the HADO staff did not have to tackle similar problems when it was responsible for enforcing the restrictions. A lament over its troubles is sung in Östberg 1986:57ff.

[21] This point could be backed up with several case studies. There is, however, no space available here for this.

Tanzanian environmentalists appears to be that the investments made, and the results achieved, should not be allowed to go down the drain. [22]

Is this what is now likely to happen? Previous experience points in that direction. Photographs taken in the area before 1973 as well as aerial photographs show a barren, dissected landscape, widespread badlands, and active gullies of horrifying dimensions. Other evidence is provided by the records of famine relief provided to the KEA before the rehabilitation program began (Östberg 1986:48ff.) as well as the experience of having to close whole areas to habitation and other uses.

Two specialists who have recently studied land quality in the area state that the land recovery "superficially apparent by the return of vegetation cover ... is illusionary". They write that for many parts of the Kondoa area the return of livestock "should not be contemplated in either the short or long term". They go on to say that the "extent of soil loss and gullying on steep hillslopes and many upper to middle pediments means that any form of agricultural land use, especially uncontrolled grazing, is unsustainable". Likewise, on gullied lower pediment slopes "livestock should be excluded, other than in zero-grazing regimes" (Shishira and Payton 1996:43). Botanists working in the area report that, while vegetative recovery is encouraging, it is still at an early stage of succession, one reason for this being that soil conditions are likely to be reversible only in a geological time perspective (Backéus et al. 1996:61).

Whereas scientists who have studied the situation hold definite opinions about what is and is not possible to do in the Kondoa area, and are worried that the situation is rapidly getting out of hand; foreign aid workers tend to be more preoccupied with whether the HADO project has been innovative in recent years and whether its institutional setup is in line with current development thinking: "local ownership" and "partnership". That the project finds itself at a standstill at present is taken lightly; this is true of many other development activities in Tanzania, we are told.

Few argue that policing the area is a long-term solution. But was it wise to give up this role overnight, which de facto is what happened? With the amount of livestock currently in the area, no easy solution is to hand. If the animals are to remain, a system of controlled grazing needs to be worked out in cooperation with all the local communities. Trials with village forest management in Tanzania may perhaps indicate a possible direction (Wily 1995, 1996). What is required in such a case is something considerably more than the acceptance of an assortment of local demands on natural resource use, namely, actually transferring authority to the villages. In the Kondoa situation, this would mean that village, not government, rules about live-

---

[22] However, it should also be noted that two senior researchers at the Institute of Resource Assessment, University of Dar es Salaam, have contributed to a report arguing that destocking has deprived farmers of access to farmyard manure and that the policy should therefore be re-examined (Dejene et al. 1997:28).

stock and pastures would become law. However, to create a climate of co-operation in which this could be achieved would in the Kondoa case be a major undertaking. An additional factor to be taken into consideration, one largely overlooked during development interventions, is that the Rangi have an economically and socially differentiated society in which those with re-sources wield considerable influence (see Mung'ong'o 1995: chaps. 4 and 7; and Kawa 1993:81ff.).

If, on the other hand, the livestock are to be removed, this would proba-bly require an unequivocal order by the political leadership of the country, forcing all government workers to give up whatever personal interests they may have in the current situation and thereby providing them with the cred-ibility required to force others to do the same.

Kondoa opinions

Within Kondoa itself, the large influx of livestock in the enclosed area is hotly debated. At a meeting of the District Council in late April 1997, one of its members alleged that permits to keep livestock in the area "have turned commercial". This was another way of saying that the village-level staff accepts money to allow free grazing. Another councillor remarked that he found it extraordinary to be asked about the presence of livestock in his constituency while there were VEOs and WEOs present whose job it is to look into such matters. The question looming large in such remarks is: who may have to bear the blame for the current state of affairs?

When the livestock question is mentioned in the course of everyday con-versation with villagers, another type of comment is encountered. Touching on the drought the northeastern part of the district experienced in early 1997, many of those we talked with emphasized how fortunate it was that livestock from such areas were able to be rescued by the lush vegetation in the enclosed area. Likewise, when encountering livestock along paths and roads people often delighted at how healthy and well fed they appeared.[23]

What such comments have in common is that the presence of livestock in the enclosed area is now taken for granted. Many villagers hold that the area is in fact rehabilitated and the problem solved. For years, restrictions have been imposed, and now the result is there to be seen. Their world has been rejuvenated. *Ardhi imerudi*, they say, "land has come back". And, indeed, who would want to see animals dying in a drought while the KEA abounds with vegetation?

People living in the KEA grew up with the badlands and the gullies. Many villagers accept them as part of an inherited landscape (cf. Dejene et

---

[23] Data from a recent survey confirm how highly livestock are valued in the area: "Even with the apparent shortage of pasture reported in all villages during field work, the majority of interviewed farmers indicated their desire to maximise herd size" (Dejene et al. 1997:34). Fifteen years earlier, Wenner (1983:16) reported similar feelings.

Livestock grazing near Haubi village inside the enclosed area. The contour ridge and agave plants tell of the HADO project's earlier activities to help conserve soil and water in this badly eroded area. As cows and goats now freely move around in the area the conservation structures are damaged. Land use appear to revert to conditions that applied before the conservation intervention.

(Photograph by Wilhelm Östberg, 1997)

al., 1997:18). However, once these hill slopes were covered with *miombo* (*Brachystegia*) woodlands (Backéus et al. 1994:327). There is no reason to believe that the bare areas and the active gullies will not return if uncontrolled grazing and tree felling are allowed. As already noted, scientists studying the area say that the land has not come back and the situation remains precarious. These, however, are the views of outsiders. Seeing the livestock moving in the landscape, another thought is foremost in many villagers' minds. They rejoice in their hearts to see how the cows shine, even if the animals do not belong to them personally. It is as if they see a life process materializing before their eyes as the starving cows gain strength.

To many livestock owners, it does not appear an illegitimate overuse of resources to graze the stands of elephant grass on the gully bottoms, at least not during the rainy period when it is green everywhere. Clearly, conservationists and most livestock owners have no common understanding of what constitutes appropriate land use in the KEA.

All this notwithstanding, it must be emphasised that most—if not all—people greatly value the improved environment. Many areas are said to benefit from an improved microclimate. They have, in local terms, become "cool". It is also common to hear worries expressed that the cows have become too numerous. Seeing headloads of green wood or charcoal being carried out of the enclosed area, people comment with dismay *"wanakata ovyo"* (they are cutting regardless) or *"wanaharibu sana"* (they are really destructive). There is no doubt that many residents of the Kondoa Eroded Area worry that active land degradation will soon be a reality again. However, this opinion finds difficulties in expressing itself both because of the prevailing power relations and because those who are worried have no primary rights to defend. The resources now being taxed are largely considered to belong to the government or alternatively to the HADO project. By rehabilitating gullies, reseeding badlands, and planting trees on dissected pediments, the responsibility for these areas has somehow been transformed to the project. From some areas, people were even physically removed. Until the former residents move back, such areas risk becoming open prey to one and all. When the HADO project became immobilized, opportunites for both entrepreneurs and locals to harvest the last twenty years' buildup of biomass and fallowed fields open up. In the areas where people return to lands they formerly cultivated, they feel they have every right to do so. This is not unique to Kondoa District but occurs also in other parts of the country that were subjected to compulsory villagization in the 1970s.

Who can act?

By the turn of the century, the HADO project was at a standstill. The policy designed to safeguard the future of the Kondoa Eroded Area had collapsed. Can anything be done? The two most obvious actors are the Forest Department and the District Council. The Forest Department has ultimate responsibility for HADO, while it typically does not involve itself in day to day management of the project. The District Council is the body that passes the bylaws governing natural resource management in the area. Neither of them has so far indicated what its intentions are. Nor is it likely that any group of local farmers will be able to force a more active policy. One group that might be thought to have a direct interest in promoting grazing restrictions is those who keep stalled animals. Their ability to get fodder for their animals is rapidly diminishing. However, their numbers are far too small to form a significant pressure group. Several of these farmers appear instead to be solving the problem by taking their own animals to the pastures.

We seem to be left with three remaining options. One is to stick to the way that land rehabilitation has been approached in Kondoa before, namely, by means of initiatives from above. A firm declaration from the central government that it will not tolerate senior staff—or any other people—keeping free-grazing animals in the area or involving themselves in other unauthorized exploitation of natural resources could pave the way for the district leadership, ensuring that the local administration will see that local bylaws are respected. Alternatively, the bylaws could be changed if dialogue were to be initiated with the local communities on how to achieve a sustainable use of land, water, and forests in the KEA in the future. The third option is to allow room for local entrepreneurs and farmers to use the available resources for their own benefit. There is a historical precedent for the expectation that this will lead to depletion of the land, which, however, may subsequently be followed by local attempts at land rehabilitation and new forms of management. Such processes of clearing, land degradation, and new forms of natural resource management, albeit at a lower level of productivity, are known from different historical periods and different parts of the world and could presumably take place also in Kondoa. Accepting such a scenario amounts to saying that the attempts during the last fifty years at soil and water conservation in Kondoa have been futile.

A visit to the area by the vice president in early 1997 raised hopes among a number of people in Kondoa that a new start could be achieved. The vice president is reported to have promised to incorporate the Kondoa Eroded Area within the concerns of an environmental division in his office. In early May, the prime minister visited Kondoa and reiterated a concern on the part of the central government about the fate of the area. He directed the Dodoma regional commissioner to look immediately into the situation with regard to the HADO project, including the alleged misuse of funds during the last few years, and to report directly to the prime minister.

Meanwhile, in February Professor I.S. Kikula, then chairman of the National Environmental Management Council of Tanzania and a board member of the National Land Use Planning Commission as well as former director of the Institute of Resource Assessment at the University of Dar es Salaam, studied the situation in Kondoa. In a strongly worded, four-page letter circulated among senior government officials and national-level politicians involved with environmental issues, he expressed his deep concern over "the pathetic situation that the HADO project in Kondoa has found itself in". Commenting on strategies outlined by the Kondoa district commissioner, Professor Kikula argued that "actions are required to be taken at a much higher level". He described charcoal pits encountered and livestock grazing freely in the area. "Already there are indications of overgrazing in many of the parts I visited. The hills behind Baura are more or less bare rocks, reminding me of what I saw in the area in 1978. I visited some of our fenced experimental plots. The difference in grass biomass between the protected plots and the surrounding areas is like day and a moon-lit night" (Kikula 1997a).

Professor Kikula's letter asks for a high-level intervention to overcome the current "glaring gap in the decision-making process concerning HADO's affairs between the Ministry, the Regional authorities, the District authorities and the grassroots".

Experiences from the extensive land rehabilitation efforts attempted during both the colonial period and the HADO era may prove that planning from the top down does not in the long run guarantee sustainable management of natural resources. But neither will a laissez faire policy do so. The gullies and the widespread badlands are overwhelming proof of that.

The responsibility of donors

The consequences of interventions in a precarious environment like that of the Kondoa Eroded Area are difficult to assess. Implementing agencies, immersed as they are in everyday practicalities, have little chance to look actively for unintended adverse consequences of their activities. Potential uncertainties easily disappear when one is hunting for spare parts for vehicles, looking after visiting dignitaries, and struggling to counteract the effects of the inevitable delays in the procurement of funds.

Perhaps it is unavoidable that projects are to a large extent governed by everyday emergencies. However, one can reasonably argue that donors are under an obligation to study continuously what happens as a consequence of their activities.[24] Since they intentionally intervene with resources in order to influence development, they also have a responsibility actively to seek out any adverse impacts their involvement may have. If they do not face up to this responsibility, they can of course harm the very people they hoped to support. An independent research initiative should therefore be linked to major projects so that their socioeconomic and environmental impacts can be studied continuously. The responsibility to document the impact of environmental interventions does not end when project support is withdrawn, as in the HADO case. The consequences of terminating support should also be studied, as should the long-term impact of the intervention.

Given the serious impact the HADO project has had on the lives of people living in the Kondoa Eroded Area, the agency that met the costs of running the operation cannot escape responsibility for the orderly development of the project. SIDA, the dominant external donor, shares with the Forestry Department the responsibility for funds being withdrawn at a stroke. SIDA may have had reasons for acting the way it did. Nevertheless it remains a fact that the investments made during the past twenty-five years and the considerable improvements achieved in productive conditions are now unprotected. SIDA is estimated to have spent some 20 million Swedish crowns (SEK) (approximately 2.7 million U.S. dollars) up to 1994 (Norén 1995:34, app.3; Erikson et al. 1995:41), while a calculation from the Tanzanian end comes to roughly 2.5 billion TSh (approximately 4 million U.S. dollars) (Kikula 1997a). The efforts of taxpayers in Sweden and Tanzania on behalf of land rehabilitation in the Kondoa Eroded Area all of a sudden appear meaningless, as do the sacrifices of the many households in the KEA that had to change their ways of life in accordance with regulations initiated by the HADO project. Now their lives are again being dramatically affected as new

---

[24] The arguments in this and the following section have been developed in somewhat more detail in an essay on the role of research in land management projects by Christiansson, Kikula and Östberg (1998).

modes of exploitation emerge. In neither situation were their opinions solicited.

Research and development projects

Apart from documenting environmental changes, there are many other important roles that research can play in natural resource management projects. One is actively to draw local communities into the research process. Researchers can stimulate negotiations and joint studies of the area, which in turn can help local communities to articulate their needs and make demands on both the authorities and development projects. When local communities become involved in studying their own area, they will, as examples from different parts of the world testify, become increasingly involved in safeguarding their resources for the future. In this way, a research input into a conservation project can help to shift the focus away from solely scientific and project needs and aims toward those of local communities and their own responsibility for caring for their natural resources.

Local groups (differentiated by ethnicity, gender, age, wealth, etc.) as well as projects and local authorities face different problems and have thus quite different demands for "knowledge". A single researcher, or even a group of researchers, cannot satisfy the demands of all, but can help to get processes going whereby knowledge is generated. This is the crux of the matter. A research program on man-land interrelations in a region can analyze major trends and provide case studies. But it can never capture "all" the processes taking place. Groups of people, specific villages, or individuals who wish to influence their situations will have to assemble the facts they need themselves. As they do so, important extra benefits will be created. Local environments will be monitored, new kinds of knowledge may be generated, and people's commitments to sustainable use of their area's resources can increase. Their identities are strengthened, and they become better equipped to defend their interests. They become more active citizens. Locally directed research thus has an importance far beyond the data it produces. But to keep it evolving and problem oriented, as well as of reasonable quality, requires a continuous dialogue with professional researchers.

When academic research becomes involved in these processes, it also benefits, as it moves closer to real life conditions. An agronomist and a farmer want different things from the same patch of land. When the researcher does not decide on his own what should be studied and how the research should be carried out, he or she has often taken an important step toward enhancing not only relevance of the research but the accuracy and the depth of the analysis.

Conclusions

Conservation in Kondoa has become politics and very much remains political. In this area of unusually severe land degradation, the administration—

whether British or independent—and local opinion have not been able to agree on what constitutes appropriate land management. The different attempts at land rehabilitation tried since the 1930s have had some success, for longer or shorter periods. The most recent example is the vegetational recovery during the dozen years or so after 1979, when a number of regulations on natural resource use were in force. As political control slackened in the early 1990s and new development ideologies favoured private initiatives, new opportunities arose. A number of political leaders and businessmen have made use of these and joined with many villagers to take advantage of the improved environment, but this stands no better chance than on earlier occasions of withstanding an unrestricted use of the land.

Characteristic of the conservation interventions in Kondoa is the way ideologies and truths produced elsewhere have for some time considerably influenced the landscape.[25] This is as true of the periods of *tsetse* clearings as it is of the conservation engineering provoked by the Western dust-bowl panic, of the *ujamaa* and villagization campaigns when a carefully planned society was to bring prosperity to the people, and of more recent ideologies of popular participation and demand-driven development. Throughout, the farmers of Kondoa have had little say. On the ground the situation today looks much the same as before: extentionists with ready recipes for a bright future vs. villagers hoping for as little outside control as possible. Meanwhile, the powerful and the daring make quick money as the administration, drained of resources, is unable to keep a check on developments. At present the gap between what is intended to be the policy and the reality is probably wider that it ever was before. The losers are the many poor who stand a better chance of making a living when reasonably clear rules govern natural resource use. The poor also benefit from a comparatively stable environment in which water, firewood, and building materials become more plentiful and general conditions improve.

References

Backéus, Ingemar, Z.K. Rulangaranga, and Jerry Skoglund. 1994. Vegetation changes on formerly overgrazed hill slopes in semi-arid central Tanzania. *Journal of Vegetation Science* 5:327–36.

———. (1996). The Dynamic interrelationship between vegetation and the environment on the formerly eroded Kondoa Irangi Hills. In *Changing environments: Research on man-land interrelations in semi-arid Tanzania*, ed. Carl Christiansson and Idris S. Kikula. Nairobi: SIDA's Regional Soil Conservation Unit.

Banyikwa, F.F., and Idris S. Kikula. 1981. Soil erosion and land degradation: The case of Kondoa Irangi Highlands, Dodoma, Tanzania. Paper prepared for the BRALUP workshop Resource Development and Environment, 22–27 June.

---

[25] It is not implied here that only outside interventions have caused land degradation in the Kondoa Hills. But land management in Kondoa is currently much influenced by policy, politics, and projects. This is a reality that all land managers (Blaikie and Brookfield 1987:74ff.) in the area must take into account.

Barr, Julian. 1994. Sustainability of agroforestry and indigenous cropping systems on recently deposited sandy soils in the Kondoa Eroded Area, Tanzania. M.S. thesis, University of Newcastle upon Tyne, Department of Agricultural and Environmental Science.

Behnke, Roy H., Jr., Ian Scoones, and Carol Kerven, eds. 1993. *Range ecology at disequilibrium: New models of natural variability and pastoral adaptation in African savannas.* London: Overseas Development Institute.

Berry, L., and J. Townshend. 1973. Soil conservation policies in the semi-arid regions of Tanzania: A historical perspective. In *Studies of soil erosion and sedimentation in Tanzania*, ed. A. Rapp, L. Berry, and P. Temple. Research Monographs, no. 1. Dar es Salaam: BRALUP

Berry, Sara. 1993. *No condition is permanent: Social dynamics of agrarian change in sub-Saharan Africa.* Madison: University of Wisconsin Press.

Blaikie, Piers. 1985. *The Political Economy of Soil Erosion in Developing Countries.* London and New York: Longman.

Blaikie, Piers and Harold Brookfield, with contributions by Bryant Allen and others. 1987. *Land degradation and society.* London and New York: Methuen.

Chambers, Robert. 1983. *Rural development: Putting the last first.* London: Longmans.

Christiansson, Carl. 1988. Degradation and rehabilitation of agropastoral land: Perspectives on environmental change in semi-arid Tanzania. *Ambio* 17(2):144–52.

Christiansson, Carl, Idris S. Kikula and Wilhelm Östberg. 1998. Research in support of soil conservation projects—or of land users? In *Soil and water conservation: Challenges and opportunities*, ed. L.S. Bhushan, I.P. Abrol, and M.S. Rama Mohan Rao. Vol. 2. Dehra Fun: Indian Association of Soil and Water Conservation.

Christiansson, Carl, Vesa-Matti Loiske, Clas Lindberg, and Wilhelm Östberg. 1992. History, society and local production systems: Crucial variables in modern soil conservation work. In *Proceeding: People protecting their land, 7th ISCO conference, Sydney, 27–30 September, 1992.* Vol. 2. Sydney: Department of Conservation and Land Management.

Dahlberg, Annika. 1996. *Interpretations of environmental change and diversity. A study from North East District, Botswana.* Doctoral dissertation no. 7, Department of Physical Geography, Stockholm University.

Dejene, Alemneh, Elieho K. Shishira, Pius Z. Yanda, and Fred H. Johnson. 1997. *Land degradation in Tanzania: Perception from the village.* World Bank Technical Papers, no. 370. Washington, DC: World Bank.

Erikson, Jan, Malcolm Douglas, Jan Lindström, and Esther Karario. 1995. Hifadhi Ardhi Dodoma (HADO): Dodoma Region Soil Conservation Project, final report by evaluation mission appointed by SIDA and Ministry of Tourism, Natural Resources, and Environment.

Eriksson, Annika. 1993. *Milk marketing based on small holder zero-grazing dairy cows systems in the HADO areas of central Tanzania: A minor field study.* Working Papers, no. 240. Uppsala: Swedish University of Agricultural Sciences, International Rural Development Centre.

Eriksson, M.G., and C. Christiansson. 1997. Accelerated soil erosion in central Tanzania during the last few hundred years. *Physics and Chemistry of the Earth* 22(3–4):315–20.

Eriksson, Mats G. 1996. Effects of tectonic activity on landform evolution in Kondoa Irangi Hills, central Tanzania. In *Changing environments: Research on man-land interrelations in semi-arid Tanzania*, ed. Carl Christiansson and Idris S. Kikula. Nairobi: SIDA's Regional Soil Conservation Unit.

——. 1998. *Landscape and soil erosion history in central Tanzania. A study based on lacustrine, colluvial and alluvial deposits.* Doctoral dissertation no 12, Department of Physical Geography, Stockholm University.

——. 1999. Influences of crustal movements on landforms, erosion and sediment deposition in the Irangi Hills, central Tanzania. In *Uplift, erosion and stability*, ed. B.J. Smith, P.A. Warke, W.B Whalley, and M. Widdowson. London: Geological Society. Special Publications, 162:157–68.

Feierman, Steven. 1990. *Peasant intellectuals: Anthropology and history in Tanzania.* Madison: University of Wisconsin Press.

Fosbrooke, Henry. 1950–51. The fight to rescue a district. In *East African Annual*, pp. 168–70.

Kangalawe, Richard Y. M. 1995. Fighting soil degradation in the Kondoa Eroded Area, Kondoa District, Tanzania: Socio-economic attributes, farmer perceptions, and nutrient balance assessment. M.S. thesis, Agricultural University of Norway.

Kawa, Ibrahim Hussein. 1993. The Dodoma Region Soil Conservation Project (HADO), Tanzania: Is it institutionally sustainable? Ph.D. thesis, School of Development Studies, University of East Anglia.

Kesby, J. 1982. *Progress and the past among the Rangi of Tanzania.* Vols. 1 and 2. New Haven, CT: HRAFlex Books.

Kikula, Idris S. 1997a. Back to office report following a visit to Kondoa. 4 March. Manuscript.

——. (1997b). *Policy implications on environment. The case of villagisation in Tanzania.* Uppsala: Nordic Africa Institute.

Kikula, Idris S., and Claude G. Mung'ong'o. 1993. *A historical review of the soil erosion problem and land reclamation in Kondoa District, central Tanzania.* Research Papers, no. 33. Dar es Salaam: University of Dar es Salaam, Institute of Resource Assessment.

Kitururu, Moses. 1997. HADO: A slowly dying project. *Daily News,* 1 March.

Lambert, J.L.M. 1957. Soil erosion of Kinyasi Ridge, Chungai Resettlement Area, Kondoa District. *Records of the Geological Survey* 1956, 3:58–61. Dar es Salaam: Government Printer.

Larsson, Christina. 1993. *Smallholder zero-grazing dairy systems in the HADO areas of central Tanzania: A minor field study.* Working Papers, no. 232. Uppsala: Swedish University of Agricultural Sciences, International Rural Development Centre.

Leach, Melissa, and Robin Mearns. 1996. *The lie of the land: Challenging received wisdom on the African continent.* London: International African Institute and James Currey.

Liwenga, Emma T., and Idris S. Kikula. 1997. *Household food insecurity and coping mechanisms in severely eroded areas: The case of Kondoa Eroded Area, Tanzania.* Research Papers, no. 39. Dar es Salaam: University of Dar es Salaam, Institute of Resource Assessment.

Lyaruu, Herbert V.M., Shadrack Eliapenda, Leonard B. Mwasumbi, and Ingvar Backéus. 1997. *The Afromontane forest at Mafai in Kondoa Irangi Hills, central Tanzania: A proposal to conserve a threatened ecosystem.* EDSU Working Papers, no. 37. Stockholm: Stockholm University, School of Geography.

Maddox, Gregory, James L. Giblin, and Isaria N. Kimambo, eds. 1996. *Custodians of the land: Ecology and culture in the history of Tanzania.* London: James Currey; Dar es Salaam: Mkuki na Nyota; Nairobi: EAEP; Athens: Ohio University Press.

Madulu, N.F. 1990. *The impact of population migration and resettlement on land degradation in the HADO areas—Kondoa District.* Research Papers, no. 26. Dar es Salaam: University of Dar es Salaam, Institute of Resource Assessment.

——. 1996. Changes in marital fertility in relation to the HADO Project activities in Kondoa District. In *Changing environments: Research on man-land interrelations in semi-arid Tanzania,* ed. Carl Christiansson and Idris S. Kikula. Nairobi: SIDA's Regional Soil Conservation Unit.

Mbegu, Alfred C. 1996. The problems of soil conservation and rehabilitation lessons from the HADO Project. In *Changing environments. research on man-land interrelations in semi-arid Tanzania,* ed. Carl Christiansson and Idris S. Kikula. Nairobi: SIDA's Regional Soil Conservation Unit.

Mbegu, Alfred C., and Wendelin C. Mlenge. 1983. *Ten years of HADO, 1973–1983.* Dar es Salaam: Ministry of Natural Resources and Tourism, Forestry Division.

Moore, Henrietta, and Megan Vaughan. 1994. *Cutting down trees: Gender, nutrition, and agricultural change in the Northern Province of Zambia, 1890–1990.* London: James Currey.

Mung'ong'o, Claude. 1995. *Social processes and ecology in the Kondoa Irangi Hills, central Tanzania.* Meddelanden series, B 93. Stockholm: Stockholm University, Department of Human Geography.

Netting, Robert M. 1993. *Smallholders, householders: Farm families and the ecology of intensive, sustainable agriculture.* Stanford: Stanford University Press.

Ngana, James. 1992. *Climatic assessment of Kondoa Eroded Area.* Research Reports, no. 80 (new series). Dar es Salaam: University of Dar es Salaam, Institute of Resource Assessment.

——. 1996. Climate and hydrology of the Kondoa Eroded Area. In *Changing environments: Research on man-land interrelations in semi-arid Tanzania,* ed. Carl Christiansson and Idris S. Kikula. Nairobi: SIDA's Regional Soil Conservation Unit.

Niemeijer, David. 1996. The dynamics of African agricultural history: Is it time for a new development paradigm? *Development and Change* 27:87–110.

Norén, Sten. 1995. *Desk study on the soil conservation project HADO in Tanzania*. Stockholm: SIDA.

Ogle, Brian. 1991. Intensive livestock-based smallholder systems for semi-arid areas of the tropics: The HADO experience. *IRDC Currents* 1:20–24.

Ogle, Brian, Hans Wiktorsson, and A.P. Masaoa. 1996. Environmentally sustainable intensive livestock systems in semi-arid central Tanzania. In *Changing environments. research on man-land interrelations in semi-arid Tanzania*, ed. Carl Christiansson and Idris S. Kikula. Nairobi: SIDA's Regional Soil Conservation Unit.

Östberg, Wilhelm. 1986. *The Kondoa transformation: Coming to grips with soil erosion in central Tanzania*. Research Report no. 76. Uppsala: Scandinavian Institute of African Studies.

———. 1995. *Land is coming up. The Burunge of central Tanzania and their environments*. Stockholm Studies in Social Anthropology, no. 34. Stockholm: Stockholm University.

Patton, M. 1971. *Dodoma region, 1929–1959: A history of famine*. BRALUP Research Reports, no. 44. Dar es Salaam: University of Dar es Salaam.

Payton, Robert W., and Elia K. Shishira. 1994. Effects of soil erosion and sedimentation on land quality: Defining pedogenetic baselines in the Kondoa District of Tanzania. In *Soil science and sustainable land management in the tropics*, ed. J.K. Syers and D.L. Rimmer. London: CAB International.

Pinner, Elise. 1996. Rainfall data analysis for Kondoa District. Kondoa Integrated Rural Development Project. Manuscript.

Redclift, Michael. 1984. *Development and the environmental crisis: Red or green alternatives?* London: Methuen.

Richards, Paul. 1985. *Indigenous agricultural revolution: Ecology and food production in West Africa*. London: Hutchinson.

Rosberg, C.G., Jr., and J. Nottingham. 1966. *The myth of "Mau Mau": Nationalism in Kenya*. New York: Praeger.

Scoones, Ian, ed. 1995. *Living with uncertainty: New directions in pastoral development in Africa*. London: Intermediate Technology Publications.

Shishira, Elia K., and Robert W. Payton. 1996. The causes and effects of accelerated soil erosion and sedimentation in the Kondoa Eroded Area: Evidence from the Lake Haubi Basin, Kondoa District, Tanzania. In *Changing environments: Research on man-land interrelations in semi-arid Tanzania*, ed. Carl Christiansson and Idris S. Kikula. Nairobi: SIDA's Regional Soil Conservation Unit.

Throup, David W. 1988. *Economic and social origins of Mau Mau*. London: James Currey; Nairobi: Heinemann; Athens: Ohio University Press.

Tignor, R.L. 1976. *The colonial transformation of Kenya: The Kamba, Kikuyu, and Maasai from 1900 to 1939*. Princeton: Princeton University Press.

Tosi, J.A., G.S. Hartshorn, and C.A. Quesada. 1982. HADO project development study and status of catchment forestry: A report to SIDA and the Forest Division, Ministry of Natural Resources and Tourism, Tanzania. San Jose: Tropical Science Center.

Ulotu, Harun Aman. 1994. An economic analysis of "zero-grazing" smallholder livestock system: A case study in the "HADO" area in Kondoa District, Tanzania. M.S. thesis. Uppsala: Swedish University of Agricultural Sciences, Department of Agricultural Economics.

Wenner, Carl Gösta. 1983. Soil conservation in Tanzania: The HADO Project in Dodoma Region, report on a visit in April–May 1983. Stockholm: SIDA. Mimeo.

Wily, Liz. 1995. Good news from Tanzania: Village forest reserves in the making—the story of Duru Haitemba. *Forest Trees and People Newsletter* 29:28–37.

———. 1996. *Collaborative forest management: Villagers and government—the case of Mgori Forest, Tanzania*. Forest, Trees, and People Working Papers. Rome: FAO.

Yanda, Pius Z. 1995. *Temporal and spatial variations of soil degradation in Mwisanga Catchment, Kondoa, Tanzania*. Dissertation Series, no. 4. Stockholm: Stockholm University, Department of Physical Geography.

Young, R.A., and Henry Fosbrooke. 1960. *Land and politics among the Luguru of Tanganyika*. London: Routledge and Kegan Paul.

"Re-claiming" Land in The Gambia: Gendered Property Rights and Environmental Intervention

Richard A. Schroeder

Introduction

There has been a long history of sweeping territorial claims in the name of natural resource conservation in Africa. Colonial powers intervened repeatedly throughout the first half of this century to try to force soil, water, forest, and wildlife conservation techniques on rural cultivators and pastoralists (Beinart 1984; Peters 1987; Anderson and Grove 1987; Bonner 1993; Bassett 1993; Neumann 1992, 1995b; Leach 1994; Hodgson 1995). These initiatives, carried out during the early stages of capitalist penetration on the continent, were frequently premised on changing natural resource tenure systems. In some areas, resident groups were pushed off the land altogether to make way for parks and protected areas or settler farms. In others, colonial administrators allowed Africans to remain on the land but encouraged them to privatize resources in the hopes of encouraging deeper market penetration. At virtually every turn, however, nature-society relations were fundamentally reorganized to accommodate conservation and development objectives.

Since the drought and famine years of the 1970s and early 1980s, managers of the United Nations, the World Bank, and wealthy bilateral donor agencies have viewed Africa's environmental problems with increasing alarm (Watts 1989). On the one hand, the perceived crisis of food production in many parts of the continent, epitomized by the Ethiopian famine of 1982–86, reinforced donor impressions that African production systems were fragile and in need of repair. On the other, the broad and deep-seated economic problems of many African states were a reflection of the failure of

I would like to acknowledge the helpful suggestions of several individuals who read earlier versions of the manuscript, including Maria Espinosa, Dorothy Hodgson, Donald Krueckeberg, Roderick Neumann, Donald Moore, Colleen O'Neill, Neil Smith, Paige West, and three anonymous reviewers for the *Annals of the Association of American Geographers*, where this essay first appeared ("Re-claiming Land in The Gambia: Gendered Property Rights and Environmental Intervention", *Annals of the Association of American Geographers* 87:3, pp. 487–508, 1077. Blackwell Publishers). I am grateful to Rutgers University's Center for the Critical Analysis of Contemporary Culture for a writing fellowship. The field research was supported by the Fulbright-Hays Doctoral Dissertation Research Award, the Social Science Research Council/American Council of Learned Societies International Doctoral Research Fellowship for Africa, the Rocca Memorial Scholarship for Advanced African Studies, the National Science Foundation Fellowship in Geography and Regional Science, and the Rutgers Research Council. Michael Siegel, of the Rutgers Cartography Lab, prepared the maps.

Map 1. *The Gambia River Basin*

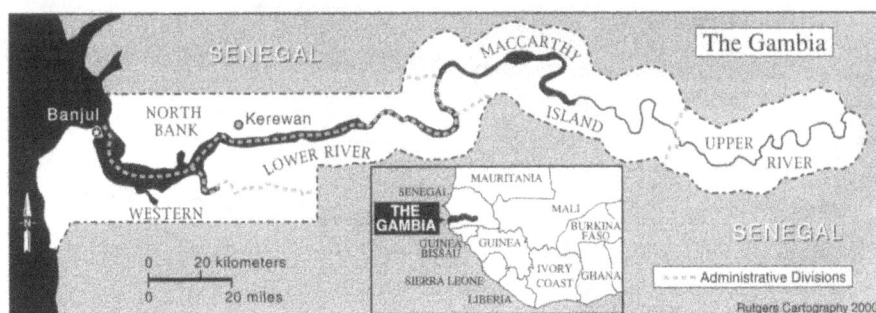

past development efforts in the region. In 1986, with the Ethiopian famine as a backdrop, the UN secretary-general was forced to acknowledge that Africa was "the only continent where standards of living ha[d] declined in the past decade" (United Nations 1986, 33).

In order to try to come to grips with this "failure"[1] and address problems of environmental degradation, donor agencies: (1) sharply increased capital investment directed at securing "sustained" economic and ecological change in the region; and (2) implemented a range of sometimes contradictory approaches to environmental rehabilitation. Between 1987 and 1993, for example, the United States Agency for International Development (USAID) committed over $350 million in support of environmental programs in Africa (USAID 1994). Over roughly the same period (1990–94), the World Bank collaborated on environmental projects in eighteen African countries totaling over $1 billion (Greve, Lampietti, and Falloux 1995). At the same time, private and nongovernmental interests greatly extended their involvement in natural resource management in the region. As of 1990, the major nongovernmental organization (NGO) engaged in environmental programs in Africa, the World Wildlife Fund, had an annual budget of $15 million for project activity on the continent (Adams and McShane 1992). Simultaneously, private business concerns became deeply involved in nature tourism. In a flurry of financial activity between 1990 and 1993, development banks and private donors committed over $80 million to tourist infrastructure development projects in Tanzania alone (Neumann 1995a). These figures provide only a partial picture of recent donor and commercial activity, yet they make it abundantly clear that environmental initiatives have become one of the major forms of foreign intervention in contemporary African affairs.

The extensive investments donors have made in natural resource management (NRM) programs have given rise to a range of new development

---

[1] Some would argue that the increasing economic marginalization of Africa was not unintentional at all but part of a deliberate logic centered on rationalizing the global capitalist system (Watts 1995; Simon et al. 1995; Mahjoub 1990).

Map 2. *The Gambia's North Bank*

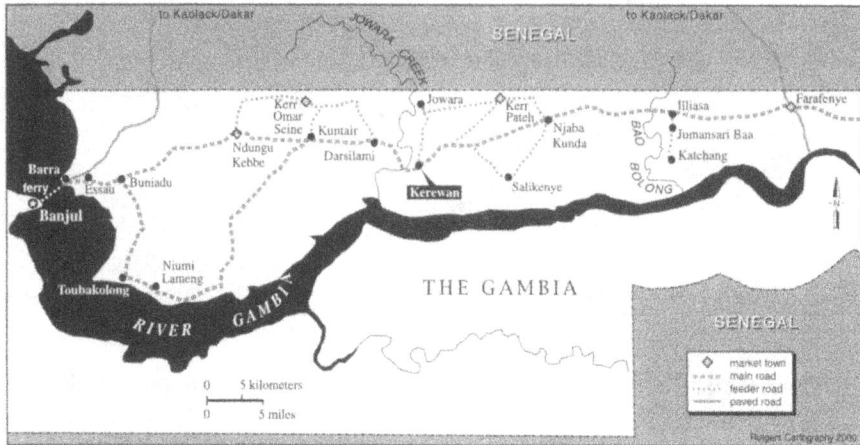

paradigms and practices, not all of which have been mutually compatible or
consistent. This essay analyzes how local groups along the Gambia River
Basin (Map 1) have absorbed and interpreted the policy shifts introduced in
conjunction with environmental initiatives over the past two decades. It explores how competing groups of potential project beneficiaries in rural
Gambian communities have vied among themselves to exploit the inconsis-
tencies of the various NRM projects at the local level and how they have
used these opportunities to open up new avenues for the accumulation of
wealth, power, and prestige. I draw particular attention to initiatives directed at "land reclamation". By definition, the renewal of soil quality
through reclamation programs renders marginally productive land resources more valuable to a broader set of users. This can open up the existing land tenure norms for renegotiation as interested parties seek to reposi-
tion themselves with respect to particular resources. Viewed from this perspective, *reclamation* has a double meaning: environmental managers
"reclaim" land resources by rehabilitating them, but in so doing they often
erase old property rights and advance new ones—they literally and figuratively "re-claim" the land.

In order to explore the implications of such shifts, I detail attempts by
women horticulturalists in The Gambia to occupy, convert, and maintain
control over low-lying land in the face of increasing local and international
pressures favoring soil and water "reclamation". I document how in the
early 1980s groups of women gardeners along The Gambia's North Bank
(Map 2) quite successfully parlayed the financial support of developers into
the acquisition of usufruct rights to valuable land and groundwater reserves. For the better part of a decade, these gardeners deepened their involvement in market gardening and gradually expanded their land use
rights, only to have male lineage heads "re-claim" the land resources in

question through NRM-related agroforestry and soil and water management projects. In sum, this is a study of the responses different community groups have made to a shifting international development agenda. The analysis lays bare both the structural constraints developers have introduced and the tactics and strategies rural Gambians have developed to manipulate these structures for personal gain.

The essay consists of five sections. In the first, I briefly describe three successive waves of development intervention growing out of two decades of political-ecological change in the Mandinka-speaking community of Kerewan (Map 3). In the second, I outline and review the basic principles of land tenure on the western half of the North Bank and present results from a 1991 survey of the transmission histories of 274 garden plots.[2] These data demonstrate the gradual erosion of male landholding privileges during a boom in women's market gardening. The third section details the backlash that took place in the mid-1980s as male landholders used the openings created by government and NGO-sponsored agroforestry projects to reclaim both some of the land they had "lost" to gardens and the initiative in controlling developmental largesse generated in the name of environmental stabilization. The fourth describes soil and water reclamation efforts focused on salt damaged rice fields, the basis of most staple food production in the area. The construction of anti-salinity dikes restored land lost to rice production for several decades and temporarily increased food production, but the projects simultaneously threatened to undermine the garden boom further when senior male community members insisted that the rice crop be prioritized

---

[2] The fieldwork for this study was conducted in three phases. The first, a seven-week trip to the North Bank in July and August 1989, included visits to eighteen communal garden sites and interviews with 127 individuals, including vegetable farmers, garden landholders, extension agents, government officials, and agricultural researchers.

During the second, from February to November 1991, I conducted a demographic and economic census of seven hundred domestic units (*dabadalu*) in 240 residential compounds (*kordalu*) in the North Bank community of Kerewan and compiled an extensive data set on a stratified random sample of one hundred women market gardeners and their families. Production surveys conducted with each woman in the data sample addressed land tenure, well construction, tree planting, cropping strategies, garden techniques, labor allocation, assistance from male family members, marketing, and changes in consumption patterns due to increased garden incomes. In-depth interviews were conducted with male landholders and female garden group leaders on the history of site development and land and tree tenure practices in each of twelve garden perimeters. These sites were mapped, measured, and inventoried as a means of assessing the threat posed by tree crops to garden enterprises. Finally, I gathered documentary evidence pertaining to horticultural policies and practices and conducted interviews with officials of several state-sponsored and nongovernmental organizations involved with horticultural projects.

The third phase of research was carried out in June and July 1995, when I made a follow-up visit to my original research sites to reinterview the market gardeners in my original research sample. I met with several development agents to discuss their shifting program priorities and gathered photographic and documentary evidence on new soil and water management projects in the area.

In addition to this formal fieldwork, I also spent two years as a Kerewan resident, from 1986 to 1988, managing food security and environmental stabilization projects for an international NGO. During this period, I provided technical and financial support to a dozen women's garden projects in North Bank communities (Kerewan was not among them).

Map 3. *Kerewan Garden District*

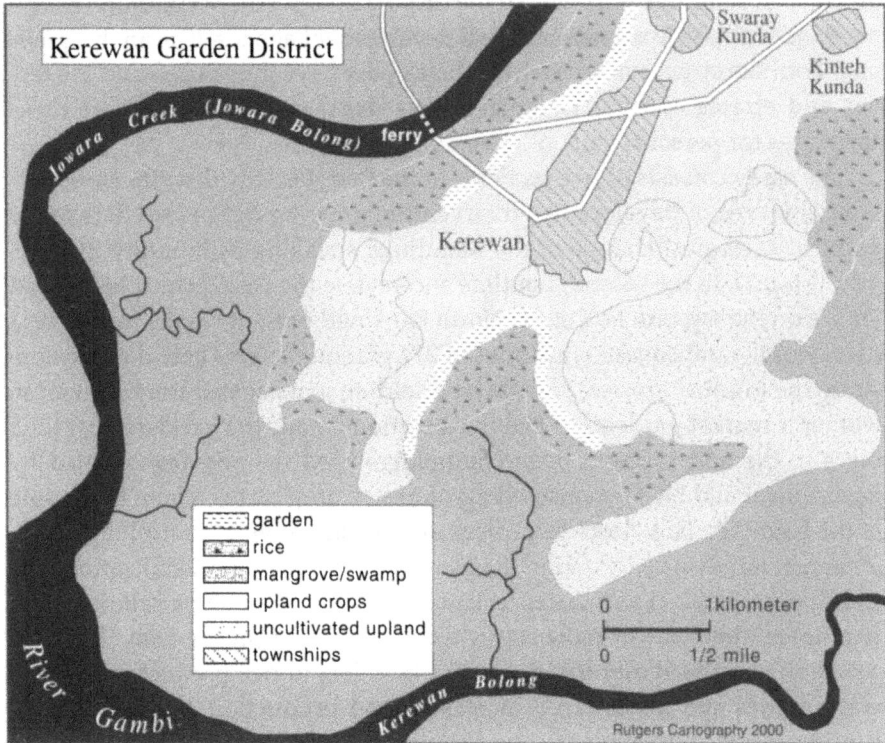

over vegetable production. In the concluding section, I inspect the seemingly well-intentioned arguments in favor of each of these land reclamation programs and show how they break down under careful political-ecological scrutiny.

Three waves of land reclamation

In the early 1980s, after a decade dominated by an average 25 percent shortfall in annual rainfall accumulations, rural families along The Gambia's North Bank found themselves in dire straits. Rice production levels were in steep decline (Kinteh 1990), imports of this principal staple were growing rapidly, and its consumer price was up sharply (Jabara 1990). The official price for imported rice doubled between 1980–81 and 1984–85 and more than doubled again in the four years following implementation of the national Economic Recovery Program (ERP) in 1985–86 (ibid.). The price for groundnuts (peanuts), the principal export crop produced in the Gambia River Basin, was extremely low due to negative terms of trade on the international vegetable oils market (ibid.; von Braun et al. 1990). At the same time, the effects of "recovery" via the ERP dramatically altered production costs for groundnut growers. Between 1984 and 1987, in real terms, fertilizer prices increased 11 percent, groundnut seed nearly doubled in cost, and the price of hired labor and draft animals rose by almost a third (Johm 1990; Puetz and von Braun 1990). The net effect of these changes was severe: a sharp decline in fertilizer usage, lower earnings for groundnut producers (primarily male farmers) on the whole, and a general withdrawal of labor from groundnut production, the source of as much as 90 percent of the country's foreign exchange throughout the period.

On the North Bank, these changes set the stage for the emergence of three competing development initiatives, one involving small-scale market gardens managed by women, a second centered on fruit tree orchards largely controlled by men, and a third emphasizing rice plots managed by women for joint family consumption purposes. The first of these to emerge in my North Bank research site, the community of Kerewan, was women's horticulture. The slumping groundnut economy in the 1980s meant that male farmers encountered increasing difficulty in meeting their customary household budget obligations. The costs of supplemental grain, clothing, and various religious and social ceremonies, once primarily the responsibility of male heads of household, were consequently shifted, by default (and at times by design—see Schroeder 1996), to women. To meet the challenge of supporting themselves and their dependents, women began converting low-lying land rendered marginally productive by drought conditions into lucrative, hand-irrigated vegetable gardens. Aided by small grants from agencies interested in "women in development" (WID) issues (Schroeder 1996) and efforts by the state, NGOs, and voluntary agencies to diversify the rural economy in response to drought, a full-scale, albeit geographically uneven, market garden boom was soon under way. In the Kerewan area alone, some fifteen to twenty projects were funded over the course of two decades beginning in the mid-1970s, while on the national scale women's garden groups in several hundred communities benefited from donor support.

The second wave of development initiatives targeting low-lying land and water resources on the North Bank came quickly on the heels of the first. Just as women before them had capitalized on funding sources generated in response to the UN's declaration of the WID decade, male landholders took advantage of a renewed developmental emphasis on environmental stabilization. Lineage heads and elders who controlled low-lying land were urged by forestry department extension agents, NGOs, missionary groups, and large-scale donors such as USAID and the European Community (EC), to establish orchards and woodlots in order to reverse deforestation and reap the economic benefits of fast-growing exotic fruit and firewood species. This approach, premised as it was on "natural" species succession and a related shift in focus from understory to tree crop, was doubly endowed with "green goodness" (Rocheleau and Ross 1995; Schroeder and Suryanata 1996) insofar as it was deemed an environmentally sustainable practice that also generated much needed cash revenues. Largely obscured from this picture, however, was the fact that projects were either superimposed on top of women's existing garden sites or established on new sites where women were only granted access to land if they watered landholders' trees and vacated their plots when the trees matured. Thus, the projects took maximum advantage of the newly productive landscapes created by the garden boom. The "natural" succession of species produced political shifts in labor claims and property rights, privileging the older claims of male landholders over the more recent and less secure usufruct claims of women gardeners. In effect, they allowed male landholders to resume control over territory they once controlled more exclusively (see further discussions in Schroeder 1993, 1995, 1999).

A third set of claims to garden lands were morally derived insofar as they sought to impose a higher "communal" good—the securing of food stocks—over and above the individual and collective needs of gardeners. A series of soil and water management projects were implemented with the aim of reversing the degrading effects of salt intrusion into rice fields through the construction of fresh water retention dikes and the liming of plots. The key actors in these projects included the Soil and Water Management Unit (SWMU), a branch of the Department of Agriculture created and funded in large part by grants from USAID; and the NGO Save the Children, which contracted the SWMU to construct the dikes in 1994. Despite being hampered by maintenance problems, these projects were quite successful in improving rice yields over the short run. They received high praise from men and women alike in the garden districts and gave rise to claims by development agencies that food self-sufficiency had been restored in several communities. The catch for women, who do virtually all of the labor on the rice crop and provided most of the labor during the intensive phase of dike construction, was the revitalization of a discourse that stressed the central importance of rice over all other forms of production, including their gardens. In effect, the promotion of new soil and water management strategies threatened to block women's access to garden land due to labor bottlenecks at key seasonal junctures just as surely as the shade canopies created by orchard owners' trees.

The nature of usufruct: ransoming garden land

The three sets of competing claims to low-lying land resources in the Mandinka-speaking community of Kerewan grew out of a tenure system that was until only recently centered on rice and groundnut/millet cultivation. The system existing prior to the garden boom (Table 1) involved a single intensive period of cultivation during the rainy season and preserved for men most of the important opportunities for earning cash from agriculture. The tenure domains were fairly sharply divided in spatial terms. Upland areas cultivated by men in a groundnut/millet rotation were known locally as *boraa banko*, or "land of the beard". So named, according to one informant, because it is "something a woman will never have", *boraa banko* was inherited along patrilineal lines. By contrast, most of the swampland lying along the main river and its tributaries was controlled by women rice growers.[3] As such, it fell under the classification of *kono banko*, "land of the [pregnant] belly", and was transferred directly from mother to daughter.

Table 1. *Division of labor before the garden boom, North Bank, The Gambia, ca. 1970*

Season	Cash crops grown by men	Food crops grown by men	Cash crops grown by women	Food crops grown by women
Rainy	Groundnuts	Millet/sorghum	—	Rice
Dry	—	—	—	Vegetables

Typically, a narrow band of low-lying land formed a boundary between these two zones. Technically *boraa banko* in most cases, the soils were at once too heavy for groundnut production and too dry to support a rice crop, especially in drought years. It was in this zone that women requested parcels of land for gardening purposes from the small group of male elders who controlled their respective lineages.[4] Before the 1970s, production on these plots rarely exceeded local demand. Most women worked single plots that were individually fenced. Outside assistance in obtaining tools, fences and wells was minimal. Seed suppliers were not yet operating on a significant scale, and petty commodity sales were largely confined to tomatoes, chili peppers, and onions. The market season, accordingly, stretched only a few weeks, and outlets on the North Bank were all but nonexistent. Most produce was sold directly to end users in the nearby villages by women who transported their fresh vegetables by horse or donkey cart and then toted them door to door on their heads *(kankulaaroo)*.

Throughout the late 1970s and early 1980s, however, economic and ecological pressures compelled women to intensify efforts to reclaim marginal land for gardening purposes, and the number of women engaged in commercial production rose precipitously. Data from Kerewan illustrate the trend: the pool of women gardeners grew from 30 selected to take part in a pilot onion project in the mid-1970s, to over 400 registered during an expansion project in 1984, and some 540 recorded in my own 1991 census. The arrival of consignments of tools and construction materials for fencing and wells in this community in 1978 initiated an expansion period that saw the area under cultivation more than triple in size, growing from 5.0 to 16.2 ha in ten years. Between 1987 and 1995, a second wave of enclosures nearly

[3] A small percentage of land in swamps was originally cleared by men on behalf of their families and passes accordingly along male lines of inheritance. Also, in cases where a woman has no female heirs her rice plots occasionally pass to her son(s).

[4] There are three lineages in my principal research site. The town chief is generally drawn from the first and the town religious leader *(imam)* from the second. The third is the remnant of a warrior clan.

doubled that area again. Meanwhile, over a dozen separate projects were funded by international NGOs and voluntary agencies, setting in place thousands of meters of fence line and dozens of irrigation wells, and aggregate community returns from vegetable sales grew to roughly $80,000, this in an area where per capita annual income was in the range of $250.[5]

In sum, between 1973 and 1991 the Kerewan garden district developed into one of the most intensive vegetable-producing enclaves in the country. In at least two key respects, the garden boom recentered the Mandinka cropping system around vegetable production, as many women began working in their gardens year round, including during the height of the rice-growing season, and as vegetable sales began yielding more cash than peanuts (the main source of income for rural men) in nearly half of all households (Field survey by the author).

The eleven communal garden sites analyzed in this essay were virtually all operated on the basis of usufruct land grants issued by senior male members of founding lineages in garden communities. Although the gardens, which ranged from a fraction of a hectare to nearly five hectares in size, were often sited on *boraa banko* land and were named after male landholders, the degree of authority these men retained with respect to land use practices in the gardens eroded markedly following the onset of the boom. Despite nominal male oversight, planning and supervision of day-to-day operations were gradually assumed by leaders of the women's groups that worked the land. It was they who organized regular maintenance functions such as fence repairs and seasonal land-clearing operations and they who levied fines against group members—for example, when a woman neglected her duty to prevent livestock from entering the gardens. The full extent of the control women asserted over their gardens on both individual and collective bases must, however, be assessed with reference to three additional aspects of garden tenure: rights of plot transmission, rights of development, and rights to tree planting.

Whereas the fence perimeters were managed communally as described above, most women worked their own land allotments (averaging three hundred square meters) individually or with a small group of female relatives. A key question, then, in understanding the tenure dynamics of the garden districts is how each woman came by her plot rights originally and whether she was free to transfer those rights to her daughter or other female relatives. In 1991, I compiled the history of tenure change in 274 plots

---

[5] Reliable income data are notoriously difficult to gather. This estimate is based on the extrapolation of 1991 sales figures. A sample of 100 vegetable growers in the village sold roughly 109,645 dalasis, worth of produce over an eighteen-week period from February to June of that year. The exchange rate at the time was approximately 7.5 Gambian dalasis to the dollar (D7.5 = $1.00). *Gross* earnings for the 540 vegetable growers in the community were accordingly on the order of D592,083 or $78,944. These figures do not include off-season or tree crop income. Net returns were lower by roughly a third. (All figures are based on field surveys by author conducted in 1991).

Table 2. *Source of women gardeners' land use rights, Kerewan, The Gambia, 1991 (n-274)*

Category	No. of Plots	%
Preexisting claims to rice land	5	2
Gifts from landholding male relatives	6	2
Temporary loans from other gardeners	17	6
Claim payments paid to landholders	128	47
Unauthorized gifts from female relatives	64	23
Plots created through site expansion	54	20
Total	274	100

Source: Field survey by the author, 1991.

located in eleven communal garden perimeters, including the origins of usufruct claims and all transfers of cultivation rights to subsequent users. I also interviewed the landholder on each site to determine what conditions, if any, were placed on access to plots. The results of these surveys (Table 2) indicate that the pattern of plot acquisition changed considerably over the course of the garden boom and that many of the plots were acquired in direct contravention of conditions stipulated by landholders. Findings show that small groups of gardeners were granted use rights to plots by virtue of preexisting claims to rice land (2 percent),[6] as gifts from landholders who were related by birth or marriage (2 percent), or as temporary loans from other gardeners (6 percent). The histories of the vast majority of plots, however, mark more substantive shifts in the nature of landholding and usufruct claims in the garden districts. Each of these plot use arrangements (land parcels acquired via claim payment [47%], unauthorized gifts from female relatives [23%], or the exploitation of openings created when existing gardens expanded [20%]) bear closer scrutiny.

The most common means of access to land was via a one-time cash payment paid directly to landholders.[7] These payments, known locally as *kumakaalu* (pl.; sing., *kumakaa;* the practice of granting *kumakaalu* is known as *kumakaaroo*), a term that once meant the ransom payments made to free family members stolen or captured into slavery or bail payments to free someone from jail,[8] were commonly assessed by landholders early in the garden boom when many of the area's horticultural perimeters were founded. The

---

[6] That is, they held prior *kono banko* rights to plots that were recognized after garden fences and wells were installed by developers.

[7] These payments ranged from D5 when the earliest gardens were established in the mid-1970s up to D30 in 1991—the rough equivalent of $1 to $3—for plots averaging one hundred square meters in size.

[8] Interestingly, none of my informants was aware of any association of the term with land allotments prior to the garden boom.

nature of the tenure hold granted under *kumakaaroo* was disputed by my informants. Most landholders saw the transfer of use rights to women under *kumakaaroo* as a temporary arrangement. As one landholder put it: "When he asked whether I sold the [garden] land to [the women] ... I replied that I *lent* the land to the people to work" (interview, Kerewan, March 1991). They maintained that the money they collected was used to help pay for fence repairs or defray other expenses incidental to the garden's upkeep. By contrast, gardeners claimed that funds paid to join garden groups were routinely diverted to the landholder's personal use and that *kumakaalu* thus constituted lease payments. In the words of one woman: "As for garden land, we *hire* that from the landholder" (interview, Kerewan, April 1991). This interpretive dispute notwithstanding, it is clear that land transactions in the early stage of the garden boom were widely monetized and that *kumakaa* payments constituted a kind of disguised rent as land was "ransomed" for gardening purposes.

The *kumakaaroo* claim system included the proviso that each time a plot was vacated due to the death, retirement, or relocation (due to marital status change) of its original occupant it had to be returned to the landholder before reallocation. This condition was set in order to give the landholder a chance to exact a new *kumakaa* payment from any prospective gardener before granting her leave to put the plot into production. In addition to the financial windfall, this resumption of plot control was meant to underscore symbolically the landholder's residual land claims. This stipulation notwithstanding, the detailed plot histories I gathered for North Bank gardens in 1991 reveal that in practice women often flouted this convention. In the survey, each plot holder was asked to indicate whether she was an "original" plot holder or whether her plot had changed hands since the site's enclosure. She was then asked whether she had been required to pay a *kumakaa* before beginning cultivation. Results were sorted by location and compared with the tenure conditions landholders claimed to be enforcing in each site. Of the 274 plots in the survey, 160 remained in the hands of their original claimants. Of the remaining 114 plots actually changing hands since the onset of the boom, only 33 reverted to the control of the landholder before being parceled out a second time and only 27 of these were ultimately reallocated on the basis of a second *kumakaa* payment (6 were awarded by landholders to family members free of charge). Thus, the vast majority of the plots changing hands (81 out of 114 or 71 percent) were passed directly from one woman to another, *without* any form of direct compensation for the exchange of use rights.

This group of plots needs to be differentiated further in order to get a clear picture of the property dynamics in play. First, many plots (17 out of 81) were allocated on a temporary basis. Women gardeners occasionally loaned plots to relatives or close friends when they themselves were pregnant, caring for a newborn, or simply too ill to work. In most cases, these were seasonal loans, but they could also be longer term, in which case the likelihood that they would eventually result in full transfers was high. A second group of plots (25 out of 81) changed hands in the one large garden site in my study where such transactions were not proscribed. This garden differed from other sites in that landholders ceded all forms of control to the women's group leaders shortly after the site was founded. Thus, the whole garden—a 3.5 ha site (over 10 percent of the total enclosed land area under garden production in Kerewan)—was removed from male control. Finally, and perhaps most telling, is the fact that 39 permanent transfers, or roughly one-third of all transfers of plots in the research sample between 1973 and 1991 (39 out of 114), took place surreptitiously, that is, without being explicitly sanctioned by the landholder in question.

That surreptitious transfers of plot rights could take place in a small community of roughly 2,500 persons might seem unlikely but for two factors. First, women's groups so thoroughly dominated gardens in the 1980s that the space enclosed within the fences became a kind of terra incognita for men. Even putative landholders were rarely seen within the fence perimeters because of the discomfort they felt in the midst of such a clearly defined women's space. Thus, the determination of who was actually cultivating a given plot was sometimes difficult to make. The situation was made more complex by the fact that as many as a third of the work units in the sample revolved around mother-daughter tandems. Younger women often assisted their mothers with their gardens and were thus in a position to gradually assume practical control over plots over the course of several years. Use rights were accordingly established on a de facto basis, *kumakaa* "ransom" payments were not required, and challenges to the younger women's claims were highly unlikely. In sum, land that was once clearly male controlled (*boraa banko* land) had been taken over by female vegetable growers and was being managed as though it were part of a *kono banko* legacy, a step that women felt was amply justified:

> A woman has the right to give her garden land to her daughter. ... Suppose I spent a lot on my garden, and after growing old, I don't have the energy to continue the work. Then in all fairness my daughter should continue to benefit from what I spent. That is no problem, since I toiled for it, and I also spent money on it, my daughter can have it. That is no problem. (Interview, Kerewan, April 1991)

The final means of establishing access to garden land involved openings created when an *existing* garden site was expanded. For several years in the 1980s, garden projects dominated the activities of NGOs and voluntary agencies working in rural Gambia. The generally industrious attitude of women's groups coupled with the prospects of simultaneously addressing goals of income generation and nutritional enhancement made gardens attractive investment targets. Not only did the gardens bear quick fruit, so to speak, but they also helped NGOs address deep-seated social inequities.

They were therefore prime targets for funds generated under the guise of WID initiatives. The intense focus on horticulture during this period meant that women's groups were often in a position to leverage successive grants to support their horticultural efforts. Whenever a new grant resulted in expanded fence perimeters, women already active in the gardens awarded themselves "expansion plots" *(lafaa rangolu)*,[9] extending their use rights *without* paying additional *kumakaalu*.[10] Such expansion came at the expense of *boraa banko* land use and accounted for 20 percent (54 out of 274) of all plots in the sample.

The lack of opposition by Kerewan landholders to such arrangements marked the fact that the community's "moral economy" (cf. Scott 1976)—the fluctuating sentiments of community members regarding notions of communal benefit and well-being—had shifted in favor of gardening. After initial resistance, most men in the garden districts had "seen the benefit" of gardens in the form of cash gifts and other financial support from their wives and became firm supporters of the garden enclaves (Schroeder 1996). Moreover, the offer of material assistance by NGOs and expatriate volunteers helped shore up the gardeners' claims. In theory, landholders might conceivably have refused individual requests from women to expand their plots, but the stakes for the community as a whole were high. Refusal to extend the women's usufruct rights would have meant denying gardeners access to additional donor assistance. Thus, for the most part the few relatively senior men who controlled low-lying land were hesitant to block expansion, despite the loss of power and prestige accompanying the loss of territorial control (but see the next section).

To sum up, the practice of *kumakaaroo*, which was nearly universal in Kerewan in the early stages of the garden boom, was only upheld in 24 percent (27 out of 114) of the land transfers that occurred during the eighteen years covered by my survey. Over 70 percent (81 out of 114) of the cases of plot transmission had been negotiated directly between women, often without the knowledge of landholders. At the same time, rights to a fifth of *all* plots under cultivation were effectively leveraged through the intercession of NGOs promoting the expansion of existing garden projects. This was a striking loss of privilege. Men who were accustomed to controlling the distribution of benefits generated by development interventions found them-

---

[9] The derivation of the term *rango*, which was used somewhat idiomatically in this community to refer to individual plots, is the English term *rank*, as in *rank and file*. Gardeners applied it to plots in recognition of the geometric grid created by extension agents, who typically provide surveying services to garden groups.

[10] New members of garden groups at the time of expansion were also required to pay *kumakaalu*. The extension of rights for existing members was thus in some sense a means of acknowledging—of *claiming*—the value they themselves had added to the garden perimeters. The assertion of such privileges by women already holding plots is evidence of hierarchical social relations *between* women gardeners, which come into play at key junctures (Schroeder 1996).

selves virtually frozen out of development altogether, and some sort of backlash was inevitable.

Branching into old territory: agroforestry reclamation

The critical break point in the tension surrounding the expansion of usufruct rights came in 1984, when a local landholder challenged the efforts of Gambian extension agents and an expatriate volunteer to secure funds to redevelop a garden site on *boraa banko* land he controlled. In spite of, or perhaps *because* of, a new alliance between state functionaries, foreign donors, and sympathetic male villagers in favor of gardening, the landholder blocked expansion. Subsequently, three female garden leaders who tried to press ahead with redevelopment plans were detained by police. This action prompted a heated public demonstration by hundreds of gardeners and an emotionally charged court case that received attention from the highest levels of the government (Schroeder 1993; Schroeder and Watts 1991). The significance of these dramatic events for my present purposes lies in what they tell us about the changing character of landholding privilege and usufruct rights during the garden boom.

Over the course of this controversial year, the landholder, a senior member of one of the town's founding lineages, was forced to claim and/or defend several specific rights to land he purportedly controlled under *boraa banko* conventions. By constantly shifting tactics, he was able to probe the pro-garden alliance, testing the resolve of the various actors to see whether he might forge alternative ties that would allow him to reassert control over the newly valuable land resources. Several of the claims he made—for example, that he should have decision-making authority over the location of wells and fences and the ability to allocate and withdraw plot use rights at will—were consistent with *boraa banko* rights as they were practiced prior to the garden boom (cf. MacKenzie 1994). That he was forced to defend them at all is indicative of the extent to which his landholding status had changed since the early stage of the garden boom. Loss of practical control over production decisions was not the only issue raised in the landholder's complaint, however. Other claims, such as the right to sign quasi-legal land use agreements with NGOs and state agencies on behalf of the gardening group, the right to store construction materials intended for use in fence repairs and well digging in his own family compound, and the ability to personally award well-digging subcontracts, reflected fears that his ability to exact real and disguised rent payments from the gardens would be greatly diminished. If he did not resist the gardeners' efforts to establish direct ties with the NGO funding community, his own ability to convert lineage rights, best understood perhaps as a kind of stewardship responsibility held on behalf of his kinsmen, into the equivalent of private property rights would be threatened. His expectation that he should be free to control development largesse, extract rent from developers and garden groups, and channel benefits to family members and friends was born of several decades of development interventions that operated in precisely that fashion (Carney and Watts 1991). When extension agents in the Department of Community Development and the expatriate volunteer coordinating USAID support for the project refused to release funds and construction materials to him directly, and proscribed many of the rent-taking mechanisms he had previously employed, the landholder simply balked and upped the ante for all concerned by threatening violence and vowing to evict the gardeners altogether (for further details, see Schroeder 1993 and 1999; and Schroeder and Watts 1991).

The courtroom deliberations that temporarily settled the dispute eventually came to rest heavily on a set of claims that were not at issue when the dispute began—the landholder's contention that women in his garden were planting trees in their plots against his will. Tree-planting rights are widely acknowledged in the tenure literature to lie *outside* the bounds of "secondary" usufruct rights such as those pertaining in Gambian garden perimeters (Freudenberger 1994; Fortmann and Bruce 1988; Raintree 1987). Landholders typically refuse to grant tree-planting rights to "secondary" tenure holders because they fear the longevity of trees will negate the possibility of alternative land uses. Part of the sensitivity of the landholder in the Kerewan case stemmed from the fact that in Mandinka areas customary law clearly maintains the partibility of tree and landholding rights (Osborn 1989).[11] While most tree planting is done by individuals on land that they control, this does not preclude entirely the prospect of planting elsewhere.

This principle was clearly illustrated in the new market gardens where, despite *boraa banko* constraints, 83 percent of the women in my research sample had incorporated trees into their crop mix.[12] Until the 1984 court case, such unilateral extension of usufruct rights went largely unchallenged in Kerewan; women were unimpeded from developing complex agroforestry practices geared toward maximizing returns from small plots and

---

[11] Gamble's anthropological account of tree tenure principles operating in the area in the 1940s confirms this claim: "So far the problems arising from trees of commercial value, e.g., fruit trees, seems not to have arisen. The people of [a nearby community] say that a man may not plant fruit trees on another man's land without his permission. If the landowner objects he can have the trees torn up. If he gives permission it amounts to a permanent alienation of the land for the granters would never claim it back. Others maintain that the original owners can claim the land, but that the trees remain the property of the planter" (Gamble 1947:22–23). Witness also the judgment of a Muslim cleric on the South Bank of the river in a dispute over tree tenure: "It has happened that when a man who has planted a tree subsequently dies, another person (possibly the owner of the land) can assume the responsibility of watering and caring for the tree. If, after some time passes, the son of the man who planted the tree claims ownership of the tree, a dispute may arise in which the caretaker claims ownership over the tree. The *imam* stated that resolution of such disputes is clear-cut—the act of watering and caring for the tree does not confer rights of ownership over the tree. The son of the tree planter inherits ownership rights to the tree" (Freudenberger and Sheehan 1994:67).

[12] These data were drawn from a sample of ninety-nine women. Interestingly, 82 percent of the women also controlled trees in upland areas, primarily in the area immediately within or surrounding family compounds in town. Of an average 11.0 trees owned in the uplands, 7.4 were inherited and 3.2 were planted by the informant herself. This suggests that recognition of treeholding rights for women was both long-standing and being continually reproduced.

extending the market season for fresh garden produce. In the 1984 court case, however, when the landholder found himself stymied on other fronts he sought to shore up his eroded land claims by asserting his right to block "unauthorized" tree planting, and the court upheld his case. Almost immediately, the landholder proceeded to uproot dozens of trees women had planted on his site. Moreover, he enlisted the technical support of the Forestry Department in planting several dozen trees on the site himself. This effort set a precedent of sorts for a practice that landholders in many communities in The Gambia would use effectively over the next several years to reclaim the initiative in developing low-lying *boraa banko* land "lost" to garden projects (see Norton-Staal 1991; Carney 1992; and Lawry 1988).

The small group of male landholders who intervened in gardens initially were primarily interested in alienating subsidies that were either directly or indirectly generated by gardeners. Garden projects founded under the auspices of WID initiatives in the 1980s (Carney 1992; Barrett and Browne 1991) increased the value of low-lying land immensely through the addition of concrete wells and wire fences. At the same time, individual gardeners made extensive soil improvements and constructed thousands of unlined "local" wells in their garden sites. As Thoma notes: "[T]he attraction of free wells, subsidized fencing, subsidized seedlings and the like [was] virtually too irresistible ... not to take advantage of, even if [male] farmers already [knew] how to plant and grow trees and women already propagate[d] fruit trees on their own" (1989, 41). Having invested hundreds of hours of personal labor in extensive soil improvements and in some cases having planted perennial tree crops, gardeners had also become more "rooted" in place. This allowed landholders to "capture" their labor to water trees, manure plots, guard against livestock incursions, repair fences, and maintain wells, all to the *end* benefit of the overlying tree crop.

The production conditions available in many garden sites were also seen as ideal from the standpoint of agroforestry extension agents. The Gambian Forestry Department was obviously interested in taking advantage of any opportunity to plant trees and see them protected against bush fire and grazing pressures from livestock. The Methodist Mission Agricultural Program, under the direction of an agricultural officer who was deeply invested in seeing tree-planting practices expand on the Sahel (Mann 1990), mounted a vigorous program of well digging, tree nursery establishment, and seedling distribution across The Gambia. At the same time, NGOs such as Save the Children and Action Aid, and voluntary organizations like the US Peace Corps and the British Voluntary Service Overseas (VSO), were in the midst of a significant shift in emphasis *away* from the gender equity programs of the early 1980s and toward environmental rehabilitation projects.

The reasons for these policy changes are not hard to trace. The specter of the Sahelian droughts of the 1970s continued to haunt the region, and most of the small development agencies working in The Gambia remained generally sympathetic toward the need to address environmental problems. Moreover, many implementing agencies derived substantial portions of their operating budgets from agencies, such as USAID and the UN programs, that had thrown their full financial weight behind environmental objectives. With so much money flowing into the environmental sector, small-scale NGOs and volunteer groups came under a great deal of pressure to demonstrate success in the area of environmental management in order to preserve some sense of institutional legitimacy in their donors' eyes (Schroeder 1995).

With the infrastructure supporting garden sites already in place and a boom in horticulture well under way, forestry extension agents very quickly made the garden sector a key site of reforestation activity. Two different spatial strategies facilitated this approach. The first involved reasserting control over *existing* garden spaces. Extension agents encouraged landholders to plant trees directly within garden plots in order to take advantage of the water women delivered to their vegetable crops, a strategy that was entirely deliberate.

As one developer put it in a recent interview, "Women are reportedly not good at watering the trees unless they are located directly in the garden and receive water indirectly when the vegetables are watered" (quoted in Norton-Staal 1991; see further discussion in Schroeder 1995 and 1999).

The second spatial strategy employed by forestry extension agents entailed founding *new* sites where gardeners would be given temporary access and their labor could be used to convert unused spaces into orchards. While the terms of access in such sites did not dispossess gardeners of previously held use rights outright, they did constitute a reassertion of landholding privilege. Most significantly, women gardeners who had successfully defended a fairly vulnerable position for over a decade were no longer able to determine unilaterally the disposition of development aid directed at reclaiming low-lying land resources.

My own discussions with landholders opening up new gardens and orchards revealed that in the dozen fence perimeters established in Kerewan between 1987 and 1995 *kumakaa* payments were eschewed altogether. In most cases, access to land was only granted under terms that required women to guarantee that they would: (1) water the tree crop as long as they stayed in the perimeter and (2) leave their plots as soon as the trees reached maturity. With a wary eye trained on the prospect of women mounting competing claims to land or trees on moral-economic grounds, the male landholders either provided fences and wells on their own account or built them with the assistance of donors interested in promoting agroforestry, many of whom had sponsored garden projects on the same sites several years earlier. In one garden destined for conversion into an orchard, a contract was signed between the landholder, the donor agency, and a garden group stipulating a five-year limit to the women's vegetable-growing rights.

In another, a project manager proposed a rule as a hedge against tenure erosion that would preclude anyone other than project participants and "one small daughter per grower" (!) from working the plot. In a third site, garden gates were padlocked in recognition of full conversion to orchard production.

Would-be orchard owners have almost invariably chosen mangoes, relatively tall trees with broad shade canopies, as the primary species to be planted in their orchards (Schroeder 1993). Consequently, the tightening of tenure restrictions has had direct and dire consequences for vegetable growers.[13] When gardeners controlled decisions over species selection, the location of trees, and rights of trimming or removal, they had less difficulty managing interspecies competition between vegetables and trees.[14] As soon as landholders reclaimed their rights to tree cropping and land development, however, these prerogatives were lost and the requirements of the vegetable crop became a secondary consideration.

It is also important to understand that dispossession of women's garden land for orchard purposes often took place without direct confrontation. An interview conducted in 1995 revealed just how such dispossession occurred. The landholder in question established his orchard site in 1992 with the assistance of free tree seedlings and extension advice provided by the Forestry Department. Women were invited to work gardens on the site, but no *kumakaa* payments were solicited and none was paid. Moreover, no assistance was required of the women for site maintenance or protection. Since the landholder spent his days at the garden himself to ensure that no livestock damaged his trees, the only obligation gardeners had in exchange for use rights was to continue irrigating the landholders' trees as they watered their vegetable plots. After the landholder proudly provided a tour of his plot, I asked him what women gardeners sharing his land would do when his trees matured and a shade canopy closed the site. Would he drive (Mandinka: *bai*) the women out of the site or would he let them stay on? His response was most telling: "*I* won't drive them. *The trees will drive them. The trees.* If not for the trees, they could stay here".

This was a remarkable admission. The landholder's claim that a "natural" species succession was responsible for driving women off the land, while perhaps consistent with a worldview that does not draw sharp distinctions between "natural" and human agencies, nonetheless masked his own explicit intention of displacing the gardeners when the orchard matured. Moreover, it obscured the all-out attempt by NGOs and donors to colonize women's gardens and reproduce what they deemed to be im-

---

[13] Mangoes are widely eaten by children during periods when other sources of food are scarce; NGOs and government agencies also imported several "improved" mango varieties at this time that fetched attractive prices on local markets in The Gambia and Senegal.

[14] Cf. Wangari, Thomas-Slayter, and Rocheleau 1996 for a contrasting case study of the incorporation of women into agroforestry initiatives.

proved land use practices and concerted efforts by Forestry Department extension agents to promote agroforestry under the direction of their superiors and donors. Indeed, the landholder's admission clearly revealed how tree cropping allowed landholders to reclaim low-lying land without *directly* evicting gardeners (cf. Rocheleau and Ross 1995). In letting trees "drive" the women, landholders and their donors/sponsors followed the contours of a moral economy wrought by the garden boom, carefully avoiding the ill will of communities grown heavily dependent on garden income. In the process, two ostensibly worthwhile development agendas, gender equity initiatives and environmental stabilization programs, were pitted against one another, undermining a market garden livelihood strategy that helped thousands of rural Gambian families adjust to financial hardship.

Soil and water management: the perils of reclamation redux?

In 1992, the scope for environmental interventions in The Gambia expanded greatly. The World Bank and its collaborators in the donor community, most notably perhaps USAID, initiated a region-wide effort to "mainstream the environment" (World Bank 1995) and rationalize environmental planning across the continent. These donors were greatly concerned that a long succession of development projects designed to reform the agricultural and natural resource management sectors in The Gambia had failed. Despite heavy programmatic investments, food production levels remained static and environmental degradation had worsened. Responding to donor concerns, the Gambian government (along with forty other African countries) developed a full-scale National Environmental Action Plan (NEAP) (GOTG 1992). USAID, in turn, launched a multiyear, 22.5 million dollar Agriculture and Natural Resources (ANR) management program (USAID 1992) designed to meet many of the objectives this plan articulated.

One of the branches of government that featured prominently in these plans was the Ministry of Agriculture's Soil and Water Management Unit. Established with USAID funding in 1978, SWMU was responsible for small-scale watershed management projects in seventy-eight Gambian communities over a twelve-year period (USAID 1992). Working toward the goal of enhancing food self-sufficiency at the village level, SWMU surveying teams constructed dozens of dike systems in low-lying areas along the River Gambia and its tributaries. These structures were designed to simultaneously halt salt intrusion into rice-growing lands, stop erosion of uplands, and capture fresh water for recharging aquifers. They were particularly welcome in the western half of the country, where riverine swamps are both tidal and saline.

In 1994, an SWMU project was designed to "reclaim" an estimated fifty hectares of rice land south of Kerewan, much of which had been lost to production for decades due to problems of salt intrusion. Better water control made possible the reintroduction of long-duration rice varieties, and this in turn brought higher yields that were universally welcomed by community members. At the same time, however, the demand for women's agricultural labor during the rainy season, which had been on the wane due to persistent drought conditions, was redoubled, and the politics of reclamation took on a new aspect centered on the seasonality of the agricultural calendar. What was not foreseen by community residents was the fact that cultivation of long-duration rice varieties would push rice harvests from September or early October into November, the height of plot-clearing and seedbed preparation in the gardens. This labor bottleneck created a dilemma for women, who were faced with the choice of completing the rice harvest or getting their vegetables in the ground in a timely manner. A delay in garden preparation would cost them, for both the yield potential and market value of their vegetable crops would decline precipitously if planting were post-poned (Daniels 1988). On the other hand, losses in the rice fields due to de-layed harvests—from bird damage or grain shattering—would cost their husbands, since it is incumbent upon *them* to buy supplementary grain for their families in the event family-grown stocks do not last the year.

It was this fear of rice harvest losses that motivated an announcement is-sued in the town mosque in late September or early October 1994. Couched in the moral-economic terms of promoting food self-sufficiency for the community as a whole, the mosque leader enjoined women from planting their gardens before completely finishing the rice harvest. The issue was then taken up by members of the quasi-official Village Development Com-mittee (VDC), who debated among themselves whether they should attempt to intervene to impose restrictions on garden preparation during the rice harvest. VDCs were established in rural communities throughout The Gam-bia in the 1980s by the Department of Community Development and the NGO community in an effort to democratize the development process. In theory, committee members worked on behalf of their communities to prior-itize community development needs and communicate them to prospective donors. Typically, these committees were comprised of between eight and fifteen members drawn from each sector of the community, including wom-en's groups. Thus, at the meeting where the Kerewan SWMU project was discussed, at least four women were present, and one of them posed a series of questions, which I paraphrase here.[15]

The issue raised by the newfound productivity of rice fields was, as she saw it, who should shoulder the responsibility for securing the rice harvest. If the question was one of selecting rice varieties appropriate to specific soil types and other microecological conditions, then the men were right to rely on the women. If it involved the timing and execution of other skilled tasks such as weeding and transplanting, that, too, was the rightful domain of

---

[15] This reconstruction is based on reports from extension agents and committee members pre-sent at the meeting.

women. It was true: women possessed the localized technical knowledge and practical experience necessary to ensure the best outcome for the crop. If, however, the question simply revolved around getting the rice in from the fields on time, then there was no a priori reason why women alone should be responsible. Harvesting rice is not a highly "skilled" task: a man can wield a sickle just as well as a woman can. Since *everyone* stood to gain from a secure supply of food, then *everyone* should mobilize around the harvest. Getting the vegetable crop in the ground was just as vital to community well-being as was the rice crop. All the concern about rice was therefore nothing more than a red herring designed to obscure the fact that men were no longer willing to work to support their families.

The lengthy discussion in the VDC meeting sheds considerable light on what has become a protracted struggle up and down the River Basin over the obligations men and women bear toward each other and their families and how these intrahousehold social relations intersect with concerns over ecological management (Carney 1988a, 1988b, 1996; Carney and Watts 1990, 1991; Barrett and Browne 1991; Norton-Staal 1991). These relations have at times become quite bitter, and women have frequently expressed considerable anger because they feel their husbands have neglected to shoulder their share of household budgetary obligations. As one garden leader put it: "Go to the hospital! You will see only women and children there who [have worked themselves] sick. You will see only women in hospital beds. You will *never* see men there. ... Most of them are lazy and useless. They do not even work hard in their fields during the *rainy* season [i.e., when the rest of the community is wholly mobilized to secure the coming year's food stocks]". In the dispute over the SWMU soil and water management project, the anger expressed by the woman VDC member forced Kerewan's male community leaders to back down on their threats to restrain the economic activities of women gardeners. At the same time, however, these men refused to take up the challenge the women VDC member issued to them to assist with the rice harvest, leaving gardeners on their own to plant and harvest both rice and vegetables according to their own schedule.[16]

The apparent resolution of the Kerewan SWMU project dispute notwithstanding, the issue of household obligation has remained central to negotiations between women and men over the opportunities and constraints embodied in development interventions. What the soil and water reclamation project accomplished in Kerewan was to enliven a discourse that prioritized rice—a crop that benefits the "household"—over gardens controlled by women. The practices of developers promoting soil and water management

---

[16] Unfortunately, this dispute may have been partially responsible for the fact that neither the various community groups in Kerewan nor the development agencies responsible for promoting the SWMU project took it upon themselves to make repairs to the dike system when they became necessary the following year. As of the middle of the 1995 rainy season, no maintenance had been performed on the dikes, and much of the reclaimed rice land was once again vulnerable to salt intrusion from tidal flows.

interventions were thus very much in line with powerful moral and political economic arguments that have been made for prioritizing food crops over cash crops (Bernstein et al. 1990). The reply of the woman VDC member, however, revealed how such arguments can obscure fundamental political realities. On The Gambia's North Bank in the early 1990s, the decision to favor rice over gardening had significant, not to say profound, implications for intrahousehold labor relations and budgetary control. There was simply no a priori reason why completion of the harvest task, which was equal in the minds of developers and male community leaders to food provisioning, should have hinged on women forgoing other economic opportunities. That the female obligation to household and family expressed in the rice-harvesting task was so readily naturalized and reinforced by environmental rhetoric favoring resource reclamation raises serious questions as to the viability of ongoing attempts to improve environmental management practices up and down the River Basin.

Conclusion

In this essay, I have inspected three successive attempts to reclaim low-lying soil and water resources in a small town on the North Bank of the River Gambia. In the first instance, rural women's groups caught up in a boom in market gardening assumed numerous rights and privileges over *boraa banko* lineage land, often without the knowledge of, and at times in the face of direct opposition by, male landholders. In the second, landholders embraced a Forest Department and development donor goal of accelerated tree planting through a garden-orchard agroforestry transition, in the process alienating land improvements and subsidies originally produced by, or intended to benefit, gardeners. In the third, the introduction of new soil and water management systems in rice-growing areas revitalized a discourse on family obligation and the moral and practical values of prioritizing staple foodstuffs for "joint household" consumption over the production of vegetable commodities for "personal" gain. The successive interventions illustrate well the dilemmas created when developers, state agents, and other interested local parties introduce the moral claims and economic mandates of "global" environmentalism into heavily politicized local development processes.

International feminist pressure in response to the egregious gender inequities of the early "development" years was a major factor in redirecting investment to The Gambia's nascent market garden sector. As women in the United States and Europe gradually worked their way into international donor agencies and fought to influence hiring decisions and redirect program objectives toward women's needs, Gambian gardeners made their own strides forward to improve production and forge crucial market connections. Consequently, by the 1980s, when nongovernmental agencies armed with capital and a definite mission to redress gender inequities sought to establish themselves in The Gambia, the increasingly viable horticultural enterprises attracted their attention. WID projects initiated by NGOs and voluntary organizations provided an ideological framework for intervention, and the negative economic circumstances created by drought and structural adjustment programs added a sense of urgency to the WID programs, as hundreds of small grants for garden wells and fences were quickly negotiated. Thus, international gender critiques helped produce real gains in rural Gambian women's incomes and enhanced women's collective power and prestige at a time of great need.

When the WID emphasis behind the garden boom of the 1970s and 1980s gave way to ecofeminist and feminist environmental critiques in the 1990s, however, many of the gains made by rural Mandinka women were threatened. In lieu of wide-ranging assistance for women during the garden boom, developers focused on environmental rehabilitation began searching for ways to enlist women in the task of producing a biologically diverse landscape and re-creating the conditions necessary for sustained food security. These goals seem unassailable, but they served dubious purposes in the context of the garden boom. Instead of supporting women in their efforts to expand land use rights, developers sought ways to tap unpaid female labor (what one agency termed "the most precious and vital local resource"— [WIF 1990] in agroforestry projects. Similarly, instead of backing women's claims to their own labor and income, developers endorsed and reified what they implicitly viewed as a static division of labor, requiring women to neglect their lucrative cash crops in order to valorize new soil and water management strategies intended to boost rice production.

The Gambian case study thus sheds a great deal of light on the political-ecological consequences of environmental interventions. It reveals quite starkly some of the structural constraints shifting development policies have created along the River Basin, many of which have been detrimental to the interests of women market gardeners. While these constraints have had their effects, the situation on the ground in Kerewan and other communities with low-lying market gardens has been far from overdetermined by the developers' actions. Intervention by WID-inspired developers into the land politics of the garden boom clearly created opportunities that women gardeners used to win expanded land rights. Similarly, the actions of NGOs, voluntary agencies, and the state forest service directed at environmental stabilization opened up opportunities that male landholders and community leaders exploited for their own social and economic purposes.

The animation of land use politics in The Gambia in connection with these policy changes underscores the contention stressed over and over in the land tenure literature that rural property systems in Africa are often quite dynamic.[17] "Traditional" or "customary" claims such as those held by

---

[17] Bassett 1993, Migot-Adholla and Bruce 1993, and Shipton and Goheen 1992 review this literature; Rocheleau, Thomas-Slayter, and Wangari 1996 focus explicitly on the literature on gendered tenure.

landholders in North Bank garden districts cannot simply be taken for granted, especially given that policy and land reforms routinely open up new opportunities for the accumulation of wealth, property, and political power (Bruce 1993; Watts 1989; de Janvry 1981). As Bruce puts it: "Often a reform is less important for its explicit objectives than for the [new] openings that [it] ... provides" (1993, 36). What the Gambian case illustrates is the fact that even "secondary" rights holders such as the North Bank women gardeners can take advantage of such opportunities whenever they present themselves.

It is worth noting as a final point that the men and women vying for position in the struggle to control The Gambia's low-lying land resources do not operate from positions of equal power. While it is clear that many of the market garden groups along the North Bank have been quite shrewd in their ability to exploit relationships with funding agencies, these advances remain more an exception than the rule. The "reclamation" efforts of male landholders fixed on orchard development have reinforced this point quite forcefully. Moreover, the labor question remains critical in explaining the outcome of development interventions in the region. As this analysis of the agroforestry and soil and water management projects has shown, the "success" of land reclamation efforts along the River Basin has hinged on capturing women's labor at every turn. Evaluation of the reclamation programs does not, therefore, simply revolve around the question of whether landholders have the "right" to plant trees on garden lands or whether one group of resource users or another has been more effective in controlling the ebb and flow of development largesse. It requires close inspection of the ways in which use rights are renegotiated and interpreted in response to shifting development paradigms and an assessment of the allocation of work obligations and the distribution of the benefits of the labor that forms the core of resource management systems. Finally, it is important to analyze carefully how critical notions of biodiversity, food security, and, not least, ideological associations between women and their environments have sometimes helped reproduce inequitable social relations rather than replace them.

References

Adams, J., and T. McShane. 1992. *The myth of wild Africa: Conservation without illusion.* New York: Norton.

Anderson, D., and R. Grove, eds. 1987. *Conservation in Africa: People, policies and practice.* New York: Cambridge University Press.

Barrett, H., and A. Browne. 1991. Environmental and economic sustainability: Women's horticultural production in The Gambia. *Geography* 76:241–48.

Bassett, T. 1993. Introduction: The land question and agricultural transformation in sub-Saharan Africa. In *Land in African agrarian systems,* ed. T. Bassett and D. Crummey. Madison: University of Wisconsin Press.

Beinart, W. 1984. Soil erosion, conservationism, and ideas about development: A southern African exploration, 1900–1960. *Journal of Southern African Studies* 11(1):52–83.

Bernstein, H., B. Crow, M. Mackintosh, and C. Martin, eds. 1990. *The food question: Profits versus people?* London: Earthscan.

Bonner, R. 1993. *At the hand of man: Peril and hope for Africa's wildlife.* New York: Vintage.

Bruce, J. 1993. Do indigenous tenure systems constrain agricultural development? In *Land in African agrarian systems*, ed. T. Bassett and D. Crummey. Madison: University of Wisconsin Press.

Carney, J. 1988a. Struggles over crop rights and labour within contract farming households in a Gambian irrigated rice project. *Journal of Peasant Studies* 15(3):334–49.

———. 1988b. Struggles over land and crops in an irrigated rice scheme: The Gambia. In *Agriculture, women, and land*, ed. J. Davison. Boulder: Westview.

———. 1992. Peasant women and economic transformation in The Gambia. *Development and Change* 23(2):67–90.

———. 1996. Converting the wetlands, engendering the environment: The intersection of gender with agrarian change in Gambia. In *Liberation ecologies: Environment, development, social movements*, ed. R. Peet and M. Watts. London: Routledge.

Carney, J., and M. Watts. 1990. Manufacturing dissent: Work, gender, and the politics of meaning in a peasant society. *Africa* 60(2):207–41.

———. 1991. Disciplining women? Rice, mechanization and the evolution of gender relations in Senegambia. *Signs* 16(4):651–81.

Daniels, L. 1988. The economics of staggered production and storage for selected horticultural crops in The Gambia. M.S. thesis, University of Wisconsin.

de Janvry, A. 1981. *The agrarian question and reformism in Latin America.* 2d ed. Baltimore: Johns Hopkins University Press.

Fortmann, L., and J. Bruce, eds. 1988. *Whose trees? Proprietary dimensions of forestry.* Boulder: Westview.

Freudenberger, M. 1994. *Tenure and natural resources in The Gambia: Summary of research findings and policy options.* Madison: University of Wisconsin Land Tenure Center.

Freudenberger, M., and N. Sheehan. 1994. *Tenure and resource management in The Gambia: A case study of the Kiang West District.* Madison: University of Wisconsin Land Tenure Center.

Gamble, D. 1947. Kerewan: A Mandinka village social structure and daily life. NAG/9/399/pp. 22–23. National Archives of The Gambia, Banjul.

GOTG (Government of The Gambia). 1992. *National Environmental Action Plan.* Banjul, The Gambia.

Greve, A., J. Lampietti, and F. Falloux. 1995. *National environmental action plans in sub-Saharan Africa.* Washington, DC: World Bank.

Hodgson, D. 1995. The politics of gender, ethnicity and "development": Images, interventions, and the reconfiguration of Maasai identities in Tanzania, 1916–1993. Ph.D. diss., University of Michigan.

Jabara, C. 1990. *Economic reform and poverty in The Gambia: A survey of pre- and post-ERP experience.* Monographs, no. 8. Ithaca: Cornell Food and Nutrition Policy Program.

Johm, K. 1990. Policy issues and options for agricultural development in The Gambia. In *Structural adjustment, agriculture and nutrition: Policy options in The Gambia*, ed. J. von Braun et al. Washington, DC: International Food Policy Research Institute.

Kinteh, S. 1990. A review of agricultural policy before and after adjustment. In *Structural adjustment, agriculture and nutrition: Policy options in The Gambia*, ed. J. von Braun et al. Washington, DC: International Food Policy Research Institute.

Lawry, S. 1988. *Report on Land Tenure Center mission to The Gambia.* Madison: University of Wisconsin Land Tenure Center.

Leach, M. 1994. *Rainforest relations: Gender and resource use among the Mende of Gola, Sierra Leone.* Washington, DC: Smithsonian Institution Press.

MacKenzie, F. 1994. A piece of land never shrinks: Reconceptualizing land tenure in a smallholding district, Kenya. In *Land in African agrarian systems*, ed. T. Bassett and D. Crummey. Madison: University of Wisconsin Press.

Mahjoub, A., ed. 1990. *Adjustment or delinking? The African experience.* Atlantic Highlands, NJ: Zed.

Mann, R. 1990. Time running out: The urgent need for tree planting in Africa. *The Ecologist* 20(2):48–53.

Migot-Adholla, S., and J. Bruce. 1993. Introduction: Are indigenous African tenure systems insecure? In *Searching for land tenure security in Africa,* ed. J. Bruce and S. Migot-Adholla. Dubuque: Kendall/Hunt.

Neumann, R. 1992. The political ecology of wildlife conservation in the Mount Meru area, Northeast Tanzania. *Land Degradation and Rehabilitation* 3(2):85–98.

——. 1995a. Local challenges to global agendas: Conservation, economic liberalization and the pastoralists' rights movement in Tanzania. *Antipode* 27(4):363–82.

——. 1995b. Ways of seeing Africa: Colonial recasting of African society and landscape in Serengeti National Park. *Ecumene* 2(2):149–69.

Norton-Staal, S. 1991. *Women and their role in the agriculture and natural resource sector in The Gambia.* Banjul: U.S. Agency for International Development.

Osborn, E. 1989. Tree tenure: The distribution of rights and responsibilities in two Mandinka villages. M.S. thesis, University of California-Berkeley.

Peters, P. 1987. Embedded systems and rooted models: The grazing lands of Botswana and the "commons" debate. In *The question of the commons: The culture and ecology of communal resources,* ed. B. McCay and J. Acheson. Tucson: University of Arizona Press.

Puetz, D., and J. von Braun. 1990. Price policy under structural adjustment: Constraints and effects. In *Structural adjustment, agriculture and nutrition: Policy options in The Gambia,* ed. J. von Braun et al. Washington, DC: International Food Policy Research Institute.

Raintree, J., ed. 1987. *Land, trees and tenure.* Nairobi, Kenya, and Madison, WI: ICRAF and the University of Wisconsin Land Tenure Center.

Rocheleau, D., and L. Ross. 1995. Trees as tools, trees as text: Struggles over resources in Zambrana Chacuey, Dominican Republic. *Antipode* 27(4):407–28.

Rocheleau, D., B. Thomas-Slayter, and E. Wangari. 1996. Gender and the environment: A feminist political ecology perspective. In *Feminist political ecology: Global issues and local experiences,* ed. D. Rocheleau, B. Thomas-Slayter, and E. Wangari. London: Routledge.

Schroeder, R. 1993. Shady practice: Gender and the political ecology of resource stabilization in Gambian garden/orchards. *Economic Geography* 69(4):349–65.

——. 1995. Contradictions along the commodity road to environmental stabilization: Foresting Gambian gardens. *Antipode* 27(4):325–42.

——. 1996. "Gone to their second husbands": Marital metaphors and conjugal contracts in The Gambia's female garden sector. *Canadian Journal of African Studies* 30(1):69–87.

——. 1999. *Shady practices: Agroforestry and Gender Politics in The Gambia.* Berkeley: University of California Press.

Schroeder, R., and K. Suryanata. 1996. Gender and class power in agroforestry: Case studies from Indonesia and West Africa. In *Liberation ecologies: Environment, development, social movements,* ed. R. Peet and M. Watts. London: Routledge.

Schroeder, R., and M. Watts. 1991. Struggling over strategies, fighting over food: Adjusting to food commercialization among Mandinka peasants in The Gambia. In *Research in Rural Sociology and Development.* Vol. 5, *Household Strategies,* ed. H. Schwarzweller and D. Clay. Greenwich, CT: JAI Press.

Scott, J. 1976. *The moral economy of the peasant: Rebellion and subsistence in Southeast Asia.* New Haven: Yale University Press.

Shipton, P., and M. Goheen. 1992. Understanding African land-holding: Power, wealth, and meaning. *Africa* 62(3):307–26.

Simon, D., W. Van Spengen, C. Dixon, and A. Narman. 1995. *Structurally adjusted Africa: Poverty, debt and basic needs.* London: Pluto.

Thoma, W. 1989. Possibilities of introducing community forestry in The Gambia, part I. Gambia-German Forestry Project, Deutsche Gesellschaft für Technische Zusammenarbeit (GTZ). Feldkirchen, Germany: Deutsche Forstservice.

United Nations. 1986. General Assembly approves call for special Assembly session on critical economic situation in Africa. *United Nations Chronicle* 23(2):32–36.

USAID (U.S. Agency for International Development). 1992. Program assistance approval document. Agriculture and Natural Resources (ANR) Program and Agriculture and Natural
Resources (ANR) Support Project. Banjul, The Gambia. Unpublished.

———. 1994. Proceedings of the USAID Natural Resources Management and Environmental Policy Conference. Banjul, The Gambia.

von Braun, J., K. Johm, S. Kinteh, and D. Puetz. 1990. *Structural adjustment, agriculture, and nutrition: Policy options in The Gambia.* Working Papers on Commercialization of Agriculture
and Nutrition, no. 4. Washington, DC: International Food Policy Research Institute.

Wangari, E., B. Thomas-Slayter, and D. Rocheleau. 1996. Gendered lessons for survival: Semi-
arid regions in Kenya. In *Feminist political ecology: Global issues and local experiences,* ed. D.
Rocheleau, B. Thomas-Slayter, and E. Wangari. London: Routledge.

Watts, M. 1989. The agrarian crisis in Africa: Debating the crisis. *Progress in Human Geography*
13(1):1–42.

———. 1995. "A new deal in emotions": Theory and practice and the crisis of development. In
Power of Development, ed. J. Crush. London: Routledge.

WIF (Worldview International Foundation). 1990. *WIF Newsletter* 3(1):4.

World Bank. 1995. *Mainstreaming the environment: The World Bank Group and the environment since
the Rio Earth Summit.* Washington, DC: World Bank.

Rethinking Migration and Indigeneity in the Sangha River Basin of Equatorial Africa

Tamara Giles-Vernick

The flora and fauna of this [Sangha basin forest] region represent a richness of great natural, scientific, economic, esthetic and cultural value that is currently severely threatened by logging exploitation, encroaching human population and subsequent heavy illegal poaching for meat and ivory. ... This reserve/park project includes a wildlife protection program, tourist development ... and efforts to ensure the cultural integrity of the BaAka Pygmies of the region.

World Wildlife Fund report, 1992

The village of Lindjombo [in the Sangha basin] is a ditch, where there is no one to see you. It's in the middle of the forest, and it's too far from large towns and markets. It's too difficult to reach it or to leave it.

Mpiemu market woman, 1993

These observations of the Sangha basin forest reflect starkly different perceptions of the forest's history, inhabitants, and value. [Map 1] In the first, an American World Wildlife Fund (WWF) employee has depicted the Sangha basin forest as a treasure of substantial ecological and cultural value threatened by population influxes and forest exploitation. Through local interventions, the conservation project sought to protect the forest's biodiversity—a vestige of the ancient past—as well as the "cultural integrity" of BaAka pygmies, an "indigenous people" for whom the forest is an "ancestral home". But the second quotation, spoken by a woman whose relatives had journeyed to the forest village of Lindjombo in the 1930s to work on coffee plantations, tells an entirely different story of the forest. While the forest had once provided African workers with access to salaries and Western consumer goods, by 1992 the plantations and nearby logging companies had gone bankrupt. And thus, this woman contended, the Sangha basin forest was a barrier preventing movement and access to wealth and power.

These strikingly different perceptions of the forest, its history, and its inhabitants underpin several questions that are central to contemporary "local" environmental interventions in Africa: how should forests be valued, who assumes responsibility for protecting these values, and who is to blame

This essay is largely based on an article entitled "We Wander Like Birds: Migration, Indigeneity and the Fabrication of Frontiers in the Sangha River Basin of Equatorial Africa", first published in *Environmental History* 4, 2(1999):168–197. Research was funded by a grant from the Social Science Research Council and the American Council of Learned Societies, with funds provided by the Rockefeller Foundation. A Fulbright doctoral dissertation research grant also contributed funding for the research. The Department of History at the University of Virginia supplied funds for additional archival research in France. I thank all of these organizations for their generous assistance. I am grateful to Rick Schroeder, Vigdis Broch-Due, and other participants in the conference The Politics of Poverty and Environmental Intervention for their comments on earlier drafts of this essay.

Map 1. *Sangha Basin*

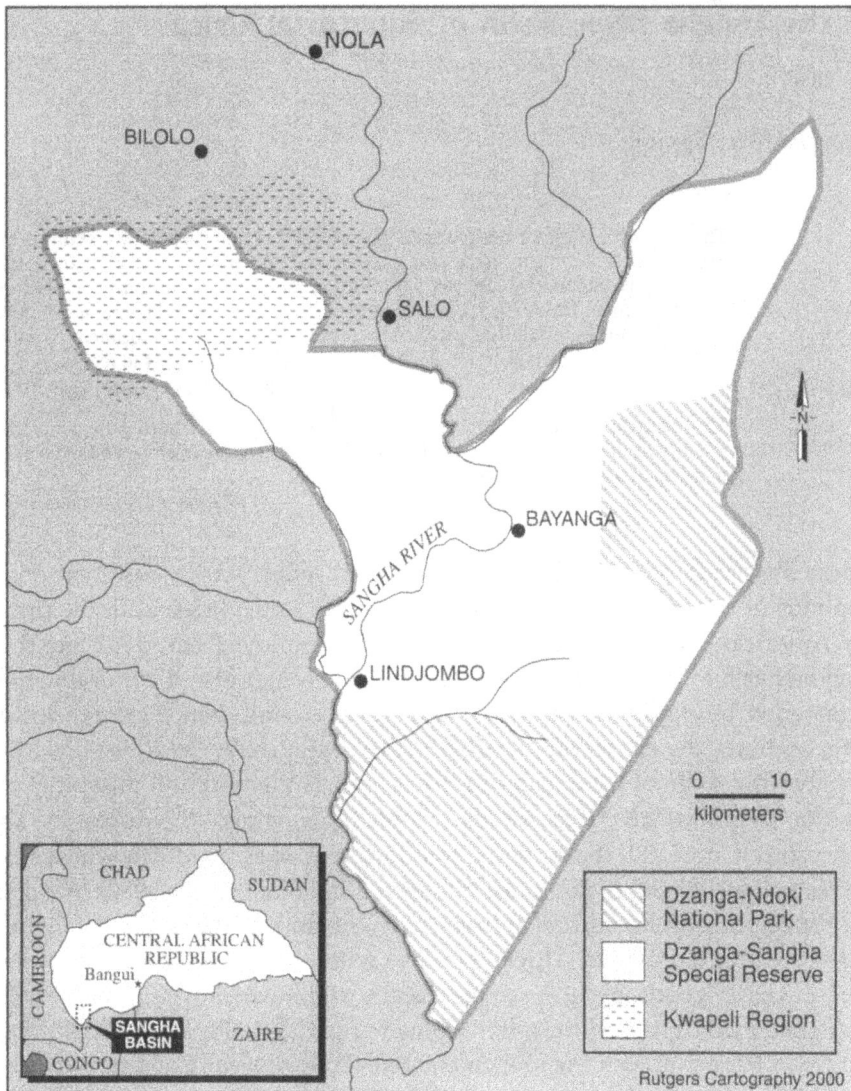

for destroying them? For conservationists who seek to intervene in patterns
of resource use and protection, these questions speak to the constitution of
"local communities" that will successfully manage the forest's resources
(Schroeder, 1999), in part by excluding migrant outsiders who would lay
waste to them. Indeed, since its arrival in the Sangha basin in 1989, the
World Wildlife Fund, in conjunction with the Central African Republic state,
has sought to resolve this question by parsing the region's heady mix of
ethnic groups into distinct categories of "migrant" and "indigenous" peoples and by allocating blame and project perquisites accordingly. For the

Mpiemu woman, her kin, friends, and acquaintances, these categories were not so clear because geographical mobility has been a crucial feature of the Sangha basin's past. Movement and stasis had long been a means of bestowing access to new people, foods, knowledge, and opportunities, but for some Mpiemu speakers in 1993 they signified disconnection with that past.

I would like to show in this essay that the categories of "migration" and "indigeneity" have long and complex histories in the Sangha basin. Both the World Wildlife Fund and Mpiemu people living in the basin articulate particular visions of mobility, indigeneity, opportunity, and environmental destruction that have developed over time and have appropriated other Africans', European explorers', and colonial administrators' historical visions of movement and stasis. This essay, then, excavates the specific historical meanings that these various historical agents have conferred on categories of migration and indigeneity. *Migration* is not a neutral term connoting the movement of a "people" or "community" across geopolitical boundaries, nor is *indigeneity* merely a description of the "original" inhabitants of a region.[1] Rather, these categories constitute various historical actors' efforts to trace past population movements and to use these histories to intervene at historical moments to build political authority and reconfigure access to and control over labor, game, and forest products.

The essay first traces the histories of population movement in the Sangha basin to demonstrate the fluidity of social, cultural, and environmental relations over the past two centuries. It then shows how Mpiemu people remember their histories of movement and the interventions of historical agents with whom they interacted, including BaAka "pygmies", early explorers, French colonial administrators, and contemporary conservationists. The Mpiemu people are a historical creation; they speak the Mpiemu language[2] and understand themselves to be a people sharing a language and similar beliefs and practices. Spatial, linguistic, cultural, social, and political exchanges among diverse people of the Sangha basin during the nineteenth and twentieth centuries have occurred, making Mpiemu a fluid category of identity. This essay examines how some Mpiemu invoke their histories of movement to describe themselves as "a dead people" bereft of history and relationships with powerful people.

The essay then assesses early-twentieth-century explorations and French colonial administrations in the Sangha basin. It illuminates the historical visions, spatial strategies, and language with which early explorers and ad-

---

[1] The voluminous borderlands literature in the United States and Mexico criticizes the spatial assumptions underpinning such concepts as culture, community, nation, and diaspora. But it does not, to my knowledge, question the centrality of state boundaries. In the Sangha basin (another geographical unit of analysis that merits questioning), geopolitical boundaries of the Central African Republic, Cameroon, and the Republic of Congo are just one set among several shifting ones (Clifford 1994; Gupta and Ferguson 1992; Derby 1994).

[2] Mpiemu is one of three dialects of the Mpo language, one of several languages spoken in southwestern Central African Republic and southeastern Cameroon.

ministrators conceptualized geographical movements, particularly those of Mpiemu speakers, and made claims to Africans' labor and forest resources such as rubber. Later, World Wildlife Fund conservationists would demonize "migrants" and sanctify "indigenous" people in their efforts to intervene in older patterns of game, timber, and diamond exploitation. The essay concludes by tracing some of the historical continuities and discontinuities in these diverse treatments of migration and indigeneity, revealing how historical participants in the Sangha basin, and particularly Mpiemu speakers, drew from all of these visions of movement, stasis, and forest histories.

Regional histories of movement

Movement has constituted a central feature of the historical dynamics of the Sangha river basin and the entire northern equatorial forest. From at least the nineteenth century, people living in the northern equatorial forest and the Sangha river basin have engaged in intense social interaction and cultural exchange. Indeed, the contemporary linguistic, cultural, and social diversity within the middle and upper Sangha basin attests to this history of movement and exchange.

Local interactions in the Sangha basin were part of two broader changes sweeping equatorial Africa. During the nineteenth century, equatorial Africa found itself increasingly incorporated into a global economy, and thus Africans in the Congo river basin began to form trading networks, exchanging manioc, palm oil and wine, slaves, and ivory for such imports as firearms, beads, and cloth (Harms 1981; G. Dupré 1982, 1985; M.C. Dupré 1995; Vansina 1990). Yet the influence of this trade in the middle and upper Sangha appears to have been indirect. Unlike some populations within the Congo basin, Mpiemu, Pomo, Bomassa, Ngundi, and Kako tellers of history mention nothing of the hallmarks of this trade in their descriptions of late-nineteenth-century life—firearms, ivory trades, and long-distance voyages. Nevertheless, all of these peoples engaged in cross-cultural interactions, which were sometimes hostile and warlike but more frequently entailed intermarriage and trade (Burnham, Copet-Rougier, and Noss 1986:87; Burnham 1980; G. Dupré 1985; Thomas 1963; Siroto 1969). Evidence of such peaceful exchanges among the various language groups of the upper and middle Sangha appear in descriptions of blood brotherhoods; trade; marriages; exchanges of different varieties of maize, manioc, and yams; and even similar forms of cultural expression (Giles-Vernick 1996; Mololi 1994; Bahuchet 1979).[3]

A second and related trend that encouraged movement throughout in the intercommunicating zone between the forest and savannas was increased enslavement. The development of Fulani kingdoms in Ngaoundere indi-

---

[3] Lindjombo Etienne, Lindjombo, 8 August 1993; Kadele Charles, Bandoka, 29 November 1993; Mabessimo Florent, Kodjimpago diamond pits, 7 December 1993.

rectly touched off movement among forest populations. As Mbum, Banda, Gbaya, and Kaka speakers took flight from Fulani raids by moving into forested regions, they created pressures among neighboring Mpiemu, Pande, Ndzimou, and Pomo speakers, among others. Oral accounts from widespread sources attest to warfare, shifting alliances, and mutual enslavement among the Mpiemu and Bakwele, Gbaya, and Kaka speakers of the northern forest.[4]

Movement continued to be a crucial dynamic of the twentieth century, as the middle and upper Sangha was increasingly incorporated into wider political economies. During this century, however, French and German colonial administrations, concessionary companies, and other, mostly European-owned, commercial enterprises provided part of the impetus for movement. Commercial enterprises sought to exploit rubber, ivory, and other natural resources of the equatorial forest but faced the formidable task of extracting labor from small, scattered populations. They tackled this challenge by tempting African workers with promises of access to new consumer goods such as cloth, salt, and alcohol.

Colonial administrations found themselves in the contradictory position of encouraging and discouraging movement across colonial boundaries. On one hand, they competed with one another and with commercial enterprises to gain control of African labor. From the 1890s to 1916, French and German administrations competed intensively to attract populations to their own territories on either side of the Sangha River (Kalck 1974). But after 1915 the French administrations sought to prevent movement, rooting people in their districts so that they could be compelled to pay taxes and labor on public works projects and to prevent the spread of sleeping sickness to other populations of the Sangha basin.

Africans participated in these commercial and administrative endeavors for a variety of reasons. Some fled from the excessive demands of colonizers and moved to Cameroon or to Catholic and Swedish Baptist missions in Berberati, Bania, and Nola. Others found the new access to such consumer goods as clothing and cooking pots appealing enough to move to workplaces in towns and the forest. Women married workers in new locations and thus moved. Kin group conflicts, frequently manifesting themselves in the form of witchcraft, also encouraged people to move. The nature of the movements was equally variable. Some groups undertook movement collectively, such as Mpiemu speakers who followed the chief Bandoka, moving from the Kodjimpago River to the west of Salo in the early twentieth century or Babenjelle (pygmies) who relocated in the 1950s from the Motaba River to the forests near Bayanga (Kretsinger 1993; Bahuchet 1985). Many other peo-

---

[4] OA (recorded oral account). 7 (Mabessimo); Kadele Charles, Bandoka, 29 November 1993; Yodjala Philippe, Lindjombo, 26 April 1993.

ple, however, moved in much smaller groups to new locations, following
the kin, inlaws, and friends who had preceded them.

Thus, continuous movement toward opportunities and away from
threats characterized at least two centuries of the middle and upper Sangha
basin's history. Various participants and observers, however, overlaid these
histories of movement with their own social, spatial categories as part of
their efforts to ground themselves historically in particular places and to
control use rights to resources in the region's forests and rivers.

Mpiemu notions of migrants and migration

Narratives of movement were a significant feature of *bi san bi doli*, "the
things of history" that Mpiemu people of the Sangha basin recalled, retold,
and enacted.[5] Mpiemu people recounted these movements to express connections with their historical identities as wanderers, forest and farming
people, broad groups with putative but untraceable kin ties (*kuli ajing*), and
Mpiemu. They detailed past wanderings across rivers and ultimately to
metego Mpiemu, or "Mpiemu soil" (Map 1), to articulate their knowledge of
the peoples, places, and resources within the forests. Environmental knowledge emphasized the benefits and dangers of cross-cultural contact and exchange. It bound tellers and listeners to past and present familial, ecological,
and ethnic identities rooted in *metego Mpiemu* (Massey 1994).[6] Some, however, recounted twentieth-century movements away from *metego Mpiemu* to
articulate their sense of loss and disconnection from historical places (forest
camps, trapping paths, fields, villages, and burial sites). These places served
as mnemonics for past people and knowledge, grounding familial, ecological, and ethnic identities (Gupta and Ferguson 1992).[7]

Of particular importance is the rhetoric that Mpiemu speakers use to depict movement, stasis, and their changing relationships to their forest environments. Mpiemu people who speak both Mpiemu and Sango do not have
words that translate as "migrant" or "migration", but they do use several
words and images that connote both movement and "sitting".[8] In Mpiemu,
tellers of history used such terms as *ntche* (to come from) or *wimbo* (to wan-

[5] *Doli* is a highly complex term that does not translate easily from Mpiemu to English. Elsewhere I have glossed this term as a distant, unchanging past; a cycle; body, and place memory;
and knowledge and practices associated with the distant past.

[6] My argument that places are historical and inhere within them various social identities draws
from Doreen Massey's work, in which she contends that space and place must be understood as
intimately connected to time and "as a configuration of social relations". She argues, moreover,
that we need to pay close attention to the dynamic struggles that constitute places.

[7] Gupta and Ferguson have suggested that this sense of disconnection and deterritorialized
identity is widespread, resulting from increased mobility. I would argue, though, that we need
to examine the specific historical processes through which this disconnection has come about.

[8] Sango is the national language of the Central African Republic. Christian missionaries are
usually credited with disseminating this language from Ubangi River populations to African
Christians in Ubangi-Shari. Mpiemu speakers may not have begun to speak Sango until the
1940s or 1950s.

der) to refer to their historical movements and *dio* (to be sitting or to live or exist in a space) to refer to times of past settlement that punctuated wanderings through forests and over rivers. In Sango, they employed similar terms: *londo* (to come from), *duti* (to sit), *ga* (to arrive), and at times *gagango*, referring to those people who had come from elsewhere.

Many tellers of *doli* recounted the movements of *bori doli* to recall a past time and people who wandered over rivers and through forests. Stories of movement in the forest had enabled *bori doli* to profit from cross-cultural exchanges with previously unknown people and to develop new ways of interacting with their environments. One tale (*saa*), for instance, explained the introduction of agriculture through the meeting of Ntchambe Mekwombo (the deity who created and commanded the forest and its people) and Ntchambe Meburi (the deity who taught people to cultivate the land).[9] Wandering through the forest in search of vines to fashion into fishing traps, Mekwombo and Mebur seized upon the same vine and thus met. They developed a fast friendship, arranging a series of marriages among their children, who produced the Mpiemu as forest dwellers and farmers. Movement—and stories about it—enabled Mpiemu to understand and reaffirm their historical, ecological identities as forest and farming peoples as well as to recall how they learned to exploit their environments in new and productive ways.

Tellers of *doli* also detailed a series of movements along rivers, which ultimately led to settlement on *metego Mpiemu*, or "Mpiemu soil". They remembered these peregrinations by invoking the place and time when *bori doli* sat. In one rendition of this narrative, Kadele Charles and a group of elder men in the village of Bandoka responded to a younger man's query about the origins of the Mpiemu people:

> Our grandparents (*besampambo*) came from many rivers, they came from many rivers. They came from Mpiemu and they came from many rivers. At that time, they crossed the Gobomon river. ... At the hour the big people (*bitumba*) sat (*dio*), they didn't know water. They didn't know the Gobomon river. They didn't know it. ... They crossed the Dja river and the Mpompo river. They crossed the Mbong Alon, the Mpiemu, they came along the Mbong Alon. ... I don't know from which river they drank before then, I only know the Dja and the Mpompo, the big rivers.[10]

---

[9] *Mekwombo* is the plural of *akwombo*, or "forest". *Meburi* is the plural of *buri*, or "fallow land". Elisabeth Copet-Rougier found one informant in Cameroon in 1980 who declared that the Mpiemu were the offspring of a man named *piemo* or *jarwa*, or "bush fire", and a woman named *akombo*, or "forest". Because Mpiemu linked bush fires with preparing fields, I take this comment to mean that the Mpiemu people were the children of the forest and the field (Copet-Rougier, personal communication, 1994).

[10] OA.21.1 (Kadele Charles). Linguistic, oral, and written evidence provides historical support for these migrations. Mpiemu in the Central African Republic speak a language, Mpo, shared by people living between Yokadouma and Moloundou in Cameroon. Their language is also very closely related to Kozime and Bajwe, spoken further west in the region between Lomie and Ngoila, and to Mezime, northwest of Yokadouma. These language groups live along the

Hence, the descendants of the first Mpiemu moved across "many rivers" and the Mbong Alon (a stout forest vine over which Mpiemu crossed the Dja River before it broke) to reach Mpiemu soil (Giles-Vernick 1996:215).[11] The Mbong Alon crossing gained significance in these narratives, for it helped to explain the widespread distribution of Mpiemu-speaking people inside and outside of the Sangha basin.

These tellers of *doli* thus employed images of cords and vines and "sitting" in stories of movement for several reasons: to evoke Mpiemu ancestors, to connect themselves to a Mpiemu past characterized by movement and settlement, and to emphasize the benefits and dangers of cross-cultural contact and exchange. "Sitting" was spatial and dynamic; once the Mpiemu ancestors reached *metego Mpiemu*, they did not remain in one place but continued to move and create large settlements such as Mpola, Adoumanjali (now Bilolo), and eventually several smaller villages.[12] Moving punctuated by "sitting" characterized the very nature of past and present Mpiemu. Hence, Kadele's remark, "Our fathers, they came, they came, they never sat in one place (*ba dio bandi woro*)", echoes one Mpiemu woman's observation, "We Mpiemu, we wander like birds".[13]

The symbols of vines and cords (domesticated vines) articulated a historical, ecological, and ethnic identity of Mpiemu as forest dwellers and farmers; it reminded tellers and listeners of the benefits of cross-cultural contact achieved through movement and learning and reiterated the movements leading "grandparents" to *metego Mpiemu*. The vines bound together Ntchambe Mekwombo and Meburi, the forest and the farm. The marriage alliances of their children ensured the continued linkages between those spaces, as each generation reproduced itself, teaching children to exploit the forests and fields. Similarly, the great forest vine, Mbong Alon, brought past Mpiemu to their contemporary home place, *metego Mpiemu*.

Twentieth-century movements away from *metego Mpiemu* took on different meanings for Mpiemu living further south living in the village of Lind-

---

migration routes described by Mpiemu tellers. Oral testimonies collected by Elisabeth Copet-Rougier in eastern Cameroon recount several Kako confrontations with Mpiemu in some locations mentioned in Mpiemu oral testimonies. These confrontations took place before 1878, when the Lamido Issa of Ngaoundere died (Grimes 1992:212; Copet-Rougier, personal communication, 1994).

[11] There are numerous variants of this telling. Some tellers located the Mpiemu along the Dja River (previously called the Mpompo or Djanampompo) and then detailed their crossings over the Mbong Alon, the Bayingi (Bangue), the Biali (Mbyali), and finally the Lekela (Ekela) and Kadei rivers, which run through Mpiemu soil. Led by chief Menkonga, Mpiemu made war and fled various peoples, including the Mpyem, Bakwele, Kako, Ngundi, Pande, and Gbaya. But Mpiemu groups also remained in some regions with these peoples, eventually identifying themselves as Bamba, Jasuwa, and Bidjuki (OA.30.1 [Mokhambo, Kadele Charles, Boluwo Adolphe]; OA.7 [Mabessimo Florent].

[12] OA.21 (Mawindi Andriane); OA.20.1(Kadele); OA.18.1(Nyambi Patrice); OA.23.1(Alouba Clotere). Mpiemu mobility displays some striking similarities with Dupré's study of Beembé mobility in the nineteenth century. (G. Dupré 1985).

[13] This woman was from Cameroon but living and trading in northern Congo (OA.20.1 [Kadele]; Atekoe Josephine, Bomassa, Congo, 5 April 1994).

jombo. According to these people, this mobility during the colonial period effectively severed the cords that bound them to past people, places, and knowledge and to present social relations through which they participated in broader political economies. Mpiemu speakers who had moved or whose grandparents and parents had relocated to Lindjombo in the 1920s through the 1960s to work for the large coffee plantations, saw these relocations as the source of their "death". Undertaking new forms and rhythms of work for wages, Mpiemu integration into a capitalist global economy dramatically produced a different kind of movement, altering their social and political relations with one another, neighboring peoples, and powerful outsiders.

Cords and vines thus illuminated the processes leading to obscurity and diminishing their participation in wider political economies. Lindjombo Mpiemu categorized and commented upon outsiders with whom they interacted. Ntchambe Mekwombo, Ntchambe Meburi, and Ntchambe Bilon crystallized historical processes by means of which white outsiders achieved political and economic dominance in Africa.[14] Ntchambe Bilon, the white deity who moved from the cleared spaces of the savanna, acquired the ability to read and write or to gain knowledge through sight. According to tellers of Mpiemu history, Bilon thus acquired authority over Mekwombo and Meburi, the deities of the forest and fallow. This story parallels Mpiemu historical visions of French colonial rule in which white people (*bimpindi*) moved from their European homes in open spaces to Africa, where they implanted their superior knowledge and cleared vast spaces from the forest.[15] Lindjombo and its inhabitants profited from this clearing. When the *patrons* lived there, hundreds of houses dotted its impeccably cleared banks along the Sangha River, and passengers on barges would exclaim at the village's size. But once these *patrons* departed tangled vines proliferated, choking the banks of the Sangha and obscuring the village from outsiders. In this context, then, Lindjombo Mpiemu could not transform these tangled vines into cords and make them useful. The vines remained undomesticated and thus prevented Mpiemu from establishing and nurturing the social relations with *patrons* that would link them to wider political economies. In this context, they bemoaned their isolation from kin in *metego Mpiemu*, who would have supported them in difficult times. These tales of movement and cross-cultural connection, then, vines and cords as ligatures and separations, articulated Lindjombo Mpiemus' sense of marginality to broader political-eco-

---

[14] Yodjala Philippe, Lindjombo, 23 November 1993; Mabessimo Florent, on the Kodjimpago River, 7 December 1993; OA.1.1 (Mpeng).

[15] This perception that powerful people lived in cleared spaces applied to Africans as well. For many Mpiemu, the 1993 election of Ange Patasse to the Central African Republic presidency confirmed that people living in the savannas were easily "seen", and thus had access to salaries, health care, transport, and consumer goods. Patasse came from the town of Paoua, in the savannas of northwestern Central African Republic, and Lindjomboans frequently referred to him as a "northerner" (*nordiste*).

nomic relations and of disconnection from knowledge of past people and places.

Thus, Mpiemu people remembered and recounted past movement and settlement in somewhat different ways. Many identified a distant past in which movement over forest rivers brought cross-cultural contact, and thus new foods and cultural practices, and propelled them to *metego Mpiemu*. For Lindjombo Mpiemu, however, twentieth-century relocations from *metego Mpiemu* resulted in temporary prosperity but eventual ruin. Although movement had bestowed access to powerful *patrons*, and thus consumer goods, it also brought "death". Mpiemu in Lindjombo found that the "cords" connecting them to kin and historical places had been severed. Movement uprooted Lindjombo Mpiemu, rendering them unable to trans-form an obscuring tangle of vines into cords linking them to networks of powerful people. They consequently envisioned themselves as living "in a ditch where no one can see you"—a place existing outside of past and future time.

How did these historical visions of past movements develop? I will argue that memories of movement served as a commentary on the past and con-temporary efforts of French explorers and administrators and Western con-servationists to cut the cords that bound them to historical processes, places, and people.

Population movements and early explorations and administrations

Early European explorers and French administrators were preoccupied with identifying Sangha populations' historical movements as a means of making claims to forest resources and labor and of establishing spatially their pres-ence in the Sangha basin. The Sangha basin was a crucial means of access to Chad during the "scramble for Africa"; the French envisioned linking their equatorial African possessions with those in North and West Africa and thus aimed to secure a solid foothold in the Sangha basin to hamper British and German expansion (Giles-Vernick 1996:41–42; Burnham 1996a:48–49). The Sangha basin explorations sought to establish a safe route to Chad, to promote trade with Africans, and to increase French influence in the region (Burnham 1996a:49). After their defeat at Fashoda in 1898, the French turned their attention to building a colonial state to exploit the riches of the Sangha basin and to protect these resources from neighboring Germans. State building, however, was a haphazard and attenuated process that made fitful headway for the first three decades of French occupation of the basin. France ruled the Sangha through concessionary companies for the first two decades of the twentieth century, setting up an incomplete and fragmentary admin-istrative structure. Concessionary companies sought to establish ill-defined, frequently contested claims to African labor, ivory, and rubber in the Sangha (Coquery-Vidrovitch 1972:315–22). Prodded by bickering companies that claimed to be losing labor and forest resources to neighbors, France and

Germany repeatedly sought to establish the boundaries separating German Cameroon from French Congo. After 1898, explorers ostensibly gathered information to settle boundary disputes and conflicting claims to Sangha peoples and commodities, but they actively sought to demonstrate knowledge of and control over these forested regions and their people as part of an erratic effort to establish state control.

Thus, in order to claim sole access[16] to forest resources and to establish French authority in the Sangha, explorers sought to learn the landscape of the Sangha basin,[17] to locate and attach names to particular peoples living there, and to trace the historical peregrinations of various linguistic groups.[18] A few explorers and administrators collected oral testimonies illuminating the historical movements of linguistic groups residing there in the early twentieth century. But they treated the populations they encountered differently. Most focused primarily on the historical wars and wanderings of "Sangha-Sangha" and other riverine people as well as savanna dwellers. BaAka (called Babingas in these early accounts), the client hunters of riverine peoples, occupied a shadowy presence in explorers' accounts but tantalized them with their intimate knowledge of the forest. And while some explorers briefly mentioned nineteenth-century Mpiemu mobility they accorded the Mpiemu a marginal place in the history of the basin, no doubt because they occupied a peripheral place along the strategic access route of the Sangha River. Mpiemu appeared in these documents as a people without history, who knew nothing of "civilization" and understood little of the riches contained in their forests. Explorers saw them as incapable of exploiting that wealth and thus in need of colonizers to "develop" forest resources and help them to acquire a history.

From their first forays into the Sangha basin in 1886, the early explorers focused on the riverine peoples, who congregated in large villages along the large navigable rivers of the northern forest (Ponel 1896; Anon. 1900a, 1900b;

---

[16] In my use of the term *access*, I am drawing from the works of Jesse Ribot, who has cogently argued that within Senegal's charcoal trade this "ability to obtain or make use of" forest resources, markets, labor, and opportunities is informed not just by legal mechanisms but more importantly by "the social and political-economic hierarchies in which extraction, production and exchange are embedded" (1998:307–41).

[17] As Benedict Anderson has observed, mapping provided a visual means by which colonizers could imagine their possessions. "European-style maps", he maintains, "worked on the basis of a totalizing classification" (1991:163–64, 173). Bassett and Porter have effectively elucidated this relationship between mapping and political control in the West African context (1991:367–413).

[18] This exploration constituted part of the process of creating ethnic identities in the Sangha basin. Tracing the creation of ethnic identities is beyond the scope of this essay, but a variety of historical interests—Mpo speakers, colonial administrators, missionaries, and other linguistic groups in the Sangha—helped to create a Mpiemu ethnic identity. The most recent studies of ethnicity have moved beyond older conceptions constructed along cultural boundaries or as a product of colonial intervention. Indeed, these studies demonstrate the complex social, cultural, and political negotiations producing these identities among and between groups (Burnham 1996b; Bravman 1998; Willis 1992; Grinker 1994; cf. Barth 1969; Vail 1991; Spear 1993).

Loyre 1909; Coquery-Vidrovitch 1972:31).[19] Most of their reports and articles mentioned something about historical movements of Sangha inhabitants. Emile Loyre, for instance, briefly discussed the origins of the Boumoali and their movements in the Sangha (406). The Mission Cottes to southern Cameroon (1905–8), a joint Franco-German expedition seeking to demarcate the boundaries of German Kamerun and French Congo, undertook a systematic exploration of local populations, dividing them into "pygmies" or "Negrilles primitifs" and the "groupe Bantou" or "groupe M'Fang". This groupe M'Fang, the mission report erroneously contended, were the offspring of pygmies and Ethiopians or "Hottentot-Boschimans" and had invaded the forests from the west (Cottes 1911). Explorer and colonial administrator Georges Bruel published two articles in which he addressed the historical movements and wars of middle Sangha populations (1909; 1910; see also Fourneau 1900).

Such ethnographic and historical treatments of riverine populations constituted an important part of learning, mapping, and reconfiguring what must have seemed a wildly complex cultural geography of the Sangha basin. Significantly, French explorers and administrators did not define these population movements as migrations. One early report, for instance, asserted that there was "practically no emigration" from the Sangha circumscription, and that the only immigration into the region came from Sierra Leoneans, "Pahouins" and "Loangos" who worked for the trading and concessionary companies (Anon. 1901).

Explorers' accounts of Mpiemu movements, however, were sparse. Mpiemu speakers, living well off these rivers, remained undocumented until their first encounter with French explorers in 1905. In that year, Commander Henri Moll undertook a two-year expedition in France's possessions in Congo and Cameroon bordering on German Kamerun. The expedition was part of an ongoing struggle between France and Germany to define boundaries between their equatorial African possessions. In an effort to secure Africans' support for the French, Moll pursued a "policy of peace", concluding treaties with various "rebel chiefs" who had apparently refused to collect rubber and ivory for the concessionary companies in the region (Muston 1933). Moll's expedition journal devotes several pages to negotiations with the Mpiemu chief N'Gombako, but it described nothing about the Mpiemu themselves.

Of Mpiemu historical movements, the Moll journal only asserted that French intervention would "civilize" N'Gombako's wandering peoples. Muston, the chronicler of the Moll expedition, reported that, after having concluded the treaty with the French, N'Gombako

---

[19] The focus on riverine peoples is not surprising. Early explorers (prior to 1898) moved along navigable rivers in the hope of linking French equatorial possessions with Sudan and the upper Nile.

was happy, for himself and his people, to be finished with an errant life and to construct a beautiful village; that he felt too old now to make war anymore; that, from the moment he accepted the peace, he would be submissive to the French like a goat, a monkey, or another beast. (1933:29)

Muston acknowledged the continual movements of the Mpiemu, but unlike so many other contemporary French reports and articles, which traced the peregrinations of riverine peoples, he never mentioned where, when, or why Mpiemu had moved. Thus, Mpiemu, in contrast to their neighbors, had no history until 1905 when, "ignorant of all civilization ... [they] came to submit themselves to the great white chief" (29). As a representative of France, the Moll mission therefore intervened in these apparently ahistorical impulses to move and make war. In settling N'Gombako and his followers, the mission "civilized" them and cleared the way for more effective control of their labor and more consistent access to forest rubber and ivory.

The Moll report was not unique in its perceptions of Mpiemu speakers as an ahistorical people. Explorers and administrators encountered plenty of mobile linguistic groups. Over the next few years, they barely mentioned Mpiemu speakers in their histories of Sangha populations. In his descriptions of Boumouali and Pomo migrations, Bruel mentioned that Mpiemu made war against various groups but confessed that he did not know to whom the Mpiemu were related. His map of linguistic/ethnic groups of the Sangha, though it identifies the Mpiemu village "Ngombako", simply subsumed Mpiemu speakers into regions with "Kaka" and "Babinga" (Bruel 1909; Cottes 1933; Faure 1937).[20]

Even the one early account of Mpiemu speakers contained nothing about a Mpiemu past. In 1907, Paul Pouperon traveled up the Sangha and into the forest near the Kadei River, where he encountered the "M'bimou", among other peoples (1908). For Pouperon, these populations, who occupied their own distinct, hermetically sealed places, could be categorized as "M'bimous, Yangheres and Kakas [who] are harvesters of rubber" and the "N'Goundis [who] paddle on the rivers Sangha Kadei, and Bandja" (4).

What is so striking about Pouperon's description of Mpiemu speakers is that he made no effort to give them a history, as so many of his colleagues had done for riverine peoples. In fact, in his description of the "M'Bimou" Pouperon took pains to blur the distinctions between the people and the forest in which they lived. "This", he began,

is the country of large monkeys: chimpanzees and gorillas. The very hilly region has a clayey soil, covered with a thick forest and traversed by several rapid waterways. One can purchase goats and chickens at an ordinary price there. If the material life of the traveler is relatively easy, his movements are rendered very difficult because of the terrain's twisted nature, its slippery soil, and its unruly inhabitants. The spirit of this population is renowned ... for its ill humor and in-

[20] Mpiemu identified the village "Ngombako" as Mpola. Kobago was a powerful Mpiemu *wani* (chief).

dependence. In the middle of this truly superb equatorial forest, in daily contact with anthropoid monkeys and beasts ... which they love to hunt, the M'bimous have adopted these animals' savage character, which renders them intractable. (1908:4–5)[21]

Most important, Pouperon identified the M'Bimou as "this magnificent country of production (rubber, ivory)", extolling its virtues as a major source of untapped riches for the French. In examining the difficulties of extracting high-quality rubber and sustaining long-term production, Pouperon worried,

> Will we hear it said again by the French that we are unfit for any sort of colonization and that we neither know how to develop our possessions nor can partake of the resources that they offer us? (6–7)

He then outlined a historical role for the French in which they would intervene in this ancient and unchanging place and people in order to exploit intensively the labor and resources that "M'Bimou" contained (Anon. 1911a, 1911b).[22] This intervention took on a real urgency for Pouperon and his colleagues in the Sangha basin, for they were engaged in a crucial struggle with Germans for access to African labor and the forest commodities of rubber, ivory, and animal skins. Hence, while French explorers and administrators in the first decade of the twentieth century took pains to historicize the wanderings and wars of riverine dwellers, they made only passing references to Mpiemu participation in these histories of movement. Mpiemu speakers, it seems, moved compulsively and without reason.

Why did French colonizers make these distinctions between Mpiemu and other Sangha populations and why did they see Mpiemu movement as ahistorical and other populations' movements as historical? French colonizers believed that instability and movement were markers of people who were "uncivilized" and without a history. Because riverine populations were more strategically located along the banks of the Sangha River, explorers surely would have had longer and more intensive contact with them. Late-nineteenth-century Mpiemu, however, lived along the peripheries of these routes of access, and they may not have been as compelling subjects of study and control. But this notion of ahistorical wanderers may not be just an artifact of French observers' ignorance or lack of concern. Explorers encountered plenty of populations that were just as mobile as the Mpiemu. Quite possibly, they perceived riverine people as "superior" to those living deep in the forests. For instance, at least one early-twentieth-century administrative report parsed Sangha basin populations by their "degrees of human evolution" (Anon. 1901). The "Bombassa" (Bomassa), who lived in larger villages

---

[21] My translation. *M'Bimou* and *Mbimou* are terms that other people used to refer to Mpiemu people but also to the colonially designated administrative region in which they resided.

[22] Concessionary company representatives, through whom the French effectively ruled during the first decades of the twentieth century, shared this opinion.

along the Sangha and had access to interior peoples and European traders, ranked higher than nomadic peoples living in the forests but also some peoples residing along smaller rivers. The report does not mention Mpiemu speakers at all, but later administrators may have appropriated this ranking when they complained in 1907–8 that the "indecisive, unstable and nomadic tribes" of the M'Bimou circumscription frequently fled to German Cameroon to escape taxes and labor demands (Congo Français 1908).

For the French, the dynamics of European territorial competition and African wanderings and wars provided ample justification to intervene in the Sangha basin and to claim access to valuable forest commodities. The French intervention would bring history to Sangha populations like the Mpiemu by settling them in newly designated *circonscriptions* and harnessing their labor to gain access to the forest's ivory, rubber, and animal skins.

Indigènes and "flight" after 1920

From the 1920s, French official interest in Mpiemu history and movement changed. The 1920s marked a more concerted effort on the part of the French to build a colonial state in the Sangha basin. This endeavor intensified following the Kongo-Wara rebellion, which swept the upper and lower Sangha basins between 1928 and 1931 and precipitated a brutal French military suppression, administrative redistricting, and extensive village resettlements, even for Africans who did not participate (Nzabakomada-Yakoma 1985; Burnham and Christensen 1983).[23] As they undertook more interventionist schemes to extract taxes and labor for roads, rubber tapping, and food cultivation, administrators increasingly defined Sangha populations' mobility (and particularly Mpiemu mobility) as a problem. Mpiemu movement hampered French control over their labor, food, and taxes. By the 1950s, some administrators would look back on the history of Mpiemu mobility and contend that it would eventually result in the disappearance of the Mpiemu altogether. From the 1920s through the 1950s, administrators sought to prevent *indigènes* (natives) from fleeing across district, circumscription, and colony boundaries. Redoubling their efforts to contain Mpiemu and other populations within colonial administrative boundaries, administrators believed that a stable, docile population would enable them to construct an administrative infrastructure and gain ready access to forest and agricultural products.[24] But to their dismay these efforts had a limited influence, as Mpiemu continued their peregrinations in response to the ravages of sleeping sickness and opportunities for paid work with expatriate commercial enterprises.

---

[23] Historians have interpreted Kongo-Wara, which translates as "The War of the Hoe Handle "in the Gbaya language, as an anticolonial and anti-Fulbé rebellion but also as one of "non-violent protection and peace" (Burnham and Christensen 1983:17).

[24] For a discussion of French administrative mapping between 1913 and 1960, see Sautter 1964:177–84.

Mpiemu wanderings grieved local administrators.[25] In 1923, Vingaras-samy, the chef of the Kadei-Sangha circumscription, complained that residents of M'Bimou Kadei were crossing the Cameroon border to sell rubber there (1923). In the same year, another administrator grumbled that the "M'Bimous" and "Kakas-Goumbes" too often practiced the "custom of easily breaking up their groups and changing the placement of their villages" (Gouvernement Général de l'Afrique Equatoriale Française 1923). A decade later, in the wake of Kongo-Wara and the French reforms, administrators were still bemoaning Mpiemu flights from "incompetent" chiefs and illegal hunting expeditions in Cameroon, though they carefully distinguished Mpiemu from participants in Kongo-Wara who had fled military reprisals and resettlement (Gouvernement Général de l'A.E.F. 1935a; Anon. 1933). Many Mpiemu moved outside of the M'Bimou circumscription to take paid jobs on coffee plantations in Salo, Lindjombo, and Berberati. Mpiemu flight intensified as sleeping sickness continued to ravage the Nola-Bilolo regions where they resided (Gouvernement Général de l'A.E.F. 1935a, 1935b; Headrick 1994:345–83; Saragba 1994:16).

The language and spatial assumptions underpinning administrators' complaints are significant. Reports never used the terms *migrant*; all Sangha basin populations were simply *indigènes* or "natives". But administrators did attach punitive references to those who moved outside of their colonially designated home districts. Indeed, these "home districts" effectively delineated the boundaries of *metego Mpiemu*. Thus, Mpiemu who refused to remain in their "places of origin" were "renegades" (*transfugés*) who "deserted", "fled", "took refuge", or undertook "exoduses" (Gouvernement de Moyen-Congo 1931; Gouverneur Général de l'A.E.F. 1935). Their "places of origin" were in fact colonially constructed districts within the M'Bimou circumscription, where chiefs appointed by the administration organized labor to gather forest rubber and cultivate manioc and other food crops for public works projects, hospitals, and prisons. The *indigènes* did not belong outside of these places—in Cameroon, towns such as Berberati, the big coffee plantations further south, or the diamond mines near Nola.

And thus French administrators struggled to confine Mpiemu to their own districts. They continually captured and returned "deserters" to their home villages, reduced labor demands on Mpiemu to prevent their flight to other regions, and patrolled all roads and paths leading to and from the M'Bimou circumscription, requiring all travelers to carry a sanitary passport (Giles-Vernick 1996:106–8). As late as 1944, they resorted to commandeering Mpiemu workers to gather rubber in the forest and to forcing the sick to cultivate manioc and other foods (Sintas 1944). And by the 1950s, though

---

[25] This vision of uprooted Africans haunted French administrators in West Africa as well. Frederick Cooper has recently argued that even as French administrators increased the demands for forced labor, through the Second World War, they simultaneously worried that labor migration fundamentally degraded African families and societies (1996:31–43; 181–82).

French administrators had abandoned coercive labor policies, they still intervened in Mpiemu mobility, attempting to "stabilize" Mpiemu families by encouraging them to cultivate coffee and oil palms in the Kwapeli and Bilolo regions in spite of the bitter opposition of plantation owners (Gouvernement Général de l'A.E.F. 1950).[26]

French administrators treated other mobile populations differently. From the early twentieth century, BaAka ("Babinga") "pygmies" had attracted the attention of administrators, who found their hunting talents fascinating and potentially lucrative, since BaAka were known for their prowess as elephant hunters. But in distinction to their heavy-handed efforts with the Mpiemu, French administrators articulated a more diplomatic approach with the "Babinga" in 1933 (Delobeau 1992:378–84). They conducted censuses among the "Babinga" but exempted them from taxes. They sought to "liberate" them from the oppressive authority of the "Bantu *patrons*" (382). BaAka participated haphazardly in such extractive and agricultural endeavors, as rubber collection in the 1930s and big coffee plantation harvests in the 1940s and 1950s. But French efforts utterly failed to contain BaAka in a single location.

The mobility of Sangha River populations hampered French efforts to gain access to forest and agricultural resources and thus to ground themselves historically in the region by arresting long-term peregrinations and transforming *indigènes* into sedentary, productive workers. Mpiemu flight remained particularly difficult for the French, who employed far more interventionist strategies to contain their movement than they did with the BaAka. According to Canal, an administrator in the Haute Sangha Region, the French colonial administration had failed completely. He compared the once populous, active M'Bimou villages to a "desert" victimized by the ravages of persistent movement and sleeping sickness (1950).

Conservationists and migration

Debates over the mobility, history, and forest resource allocation patterns of Sangha basin inhabitants continued during the postcolonial period. Much like its colonial predecessors, the cash-strapped independent Central African Republic state turned to its forests to produce export earnings. In 1963–64, French foresters conducted the first forest inventories of the Sangha basin, identifying several exploitable timbers, including *sapeli* (*Entandophragma cylindricum*) and *sipo* (*E. utile*), high-quality hardwoods used for veneers and fine furniture, and less valuable whitewoods used for plywood such as *ayous* (*Triplochiton scleroxylon*) and *limba* (*Terminalia superba*) (Lanly 1966a, 1966b). In the early 1970s, a rapid succession of mostly European-controlled enterprises acquired logging concessions in the region. Logging enterprises rarely remained commercially viable for long since they were constantly beset by

---

[26] Pere M. Morel, in Chambèry, France, 14 May 1998. This policy seems to have been part of a broader French effort in Africa to stabilize laborers' families (ibid.:198, 262).

poor management, high transportation costs, and shifting world prices. Yet the state's need for concession rents and stumpage fees kept it sympathetic to the plight of these logging concerns. In 1986, the Central African Republic state found new reasons to support European logging companies. Under pressure from the World Bank and International Monetary Fund to adhere to a structural adjustment program (which was to reverse the growth of government spending, liberalize prices, encourage a more open investment code, and provide incentives to agriculture and forestry), the state continued to welcome foreign-owned logging companies (U.S. Department of State 1989:5). Hence, from 1971 through the 1990s logging concerns set up shop in the forest, exploited their concessions, went bankrupt, and resurrected themselves in Lazarus-like fashion, repurchasing the same concessions from the state and recommencing timber harvesting.

New logging enterprises, as well as increased diamond prospecting, altered the composition and destinations of mobile peoples in the Sangha basin. People from more distant locations moved into the basin to find wage labor as loggers and sawmill employees or to seek their fortunes in the diamond pits. But some of the major logging enterprises were located in or near many of the old coffee-producing centers. Children of Mpiemu and other coffee workers moved into logging, sawmill, and diamond-digging jobs, thus sustaining twentieth-century patterns of movement between *metego Mpiemu* and such towns as Bayanga and Lindjombo.

In the 1980s, American researchers under the auspices of the World Wildlife Fund expressed increasing concern that logging companies had harvested too much timber, failed to allow sufficient time for regeneration, and opened up the forest to make hunting easier. In 1988, the WWF set up a project,[27] which in 1990 established the Dzanga-Ndoki Park, 1,220 square kilometers of tropical rain forest, allowing only approved research and regulated tourism, and the Dzanga-Sangha Special Reserve, a 3,359 square kilometer buffer zone around the park that permitted hunting with "traditional" weapons (nets, crossbows, and spears) and registered guns as well as food and medicinal plant gathering. With funding from its U.S. branch, various agencies of the American government, GTZ (Gesellschaft für Technische Zusammemarbeit), the World Bank, and the Central African Ministry of Water and Forests, the WWF has since managed the park and reserve, seeking to promote ecologically sustainable rural development, improve roads and construct camps, and set up a research program to ensure sound forest management.

---

[27] Clearly, many different people from international conservation agencies and Central African ministries make up the project, and it is impossible to represent the diverse opinions of all people who have worked on it. In this essay, *project conservationists* and *personnel* refer primarily to U.S., European, and Central African conservationists who hold positions of authority within the project.

In order to bolster its activities, the project has developed a historical vision of population mobility and resource use in the forest. Invoking the categories of "migrant" and "indigenous" people, project personnel have been just as concerned with justifying their interventions in a history (though one of "uncontrolled" environmental exploitation and degradation) as French explorers and administrators were earlier in this century. This time, however, the spatial frontiers underpinning migrant and indigenous categories have vacillated substantially. It is instructive to explore how Mpiemu fared in project efforts to parse inhabitants of the Sangha into distinct categories.

Over the years, the project's goals have remained relatively constant: the "long-term conservation of the natural ecosystems" in the area and the preservation of "the forest as a viable home and source of livelihood for the Aka people" (Fay 1993; Telesis 1991; Carroll 1988:20). In an effort to spur what it calls "ecologically sustainable rural development", the project has sought to provide some jobs and services to particular groups living in the reserve. It has attempted to enforce the Central African Republic's laws restricting hunting and gathering in the park, outlawing widespread snare trapping, requiring various permits for hunting certain kinds of game with shotguns, and substantially limiting the bush meat trade.

Since its inception, the project has sought to provide services and jobs to some people of the Dzanga-Sangha reserve's highly mixed populations. It recruited antipoaching guards, tour guides, and construction workers from various ethnic groups in the region, though it appeared that most workers came from Bayanga, the town where the project headquarters existed. But the project allocated health care services on different grounds. Arguing that the "indigenous" BaAka should not shoulder the entire burden of conservation, it provided free health clinics for BaAka but not for most other reserve inhabitants.[28] In part, the project personnel based the allocation of such services on the grounds that the BaAka were "owners of the forest" because of their status as the "first" inhabitants there. This definition, of course, holds up poorly in a larger historical context of continual geographical mobility in the northern equatorial forest; many people, including the BaAka, spoke of arriving in the Bayanga area from another part of the forest (Giles-Vernick 1996:126–27; Kretsinger 1993:1; Bahuchet 1985).

Other project practices allowed BaAka greater access to forest resources, particularly game. The project enforced state laws governing hunting practices in the reserve, which effectively favored BaAka over other hunters. Reserve laws allowed hunting with shotguns (though these guns required expensive permits) as well as hunting with "traditional" weapons, including nets and crossbows. Only the BaAka now use these "traditional" weapons; others must take their chances trapping with illegal snares or hunting with

---

[28] Currently, however, in some smaller villages nonpygmies receive free health care.

unregistered guns or amass enough money to pay for gun permits (Giles-Vernick 1996:66–68).

More significantly, project personnel invoked the terms *migrant* and *indigenous* to support a particular historical vision of the Sangha basin.[29] This history assigned blame for environmental destruction of the forest and supported the project's mission to prevent destructive hunting, meat trading, and diamond mining and promote "ecologically sustainable" activities. In this historical vision, "indigenous" people were vulnerable forest people who required protection to continue their "ancient" way of life in the forest (Carroll 1986:38; 1988:320–21). While project personnel at first depicted the BaAka as "noble savages" struggling to retain their hold on older ways of living in the forest, more recently they have portrayed them as an "acculturated" people seduced by Western capitalism (Carroll 1986:40–41, 1988:320; Telesis 1991; Linden 1992).

"Migrants", however, have served as sources of destruction in this fragile landscape, an unspoiled remnant of ancient times (Taylor and Garcia-Barrios 1995:5–7). In the early twentieth century, economic opportunities for forest resource exploitation encouraged human migration into the area. This migration brought about "uncontrolled exploitation" and exerted increased demographic pressure on available resources. Indeed, this exploitation was and had remained uncontrollable because the migrants' propensity to exploit was "ingrained in their psyches" (Fay 1993). Project conservationists insisted that their purpose in the Dzanga-Sangha reserve and park was to stop this "uncontrolled exploitation", discourage migration into the reserve, and promote less ecologically destructive ways of making a living.

Who was considered a migrant? The project's units of analysis varied considerably over time. In 1986, one of the park's founders effectively defined a migrant as anyone who was not BaAka (Carroll 1986:40–41). This early definition of *migrant* probably resulted from the fact that the two project founders began their work as botanists in the region and employed primarily BaAka in their fieldwork in the forest. Only later did they come into contact with other peoples in the region.[30] By 1993, however, the definition had contracted somewhat; project personnel referred to "migrants" as those who were neither BaAka nor "Sangha Sangha" (Kako, Pomo, Bomassa) (Projet Dzanga-Sangha 1993). Subsequently, project personnel defined "migrants" more variably: they had either originally come from outside the Dzanga-Sangha reserve's boundaries or they came from areas inside the reserve but outside of the project's effective reach—beyond Bayanga and its

---

[29] The definition of *indigenous peoples* has varied considerably throughout Africa and the world. Some countries have adopted legal definitions to connote the descendants of "original" populations. Indigenous people are usually a numerical minority possessing social, cultural, and linguistic characteristics that distinguish them from most other populations in the country. To my knowledge, however, the Central African Republic has not defined its pygmy populations as indigenous (Hitchcock 1994; Gray 1991).

[30] Hardin, Rebecca. 1996. Personal communication.

surrounding villages and forests (Fay 1993).[31] Now, even as some WWF personnel have reconsidered their long-standing perceptions of migrants in Sangha basin history, they have retained the category. It is not clear, though, whether they have altered their definition of *migrants*.

Mpiemu have occupied a variable place in these changing categorizations. Although part of the Dzanga-Sangha reserve extends into Kwapeli, one of the "home places" that Mpiemu have occupied since at least the early twentieth century, most project documents since the mid-1980s have identified Mpiemu as migrants, not indigenous peoples (Fay 1993; cf. Carroll 1993).[32]

What inspired the changes among these categories? At this point, I can only hypothesize the concerns that shaped these shifting categories. Early on, both Western and Central African conservationists uncritically accepted exoticizing images of BaAka as quintessentially "indigenous" forest people. The category "migrants" served as a means by which they could articulate what indigenous people were not (see, e.g., Goyemide 1984). Moreover, it is important to remember that these WWF workers constituted part of an international network of conservationists who read much the same literature, attended the same conferences, and ostensibly dealt with similar concerns. It is likely that they imported many of the assumptions shaping debates over migrants and conservation in the Amazon river basin (Schmink and Wood 1992). Both of these concerns helped early on to define stark boundaries between migrant and indigenous peoples.

In addition, since the project's inception, personnel have limited their activities to Bayanga and its surrounding villages. As a result, they have perceived twentieth-century social dynamics in the Sangha basin from a "Bayanga-centric" perspective. In effect, because the project didn't work outside of Bayanga and nearby villages it neglected the broader region, its populations, and its historical dynamics. It thus premised its stark distinctions between migrant and indigenous peoples on a highly limited geographical space and understanding of the Sangha basin.

Thus, the WWF has sought to categorize migrant and indigenous peoples and demonize a historical and present "migration". These efforts have underpinned an attempt to allocate certain rights to project and forest resources, cementing the project's role as arbiter over access to forest resources and establishing its place in reversing historical trends of forest exploitation.

Conclusion: Historical continuities and discontinuities

This essay has demonstrated that the changing categories and historical visions of "migration" and "indigeneity" are part of broader, long-term debates over movement and settlement, history, state building, and access to

---

[31] U. Ngatoua and P. Yangoui, Bangui, 18 December 1993.

[32] A. Blom, Bangui, 19 May 1993.

and control over productive resources. In questioning the spatial, historical assumptions buttressing these categories, this essay has argued, too, that the frontiers underpinning these categories have been not been fixed but have resulted from dynamic struggles among Africans and powerful outsiders to forge, both spatially and temporally, a particular vision of the Sangha basin's history. Thus, it seems more useful to focus not on the frontiers that these historical actors created but on the political and economic struggles leading some of them to claim that these boundaries were meaningful ones. In locations where populations have used spatial mobility as a strategy to create and cope with change and have woven it into their historical imaginations, these struggles illuminate just how treacherous it can be to identify a "local community" by creating artificial, ahistorical distinctions between "destructive migrants" and "threatened indigenous people".

I would like to conclude by addressing one of the volume's themes concerning historical continuities and discontinuities in environmental interventions, particularly those between the colonial and postcolonial periods. The historical interests in the Sangha basin did not develop their notions of movement and stasis in a vacuum but instead actively disputed and appropriated from one another. Several scholars of environmental interventions in Africa have underscored the continuities between the activities of contemporary conservationists and colonial officials (Fairhead and Leach 1996; Neumann, 1998; Anderson and Grove 1987). And in the Sangha basin, it is true that both French colonials and conservationists were powerful outsiders with a direct interest in access to Sangha basin resources. But their interests differed substantially, and hence the WWF did not simply resurrect colonial visions of the forest and Mpiemu migrations. French colonial officials and concessionaires sought to marshal African labor so as to exploit Sangha basin resources; WWF conservationists, however, have attempted to "protect" these resources from human exploitation. Thus, these distinct goals for resource use produced strikingly different approaches to and visions of a Mpiemu past. The French perceived the age-old Mpiemu practice of migration as a problem because it denied them control over their labor. Accordingly, Mpiemu people needed to "forget" their past practice of moving and settle docilely in their home places of *metego Mpiemu*. But the WWF has seen this history of migration as problematic because "migrants" were responsible for the overexploitation of natural resources. Conservationists therefore encouraged Mpiemu spatial mobility in one direction—out of the Dzanga-Ndoki park and Dzanga-Sangha reserve.

Mpiemu speakers, too, interpreted and reappropriated these colonizers' and conservationists' histories of mobility and stasis and thus illuminate both continuities and discontinuities in their century-long responses to environmental interventions. Whereas some living in Kwapeli, part of a colonially constructed *metego Mpiemu*, embraced the advantages of mobility, others in Lindjombo recalled their past movements and settlements in order to reject, not embrace, the past.

Prior to independence, Mpiemu people evidently gave little credence to French official insistence that they settle inside colonially designated districts, and thus they persisted in older historical practices of mobility. Between 1920 and 1959, Mpiemu responded to French attempts to stabilize them by continuing to move. Movement permitted escape from sleeping sickness, heavy taxes, and insupportable labor demands, but it also propelled Mpiemu toward new opportunities fostered by European *patrons*. Indeed, for those who had left *metego Mpiemu* during these decades and subsequently returned, this movement constituted part of a long history of wandering and settlement. It reinforced the perception that mobility could be beneficial since it conferred access to new foods and consumer goods and provided a measure of autonomy from onerous French demands.

For those who left *metego Mpiemu* and had never returned, historical visions of movement differed substantially. Lindjombo Mpiemus' ambivalence toward historical spatial mobility seems to have resulted from the establishment of the Dzanga-Ndoki park and Dzanga-Sangha reserve in 1990. The establishment of the park and reserve profoundly affected Mpiemu residents of Lindjombo, sending a very clear message to Mpiemu and other "migrants" that they were not free to wander through the forests, to exploit resources within the park, to mine diamonds, to hunt protected species, or to use affordable (but ecologically destructive) snare traps. It is no surprise that Lindjombo Mpiemu, finding their old strategies of spatial mobility criminalized, their access to places in the forest curtailed, and their opportunities for taking advantage of the forest's wealth hampered, would perceive themselves as a "dead people". They thus remembered their past peregrinations to reject them, to place themselves outside of earlier historical dynamics, and to protest their lost access to the wealth produced in a contemporary global economy.

References

Anderson, B. 1991. *Imagined communities: Reflections on the origin and spread of nationalism.* Revised ed. London and New York: Verso.

Anderson, D., and R. Grove, eds. 1987. *Conservation in Africa: People, policies, and practice.* Cambridge: Cambridge University Press.

Anon. 1900a. L'expedition du Dr Plehn au Sud-Kamerun. *Le Mouvement Géographique* 7:85–88.

———. 1900b. La Règion des Concessions dans le Bassin de la Sangha. *Le Mouvement Géographique* 7:85–88.

——— 1901. Questionnaire. Centre des Archives d'Outre Mer (CAOM), 4(2)D1.

———. 1909. Les Basse vallées de l'Oubangui et de la Sangha. *La Géographie: Bulletin de la Société de Géographie* 19:353–66.

——— 1911a. Rapport de la Compagnie Forestière de la Sangha Oubangui. CAOM, A.E.F. 8Q30.

———. 1911b. Rapport de la Compagnie de la N'Goko Sangha. CAOM, A.E.F. 8Q20.

——— 1933. Rapport de tournée effectuée par le Chef de la Subdivision de Nola dans la région M'Bimou, du 8 au 14 Decembre. CAOM, A.E.F. 4(2)D56 bis (Tournées 1933).

Bahuchet, S. 1985. *Les Pygmées Aka et la Forêt Centrafricaine: Ethnologie ecologique*. Paris: SELAF.

Barth, F. 1969. *Ethnic groups and boundaries*. London: Allen and Unwin.

Bassett, T., and P. Porter. 1991. From the best of authorities: The mountains of Kong in the cartography of West Africa. *Journal of African History* 32:367–416.

Bravman, B. 1998. *Making ethnic ways: Communities and their transformations in Taita, Kenya, 1800–1950*. Portsmouth, NH: Heinemann.

Bruel, G. 1909. Les basses vallées de l'Oubangui et de la Sanga. *La Géographie: Bulletin de la Société* 19:353–66.

———. 1910. Les populations de la Moyenne Sanga. Les Pomo et Les Boumali. *Revue d'Ethnographie et Sociologie* 1:3–16.

Burnham, P. 1980. *Opportunity and constraint in a savanna society: The Gbaya of Meiganga*. London: Academic Press.

———. 1996a. Political relationships on the eastern marches of Adamawa in the late nineteenth century: A problem of interpretation. In *African crossroads: Intersections between history and anthropology in Cameroon*, ed. I. Fowler and D. Zeitlyn. Cameroon Studies, vol. 2. Oxford: Berghahn Press.

———. 1996b. *The politics of ethnic identity in northern Cameroon*. Washington, DC: Smithsonian.

———, and T. Christensen. 1983. Karnu's message and the "War of the Hoe Handle": Interpreting a Central African resistance movement. *Africa* 53(4):3–22.

———, E. Copet-Rougier, and P. Noss. 1986. Gbaya et Mkako: Contribution ethno-linguistique à l'histoire de l'Est Cameroun. *Paideuma* 32:87–128.

Canal, A. 1950. Rapport économique. CAOM. A.E.F. 4(3)D61.

Carroll, R. 1986. Status of the lowland gorilla and other wildlife in the Dzanga-Sangha region of southwestern Central African Republic. *Primate Conservation* 7:38–44.

———. 1988. Relative density, range extension, and conservation potential of the lowland gorilla (*Gorilla gorilla gorilla*) in the Dzanga-Sangha region of the southwestern Central African Republic. *Mammalia* 52:309–23.

———. 1993. The development, protectioin, and management of the Dzanga-Sangha Dense Forest Reserve and the Dzanga-Ndoki National Part in southwestern Central African Republic. Unpublished report.

Chadwick, D. 1995. Ndoki: Last place on earth. *National Geographic* 188:2–45.

Clifford, J. 1994. Diaspora. *Cultural Anthropology* 9:302–38.

Congo Français et Dépendances, Colonie du Moyen Congo. 1908. Rapport du 2ème trimestre, 29 Août 1908. CAOM, A.E.F. 4(2)D2.

Cooper, F. 1996. *Decolonization and African society: The labor question in French and British Africa*. Cambridge and New York: Cambridge University Press.

Coquery-Vidrovitch, Catherine. 1972. *Le Congo au temps des grandes compagnies concessionnaires, 1898–1930*. Paris: Mouton.

Cottes, C.A. 1911. *La Mission Cottes au Sud-Cameroun (1905–1908): Exposé des resultats scientifiques s'aprés des travaux de divers membres de la section française de la Commission de Délimitation entre le Congo Français et le Cameroun (frontière meridionale) et les documents Etudiés au Museum d'Histoire Naturelle*. Paris: Ernest Leroux,

Delobeau, J.-M. 1992. La colonisation française et une minorité ethnique de l'Afrique Centrale: Les Pygmées. In *Histoires d'Outre-Mer: Mélanges en l'honneur de Jean-Louis Miège*, offerts par l'Institut d'histoire de Pays d'Outre-Mer, vol. 2. Aix-en-Provence: Publications de l'Université de Provence.

Derby, L. 1994. Haitians, magic, and money: Raza and society in the Haitian-Dominican borderlands, 1900–1937. *Comparative Studies in Society and History* 36:488–526.

Dupré, G. 1982. *Un ordre et sa destruction*. Paris: ORSTOM.

———. 1985. *Les naissances d'une société: Espace et historicité chez les Beembé du Congo*. Paris: ORSTOM.

Dupré, M.C. 1995. Raphia monies among the Teke: Their origin and control. In *Money matters: Instability, values and social payments in the modern history of West African communities*, ed. J.I. Guyer. Portsmouth, NH: Heinemann.

Fairhead, J., and M. Leach. 1996. *Misreading the African landscape: Society and ecology in a forest-savanna mosaic*. Cambridge: Cambridge University Press.

Faure, H. 1937. Notes sur l'exploration de la Haute-Sangha. *Bulletin de la Société des Recherches Congolaises* 24:114–16.

Fay, J. Michael. 1993. Ecological and conservation implications of development options for the Dzanga-Sangha Special Reserve and the Dzanga-Ndoki National Park, Yobe-Sangha, Central African Republic. Report to GTZ, Mission Forestiere Allemande, Cooperation Technique Allemande.

Fourneau, L. 1900. Mission Fourneau: Rapport anecdotique. *Revue Coloniale* 6:1341–42.

Fowler, I., and D. Zeitlyn, eds. 1996. *African crossroads: Intersections between history and anthropology in Cameroon*. Cameroon Studies, vol. 2. Oxford: Berghahn Press.

Giles-Vernick, Tamara. 1996. "A Dead People"? Migrants, land, and history in the rainforests of the Central African Republic. Ph.D. diss., Johns Hopkins University.

Gouvernement Général de l'A.E.F. 1935a. Rapport trimestriel, 1er trimestre, 1935. CAOM, A.E.F. 4(2)D66.

———. 1935b. Rapport trimestriel, 2e trimestre. CAOM, A.E.F. 4(2)D66.

———. 1950. Rapport économique, 1950. CAOM, A.E.F. 4(3)D61.

Gouvernement Général de l'Afrique Equatoriale Française, Colonie du Moyen Congo, Circonscription de la Haute-Sangha, Subdivisions de Nola et Berberati (ancienne Kadei-Sangha). 1923. Rapport mensuel, année 1923, mois de Novembre. CAOM, A.E.F. 4(2)D37.

Gouvernement Général de Moyen-Congo, Haute Sanga. 1931. Rapport du 2e trimestre, 1931. CAOM, A.E.F. 4(2)D52.

———. 1935a. Rapport Trimestriel, 1er trimestre, 1935. CAOM, A.E.F. 4(2)D66.

———. 1935b. Rapport Timestriel, Année 1935, 2e trimestre. CAOM. A.E.F. 4(2)D66.

Gouverneur Général de l'A.E.F. 1935. Lettre au M. le Chef de Department de la Sangha. CAOM, A.E.F. 4(2)D66.

Goyemide, E. 1984. *Le silence de la forêt*. Paris: Hatier.

Gray, A. 1991. *Between the spice of life and the melting pot: Biodiversity conservation and its impact on indigenous peoples*. Document 70. Copenhagen: IWGIA.

Grimes, B., ed. 1992. *Ethnologue: Languages of the world*. 12th ed. Dallas: Summer Institute of Linguistics and University of Texas at Arlington.

Grinker, R. 1994. *Houses in the rainforest: Ethnicity and inequality among farmers and foragers in Central Africa*. Berkeley: University of California Press.

Gupta, A, and J. Ferguson. 1992. Beyond "culture": Space, identity, and the politics of difference. *Cultural Anthropology* 7:6–23.

Guyer, J.I., ed. 1995. *Money matters: Instability, values and social payments in the modern history of West African Communities*. Portsmouth, NH: Heinemann.

Harms, Robert. 1981. *River of wealth, river of sorrow: The central Zaire basin in the era of the slave and ivory trade, 1500–1891*. New Haven and London: Yale University Press.

———. 1987. *Games against nature: An eco-cultural history of the Nunu of Equatorial Africa*. Cambridge: Cambridge University Press.

Headrick, R. 1994. *Colonialism, health, and illness in French Equatorial Africa, 1885–1935*, ed. Daniel R. Headrick. Atlanta: African Studies Association.

Hitchcock, R. K. 1994. International human rights, the environment, and indigenous peoples. *Colorado Journal of International Environmental Law and Policy* 5(1)1–10.

Kalck, P. 1974. *Histoire de la République Centrafricaine des origines préhistoriques à nos jours*. Paris: Èditions Berger-Levrault.

Kretsinger, A. 1993. Recommendations for further integration of BaAka interests in project policy, Dzanga-Sangha Dense Forest Reserve. Manuscript.

Lanly, J. 1966. Inventaire forestier en République Centrafricaine. *Revue Bois et Forêts Tropiques* 107:33–56.

———. 1966. La forêt dense Centrafricaine. *Revue Bois et Forêts Tropiques* 108:43–55.

Linden, E. 1992. The Last Eden. *Time* Vol. 140 (July 13):42–48.

Loyre, E. 1909. Les populations de la Moyenne-Sangha. *Questions diplomatiques et coloniales,* 28:406–20.

Massey, D. 1994. *Space, place, and gender.* Minneapolis: University of Minnesota Press.

Mololi, A. 1994. Rapport de l'étude sociolinguistique comparée de deux langues minoritaires bantu: Le Kaka ou Linzali et le Mpiemo. Manuscript.

Muston, E. 1933. Petit journal de la mission de délimitation Congo-Cameroun, 1905–1907, *Bulletin de la Société des Recherches Congolaises* 19:7–155.

Neumann, R.E. 1998. *Imposing wilderness: Struggles over livelihood and nature preservation in Africa.* Berkeley, Los Angeles and London: University of California.

Nzabakomada-Yakoma, Raphael. 1985. *L'Afrique Centrale insurgée: La guerre du Kongo-Wara, 1928–1931.* Paris: L'Harmattan.

Ponel, E. 1896. La Haute Sangha. *Bulletin de la Société Géographique* 17:188–211.

Pouperon, P. 1908. *Deuxième mission de P. Pouperon dans la Sangha.* Alger: IMP, Typo-Lithographique S. Léon.

Projet Dzanga-Sangha WWF-US/RCA. 1993. Bienvenue à la Reserve Spéciale de Forêt Dense de Dzanga-Sangha et au Parc National de Czanga-Ndoki.

Ribot, J. 1998. Theorizing access: Forest profits along Senegal's charcoal commodity chain. *Development and Change.*

Saragba, M. 1994. La trypanosomiase humaine en Oubangui-Chari: Son extension pendant la periode coloniale. *Ultramarines* 9:11–18.

Sautter, G. 1964. *De l'Atlantique au Fleuve Congo: Un géographie du sous-peuplement.* Paris: Mouton.

Schmink, M., and C.H. Wood. 1992. *Contested frontiers in Amazonia.* New York: Columbia University Press.

Schroeder, Richard. 1999. Geographies of environmental intervention in Africa. *Progress in Human Geography* 23:359–78.

Sintas, P. d'A. 1944. Extraits du rapport de visite pastorale de Mgr Sintas en pays Mbimou du 8 au 16 octobre 1944. Personal archives of Père Ghislaine de Banville, Congrégation Spiritaine, Chevilly-la-rue, France.

Siroto, L. 1969. Masks and social organization among the Bakwele people of Western Equatorial Africa. Ph.D. diss., Columbia University.

Spear, T. 1993. Introduction to *Being Maasai: Ethnicity and identity in East Africa,* ed. T. Spear and R. Waller. London: J. Currey; Athens: Ohio University Press.

———. and R. Waller, eds. 1993. *Being Maasai: Ethnicity and identity in East Africa.* London: J. Currey; Athens: Ohio University Press.

Taylor, P., and R. Garcia-Barrios. 1995. The social analysis of ecological change: From systems to intersecting processes. *Social Science Information* 34:5–30.

Telesis, USA, Inc. 1991. Sustainable economic development options for the Dzanga-Sangha Reserve, Central African Republic. Final report prepared for the World Wildlife Fund and the PVO-NGO/NRMAS Project, Washington, DC.

Territoire de l'Oubangui Chari, Région de la Haute Sangha, District de Nola. 1950. Rapport économique, 1950. CAOM, A.E.F. 4(3)D61.

Thomas, J.M.C. 1963. *Les Ngbakas de la Lobaye: Le Dépeuplement rural chez une population forestière de la République Centrafricaine.* Paris: Ecole Pratique des Hautes Etudes.

U.S. Department of State, Bureau of Public Affairs. 1989. *Background notes, Central African Republic.* Dept. of State Publ. 7970, Background Notes Series. Washington, DC. GPO.

Vail, L. ed. 1991. *The creation of tribalism in Southern Africa.* Berkeley and Los Angeles: University of California Press.

Vansina, J. 1990. *Paths in the rainforest: Toward a history of political tradition in equatorial Africa.* Madison: University of Wisconsin Press.

Vingarassamy, L. 1923. Lettre au monsieur le lieutenant-gouverneur du Moyen Congo (affaires civiles). CAOM, A.E.F. 4(2)D37.

Willis, J. 1992. The makings of a tribe: Bondei identities and histories. *Journal of African History* 33:191–208.

Fueling War: A Political-Ecology of Poverty and Deforestation in Sudan

Cindi Katz

Civil war has resumed in Sudan with a vengeance. In 1983, following an eleven-year hiatus, a series of aggressive acts by the increasingly desperate Nimeiri government made wastepaper of the 1972 Addis Ababa peace accords and recommenced the war that had simmered and raged since 1955, the year just prior to Sudan's independence from Britain. Over the last two decades, the war has resulted in the deaths of nearly two million—mostly southern—Sudanese people and forced the dislocation of upward of four million southerners. Conservative estimates are that the civil war costs the drastically indebted government of Sudan more than a million dollars a day. The war has made Sudan an international pariah, earning its government the condemnation of Amnesty International, scores of humanitarian aid groups, and virtually every relief worker on the ground. In response to the war, and the broader politics of the Islamic fundamentalist Sudanese state, the United States, Canada, and most European governments have withdrawn almost all "development" assistance and investment apart from relief. The International Monetary Fund (IMF) and World Bank remain on the scene, however, as they continue to try to whip Sudan into line with ongoing structural adjustment measures that have taken a punitive toll since the late 1970s.

The environmental effects of the civil war have been well documented in the popular press and elsewhere. Some places have suffered the effects of a scorched earth policy—towns, villages, fields, and pastures have been bombed, burned, and devegetated, while elsewhere serious environmental degradation has resulted from fighting, the abandonment of settlements, and the wholesale displacement of hundreds of thousands of people and animals, resulting in the disruption of migratory, grazing, and agricultural patterns. All of these problems have been exacerbated by recurrent drought and governmental restrictions on the interventions of relief organizations. My concern here, however, is on a less extreme arena of environmental

Thanks are due to Mohamed Salih and Rick Schroeder for their helpful comments on an earlier draft of this essay, and to Dusana Podlucka for her energetic and careful research assistance. I am grateful to PSC-CUNY for a faculty research grant that enabled me to return to Sudan in 1995, and to el Haj Bilal Omer, who graciously hosted me and astutely informed me about the situation in Sudan; together, they made possible the field research that informs this essay.

degradation that nevertheless has a peculiar relationship with the prosecution of Sudan's civil war. I will focus on two sites, one a village I call Howa, located on the Dinder River in northern Sudan, where I have done fieldwork intermittently since 1980[1] and the other the area around the White Nile town of Renk in the Upper Nile Province of southern Sudan, which I have never seen but which has come to figure prominently in the economic life of Howa. The connections between these disparate and distant sites (figure 1) are political, economic, and environmental. Unraveling them, as I will do here, reveals the northern government's hand and interest in the felling of old growth and other forests around Renk and a sinister dovetailing between these practices and more local interests elsewhere, including Howa and villages like it and Sudan's growing urban centers.

Dispersing destruction

The deforestation of the area around Renk suggests a perverse production of nature twinned with the enduring production of poverty, north and south. This production of nature and poverty are the harvest of misdevelopment, the ongoing war, and the extraordinary toll of two decades of structural adjustment measures imposed upon Sudan. The problems that concern me here are rooted in part in the unfolding of a state-sponsored, multilaterally financed, agricultural development project that incorporated Howa and its surroundings in 1971. The Suki Agricultural Development Project was an environmental intervention in the obvious sense of its provoking dramatic changes of land use in the 35,700 hectares that it embraced. But these very changes—which, among other things, circumscribed the area of woodland and pasture in the vicinity both in size and quality and forced the alteration of agricultural practices—also led to pronounced environmental shifts because they propelled the commodification of formerly free or commonly held goods and required new allocations of household labor. These shifts, in

---

[1] This chapter draws on field research I conducted in Sudan in 1979, 1980–81, 1983, 1984, and 1995. There are a number of reasons for the long hiatus between field visits. At first, I did not return because I had reoriented my work to look at the questions of social reproduction that concerned me in Sudan in New York City for political reasons and comparative purposes. By the time I was ready to return to Sudan, the government of 'Umar al-Bashir, in partnership with the National Islamic Front under the leadership of Hasan al-Turabi, had come to power. This regime, which has remained in power since 1989, spent its first few years consolidating power in a heavy-handed way, including a visible military presence on the streets of the capital, the intensification of the war against the south, the development of an extraordinary surveillance and policing apparatus, and, on a personal note, the imprisonment, oppression, and eventual exile of a number of my friends and colleagues. Under these circumstances, I could not in good conscience visit Sudan. When some of the worst excesses appeared to have abated and I learned that my friend, the human rights activist 'Ushari Ahmad Mahmud had been released from prison, I made plans to return. Thus, this work represents findings of research I conducted between 1979 and 1984 and in 1995. What is reported for the interval when I was not visiting Sudan is based on what people in the village, the agricultural project, and Khartoum University told me as well as, of course, secondary sources.

Figure 1. *Map of Sudan*

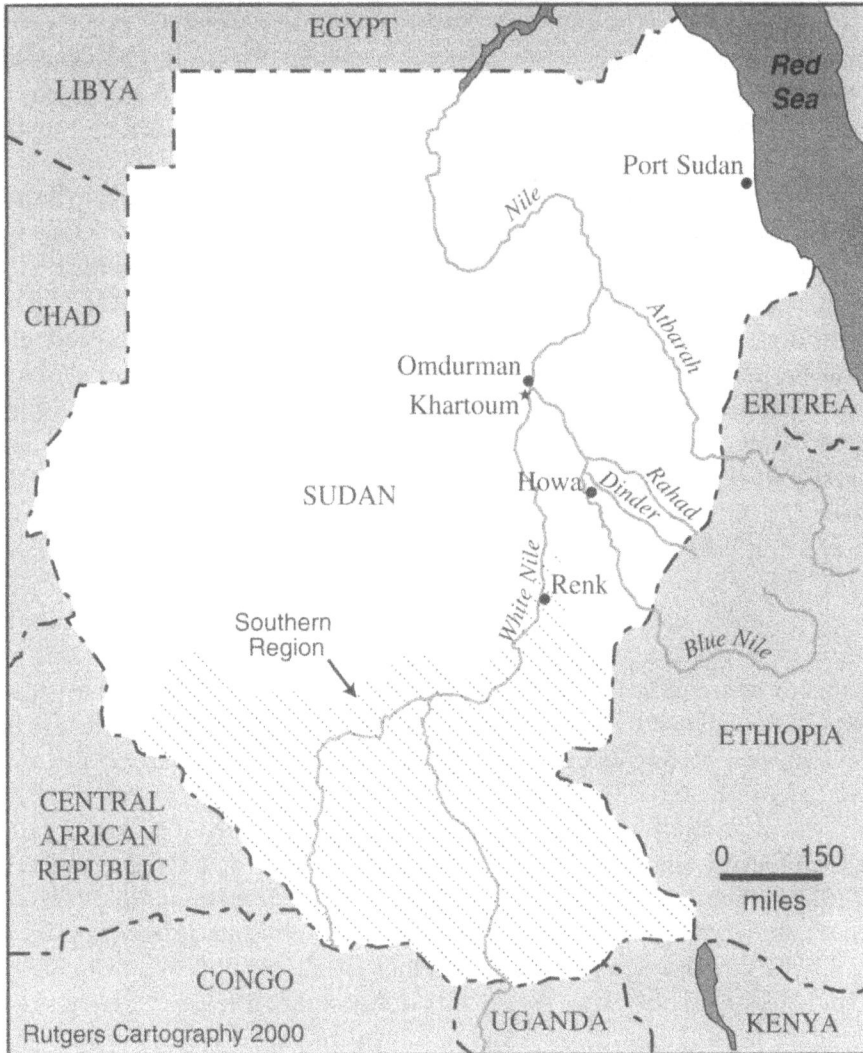

turn, were complemented and exacerbated by the intensification of the local area's articulation with the cash economy, which was, of course, part of the implicit intent of the project. In other words, the Suki Project provoked a political-economic and environmental squeeze on tenants and other residents in its area, which was redressed by practices that as often as not exacerbated the very environmental problems that had engendered them.

It comes as no surprise that under these conditions the time-space of everyday life has been reworked. In Howa, for instance, the talked about world has expanded dramatically in the last two decades. Whereas before my friends there expressed little interest in the fine points of my goings and comings in Khartoum, when I returned in 1995 they wanted to know what street the university guest house was on and got into the intricacies of how to locate the home of my research assistant, 'Intisar. More striking, in a year of living there in 1980–81 and in subsequent visits in 1983 and 1984, I cannot recall a single person mentioning, let alone visiting, the southern Sudanese town of Renk. By 1995, at least a quarter of the men in the village were spending much of the dry season engaged in the energetic charcoal trade around this White Nile town more than two hundred kilometers away. Because of extensive deforestation in the area around Howa, men there found it necessary, and even lucrative, to pursue charcoal production and trade in the relatively well forested area around Renk. This practice, which had many troubling elements undergirding it, reflected a relatively benign form of dislocation provoked by the environmental interventions of the agricultural project and its entailments. By going to Renk, men in and around Howa were able to preserve much of the seasonal labor pattern that had sustained them for decades—rainy season cultivation complemented by forestry work during the dry months following the harvest. In this expanded field, the landscape, though not the practices of production and reproduction, had been refigured entirely in under fifteen years. The trek to Renk was not the only vector along which the geography of Howa expanded, however. Many more households than before relied on the remittances of members working seasonally or permanently in the towns of Sudan or internationally. This sort of labor migration, long a staple of other parts of Sudan, had not featured in Howa much until the late 1980s, and in the span of a decade its importance to the local political economy had become compelling.

The resumption of war in Sudan was provoked by the discovery in the early 1980s of significant oil reserves in the south and the plan of the northern-dominated government to build a refinery in the north. This affront to the south was exacerbated in 1983 by Nimeiri's desperate imposition of Shari'a law throughout Sudan, whether Islamic or not. The historically uneven relationship between Sudan's north and south has been fissured along the lines of race, religion, and resources, and the discovery of oil in the area around Bentiu promised yet another galling instance of the resources of the south being exploited for northern gain. Stymied in their intent to construct a refinery in the north, the government decided to build a nearly 1,500 kilometer pipeline from the Bentiu oil fields to Port Sudan and export crude oil. The civil war halted any development of Sudan's oil capacity, and instead of at least achieving self-sufficiency Sudan has remained dependent on imported oil and petroleum products. According to available figures, oil imports absorbed approximately 72 percent of Sudan's foreign exchange budget in 1995, and Sudan's dependence upon imported petroleum products has led to periodic fuel shortages and chronic difficulties with procuring basic commodities such as cooking fuels, electricity, and gasoline. Even the wealthy have had enormous difficulties in procuring reliable household energy supplies. Urban frustration often ran high in Sudan, fueled by such things as chronic power cuts, lack of supplies, and inadequate infrastructure. Keeping such frustration in check, if not entirely appeased, was obviously an important goal of any government hoping to stay in power.[2]

The war, the reworked geographies of everyday life I found in Howa, and the pervasive urban frustration that has characterized Sudan since the late 1970s are connected. They are pieces of a politicized ecology of deforestation that I want to pull apart a bit more and then reassemble. Along the way, I hope to demonstrate how the everyday social, economic, and environmental practices of differently placed social actors work in concert and operate at cross-purposes to devegetate particular and strategic forests in southern Sudan and examine the critical intersections of these practices across scale. I will argue that the confluence of the interests of the state, the rural poor, and the urban elite has conspired to destroy the environment around Renk in potentially disastrous ways. Analysis of this assemblage offers an illuminating perspective on how environmental intervention and socioeconomic displacement are intertwined and are as much produced by as productive of poverty and underdevelopment.

"Development" and the shifting terrain of production

The Suki Agricultural Development Project, launched during 1971, was intended to incorporate a number of villages in the Blue Nile Province more firmly into the cash economy through the contract farming of cotton and groundnuts for the world market. It was also a vehicle for resettling people from western Sudan and sedentarizing various nomadic groups in the area; the former was more successful than the latter. It was the first agricultural project of the May Revolution, as the regime of Ga'afar Nimeiri (1969–85) was known, and as such was motivated more politically than ecologically if I can separate the two. According to the geographer Salih el Arifi, the Suki Project was a "rush job" intended to show the Blue Nile population that the new government cared about them. While the extension of irrigated cultivation had been on the books for this and other parts of the Blue Nile basin since prior to independence, project planners drew on a six-year-old survey of the area by the British consulting firm Hunting Technical Services to demarcate the project's boundaries without recognizing or compensating for

---

[2] A brief anecdote might give this "urban frustration" some texture. In 1982, I received an amusing but agitated letter from a friend at Khartoum University recounting the trials and tribulations of filling his *ambuba* or butagas tank. He began at 4:20 in the morning just to reserve a place in the queue and was told he had to stay and "guard" his tank until the lorry came to take it to the depot and then wait until it came back. The grating thing was that nobody knew when any of these operations would take place, and two days were wasted in the process. This was a scholar who published his work internationally, taught full time, ran a linguistics institute, and was a single parent of a young child. His harried academic life was compounded by a daily grind replete with almost daily electricity cuts, constant fuel shortages, intermittent and unreliable telephone service, and frustrations like the *ambuba* line. And that was in 1982, before things in Sudan got really bad.

its errors in land classification. Hunting had identified large tracts as "vacant" when they were important seasonal grazing areas for a number of pastoral groups that happened to have been elsewhere when the survey was conducted (cf. MacDonald et al. 1964). Among other things, the project cut through one of the key migratory routes of the Rufa'a el Sharq, whose winter grazing pattern east of the Blue Nile took them southward through the area. Besides building land use conflict with great potential for environmental degradation into the Suki Project's foundation, the Ministry of Agriculture, in its haste to make demonstrable progress, overlooked a contouring job so shoddy that the size of the project had to be reduced from 50,400 cultivated hectares to 33,600 (el Arifi, personal communication, 1979). Even at that, there remained drainage and irrigation problems in many parts of the project because in an area noted for poor drainage it was inappropriately leveled and graded at the outset (cf. Barbour 1961).

The political ecology of Howa and its surroundings was altered dramatically in the process. One thousand and fifty hectares of land that had been used in a mixture of dryland cultivation, grazing, and forestry were cleared and graded for the irrigated cultivation of cotton and groundnuts as cash crops in 100 percent annual rotation. The lack of any fallow period increased the dependence upon imported chemical fertilizers, and by precluding the cultivation of any food crops, particularly the staple sorghum, the project management rendered the participating population increasingly vulnerable to an inflated and unreliable grain market. The Suki Project also ushered in a new form of accounting with striking parallels to what has become known as post-Fordism. Unlike any other agricultural project in Sudan at that time, the Suki Project pioneered the "individual account system", which placed almost all of the risk and expenses on the individual tenant, rather than the tenants as a group or the state. The accounting system was intended to encourage harder work. But, despite the tenants' extraordinarily hard work, the project and its accounting system led to fantastic levels of immiseration and exacerbated socioeconomic differentiation in and around Howa.

The problems facing tenants were quintessentially political-ecologic. The state-sponsored agricultural project altered customary land rights, and the environmental changes it engendered created shifts in access to local environmental resources and altered the nature and quality of those resources. The new landscape of production and reproduction required increased labor time of households with diminishing results, among other things. As I have discussed elsewhere, the impress of the state and market through the aegis of the Suki Project also resulted in a reworking of the allocation of household labor along gendered and age-related lines, the redefinition and heightened commodification of environmental resources, the sharpening of existing socioeconomic differences, and the production of new forms of differentiation within the village (e.g., Katz 1991a, 1991b, 1994). This essay examines the intersections between state, local, and urban interests in the ex-

Figure 2. *Photograph of Dinder River showing absence of trees*

(Photo: Cindi Katz)

pansion of both the geography of charcoal production and the extent of local people's involvement in it as a means of coping with the altered ecologies and political economy of their everyday lives in Howa.

The transfer of more than a thousand of hectares of the best land in the immediate area of the village from mixed land use to intensive cultivation exacerbated pressure on the woodlands and pastures that remained around the village. While the best available estimates put the annual rate of defor-estation in Sudan at about 1.1 percent of total forests, suggesting a wide-spread and serious problem, the gravity of the situation around Howa was relatively new (Sudan Update 1995a, 1995b). The Hunting Technical Services survey had shown "thick" woodlands along the rivers in the area and even the nearby *kerrib* (gullied clay) lands were depicted as forested (MacDonald et al. 1964). Likewise, a 1970 study in the vicinity by the geographer Gelal el Din al Tayib (1970) had indicated that the local population used wood felled within a day's walk of their villages from "naturally regenerated" wood-lands. Perhaps even more striking, a description of the area in the late 1950s by the British geographer K.M. Barbour (1961) indicated that areas reserved for charcoal production in the "peninsula" between the Dinder and Rahad Rivers were little used and that there was "no use being made at present of the *Acacia nilotica* belts along the water courses, though they encourage so many birds to roost during the summer that cultivation near them has been almost abandoned" (193; cf. figure 2). As might be suggested by the photo-graph in figure 2, by the time of my stay in 1981—only ten years after the inception of the Suki Project—local woodlands were severely deteriorated. Local residents remarked on this all the time. They had begun cutting immature and less desirable trees for their household needs and going further from home or using poorer quality species for charcoal production (cf. Katz 1991b).

This situation was increasingly common throughout Sudan. During that same period, for instance, there were a number of agricultural development projects established, particularly in northern Sudan. According to Kamal Hassan Badi, then director of Sudan's Forests Administration, virtually all of the substantial expansion of land in agriculture between 1967 and 1981 (from 1.3 percent of Sudan's total land area in 1967 to 3.4 percent in 1981) was at the expense of forests. Over 50,000 square kilometers Sudan's "natural" forests were lost during that period (Madut 1983, 9–10). What has ensued since that time is even more serious. While accurate figures are difficult to obtain, the situation was clearly serious. According to one estimate, about 25 percent of Sudan's total land area was forested in 1982 whereas by 1992 the area forested was said to be only 18.6 percent (Planting Trees 1982; Sudan Update 1995b). These figures suggest a loss of over 142,000 square kilometers of forest and woodland in a decade. According to the 1994 *World Development Report*, the forested areas of Sudan had declined from 478,000 square kilometers in 1980 to 430,000 a decade later. This puts the total loss over the decade at a more modest, but still huge, 48,000 square kilometers, with a rate of deforestation of 4,800 a year (Sudan Update 1995a). The declines were almost entirely the result of land use practices that varied geographically such as the expansion of cultivated areas; the increase in herd sizes and intensification of grazing in certain areas as a result of altered migratory routes, elimination of pasture areas, or drought conditions elsewhere; and increased firewood and charcoal consumption (Khairi 1985; Sudan Update 1995b; cf. Schlesinger and Gramenopoulos 1996). Figures from the United Nations Food and Agricultural Organization suggest that per capita firewood consumption increased by over 10 percent between 1967 and 1981, and the Sudan Forest Administration estimated that charcoal consumption increased eightfold between 1963 and 1977 (Planting Trees 1982). Since then, increases have been even more dramatic. For instance, according to the World Bank, between 1987 and 1988 charcoal consumption increased in Sudan from 32 to 46 million cubic meters (Sudan Update 1995b). This astonishing shift was largely the result of Sudan's inability to afford the oil imports necessary to provide cooking and other fuel for its increasingly urbanized population. Under these conditions, which were the fallout of enormous debt, structural adjustment, and the renewed civil war, more than 80 percent of the population was said to rely on wood fuels for cooking by the early 1990s (Sudan Update 1995b). The continuation of these conditions called for investments in reforestation, which, because of the same constellation of conditions, were not forthcoming.

Grazing lands in the area of Howa and elsewhere had also declined. Research indicates that even before the Suki Project dry season pasture areas were inadequate to sustain the local livestock (e.g., Salih 1989). Large stockholders often farmed animals out to nomadic relatives or a hired herder during the dry season prior to the establishment of the project. But, according to local accounts, the problems were worsened by the agricultural project when pastures were incorporated in the cultivated area and wastelands were made to serve as pastures (cf. Koch and Bischof 1982). The limited and poor-quality pasture left in the vicinity after more than a thousand hectares were appropriated for the Suki Project spurred one of the more striking forms of resistance I found in the course of my research: encouraging animals to graze on the ripening cotton. The reasons for this practice were several; most tenants' cotton yields were inadequate to see any profit after the costs for all expenses were deducted under the individual account system, which reckoned all production costs—whether for cotton or groundnuts—from the cotton account. Allowing one's animals to eat the crop was a conscious attempt to ensure that household members' farming efforts went to their own benefit rather than the project or the government's cotton marketing board. The practice was a means of redirecting the value of their labor and land to serve their own interests and was an effective resistance strategy to the project (cf. Ali 1985a). Not only were resident animals better off from browsing this high-quality fodder, but the project administration had to deploy guards in the fields to protect the ripening cotton from the very farmers who cultivated it. This strategy notwithstanding, owners complained that their animals were ailing and dying in unprecedented numbers as a result of grazing on poor-quality and even toxic plants because of the constricted pasture area that resulted from the project. While precise labor accounts are unavailable, it was apparent that herdboys had to work harder and longer than their counterparts had in the past just to get their flocks adequate nourishment throughout the year. The days were particularly long by the end of the dry season, when the landscape was parched, crop residues all but consumed, and the humid heat most oppressive. Brothers often spelled one another during this period so that the animals could be grazed as long and as far as possible without endangering the well-being of herdboys.

The declines in woodlands and grazing areas were interconnected, of course, and "fed" on each other, as it were—with fewer acacia trees in the vicinity, the animals lost a high-quality source of fodder and grazed more intensively elsewhere. By the same token, shepherds thought nothing of letting their goats browse on what saplings there were, and by the end of the dry season in April and May they even climbed trees to cut what few branches had eluded woodcutters and hungry goats. This situation, provoked largely by the shifts in land use associated with the agricultural project and the influx of new people to work in or become tenants in the area— western Sudanese were resettled there to take up tenancies in the Suki Project and Ethiopian and Eritrean refugees drifted in from their settlement camps nearby in search of work (and wood)—made the local ecology susceptible to the prolonged drought of the 1980s. Many households in Howa, as elsewhere in Sudan, lost their entire flocks during the most dire year of drought in the 1984–85 season (cf. Mahran 1995). These environmental problems led to and were compounded by the commodification of formerly free or commonly held goods such as wood and other tree resources and, to the horror of locals, milk. As households' need for cash increased in the face of widespread commodification, woodcutting and charcoal production became ready means of earning money, and thus the pace and intensity of cutting increased (Katz 1991a, 1991b).

This situation continued through the 1980s and early 1990s with various ups and downs. Village leaders told me that following the 1984–85 drought, when the coalition government of Sadiq al-Mahdi came to power, the state's presence in the local area was nearly totally absent for a number of years while the Mahdi government struggled to stay in power and prosecuted the war in the South and then the Bashir government came to power. Mohamed, a farm tenant and community leader with strong ties to the original project management under Nimeiri's Sudanese Socialist Party, and one of my key informants, told me that in the years after Nimeiri fell and before Bashir consolidated power for the National Islamic Front, the Suki Agricultural Project was essentially left to its own devices with few state inputs and virtually no staff. These circumstances enabled farmers to divert irrigation water as they wished and to grow sorghum much more widely in an agricultural free-for-all that lasted several years. In addition, as local residents struggled to restock after the drought and famine of the mid-1980s, the area was hit by a series of diseases that by several accounts wiped out much of the local livestock in the early 1990s. The high, and often staggering, rates of inflation throughout the period coupled with the loss of food subsidies and the imposition of other structural adjustments by the IMF, which began in the late 1970s and has lasted through the 1990s, ensured that local residents had to redouble their efforts to earn cash. Reliance on charcoal production and sales remained a key and stable way to earn a living throughout the dry season. The testimony of local residents suggests that it took on special importance in years when the inputs, support, and marketing infrastructure for cash crop production were unreliable or absent and when domestic animals were failing. These were, of course, the same years that urban demand for charcoal was surging.

When I returned in 1995, the agricultural project was back on (a new) track, the animal population had not recuperated—local large holders who had managed to restock kept their animals almost permanently out of the vicinity with family or hired herders—and an increasing number of men were relying on charcoal production to earn money. Most striking to me, caught up with documenting and analyzing the shifting political ecology of production and reproduction in Howa, were the ways that many of the everyday activities of subsistence procurement and income generation remained the same as before *but* required a tremendously expanded geography to be viable. Most adults in the village continued to engage in the old mix of agriculture, animal husbandry, and forest work (albeit with important supplements and changes), but the space of its enactment had expanded more than tenfold. Where people had walked or ridden a donkey for no more than a day to cut wood for fuel or charcoal production in the early 1980s, by 1995 they traveled 225 kilometers over poor roads to do the same work. Animals, when they were kept at all, were grazed increasingly for extended periods away from the village. As before, many young men worked as agricultural laborers in the project and further afield, but as a growing percentage came of age without access to a tenancy or other productive resources more young men worked in other agricultural projects and in the rain-fed sesame areas to the south. This constellation of activities relied upon and produced an expanded geography of everyday life that kept a growing number of young men away from the village for weeks if not months at a time. Yet for many people in Howa these alternatives were preferable to making their way permanently in the towns and cities of Sudan, which under the prevailing political-economic conditions offered them little. Nevertheless, the expanded geography increasingly included migration—both seasonal and more permanent—not only to Sudan's towns and cities but to the Gulf States, Libya, and Saudi Arabia as well. These sociospatial strategies, which for many households included the return of young men to assist with agricultural work during the rainy season, enabled the endurance of familiar patterns of work and had the tangential effect of reducing some of the potential pressure upon urban areas that might have ensued from the influx of growing numbers of disenfranchised rural residents. These issues will be addressed below.

War and the shifting terrain of production

For now, I want to turn to the interests of the Sudanese state, which is waging an erosive war on the South. The state, as I have mentioned, has been beset by crisis—internal and external—for the past two decades. Sudan, like most of Africa and elsewhere in the so-called Third World, has been pummeled by the punitive effects of IMF/World Bank-imposed structural adjustments since the end of the 1970s. Its economy has been in disarray (cf. Ali 1985b). Among other things, the Sudanese political economy has been unable to cope with the proletarianized people it has produced in the countryside more effectively than cash crops. Like many countries in Africa, the Sudanese economy has neither the employment possibilities nor the infrastructure to absorb its rapidly urbanizing population. Its problems have been compounded by the costs of its protracted civil war. The government's own "austerities" have included the reduction of imports to the bare essentials, a striking withdrawal of capital from virtually all productive activities, and exorbitant rates and new forms of taxation (as well as its more vigorous collection). Yet, the war's costs, estimated conservatively at a million dollars a day, have required the infusion of funds from a consortium of states and private interests supportive of the Sudan government's hard-line Islamic fundamentalism. The wastage is tragic in every way imaginable.

On this stage, the government has seen fit to authorize and encourage the cutting of old growth and other forests in the Upper Nile Province around Renk, among other places (figure 1). The policy and practice cover a number of important bases. First, the taxes and fees imposed—and collected—on charcoal production add significantly to the government's coffers. Second, the opportunity to cut wood for charcoal enables marginalized rural dwellers to remain in the countryside and out of the already overburdened towns. Third, the forests being cut are in "the South", and thus their destruction has been at the expense of the South and for the benefit of the North. All of these practices contributed directly to propping up the National Islamic Front government of 'Umar al Bashir and the cleric Hasan al-Turabi. On a more local level, Renk literally was "not on the map" for residents of Howa as recently as the mid-1980s. Yet when I returned in 1995 I found at least a quarter of the male population between the ages of eighteen and fifty going there regularly to cut wood, burn charcoal, or trade. This "new world" has helped enable the old to persist.

The exploitation of southern energy sources for the benefit of the North is an old and painful story in Sudan. One of the sparks that renewed the fighting in 1983 was the discovery of substantial reserves of oil (over 950 million barrels) in the South and the outrageous (Nimeiri) government decision to locate the refinery more than three hundred kilometers away in the northern White Nile town of Kosti. When this plan was shelved, it was not replaced with plans for a southern refinery but with plans to construct a pipeline for crude oil to the northern city of Port Sudan. By the end of 1999 the latter had (precariously) come to pass. The southern leadership and population remained adamantly in favor of constructing the refinery in the southern town of Bentiu near the fields, as Chevron originally had proposed.

In a similar vein, most of the best forest reserves and resources were in the southern Sudan, where the majority of the economically exploitable stands of valuable tropical hardwoods such as mahogany and teak were located. While these forests were cut to meet local needs, they came under more significant exploitation beginning in the early twentieth century by the colonial government for construction timber, furniture, the railroad, and fuel for Nile steamers. The demands upon forest resources intensified during the Second World War, when new areas were opened up for cutting and conservation for the war effort (Wani 1983; Madut 1983; cf. Cline-Cole 1993). This use and "reservation" of southern forests for national (and imperial) interests set in motion a pattern that was naturalized and continued by colo-

Figure 3. *Photography of charcoal lorry*

(Photo: Cindi Katz)

nial and postcolonial governments alike. Numerous environmental analyses
over the years have pointed out that the South has a "surplus" of trees relative to its population, whereas the more arid and more populous North has
a deficit (cf. Whitney 1987). Thus, cutting in the North has routinely exceeded replacement rates, or the "allowable cut", for years. It was "natural",
then, in the current hostile climate, that southern forests were constructed as
an ideal site for resource accumulation at the national (i.e., northern) level.
The area around Renk, for instance, which has sandier soils than most of the
rest of the Upper Nile region, was one of the few areas of the region that remained well forested, and thus available for significant exploitation, in recent decades (Madut 1983). While southerners surely engaged in cutting
around Renk, the commercial development of the forests in this district
appeared to be securely in northern hands. The commercial operations included procuring charcoal from itinerant and local producers, securing the
permits to transport and sell it, transporting it to towns in the North (most
commonly the capital district), storing and warehousing it, and, finally,
selling the product in urban areas. Thanks to their access to capital—both
money and trucks—and to urban markets, most of these operations were
run by lorry-owning traders from the North (cf. Ribot 1998). Several such
traders from Howa participated vigorously and profitably in the process. In
addition, judging from what people in Howa and elsewhere told me, the
better part of the cutting and burning around Renk was also done by itiner-

Figure 4. *'Charconomics' for charcoal produced near Renk, Sudan in 1995*

Income per 50kg sack for member charcoal work gang	150[*]
Expenses per 50kg sack for charcoal trader	
Price paid to jobber	1,000
Government permit	1,000
Sack	150
Loading on Lorry	120
Misc. bribes and other expenses	180
Transport	1,000
Total	3,450
Charcoal sold by wholesaler in Khartoum at mark-up of 10% minimum	
Price received	3,800
Percent of total wholesale price received by charcoal gang worker	4%

[*] All figures in Sudanese Pounds (500£S = US$1.00 at 1995 exchange rates)

ant northerners who worked in gangs or independently. Still others went south during the season to work as lorry loaders and charcoal handlers.

The economics is compelling. Woodcutters, who worked in gangs, earned about 1,000 Sudanese Pounds (£S) (just under U.S.$2 at 1995 exchange rates) per fifty kilogram sack of charcoal produced. An individual could produce an average of one to two sacks of charcoal a day depending on conditions. Work gangs were usually paid by a trader who sold charcoal from Renk at a 100 percent markup on labor costs (i.e., the trader added 1,000£S per sack to the price paid to charcoal makers). (Charcoal from near the northern town of Dinder, by contrast, was sold by traders with a markup of 500£S.) More often than not, the trader owned a lorry and transported the charcoal—approximately two hundred sacks each run—to urban areas in the North, most commonly Khartoum, over four hundred kilometers away (figure 3). The trader paid a fee of 1,000£S per sack for a government permit, and, whereas in the past there was lax enforcement of all restrictions on cutting, the permits were checked at several points on the way to Khartoum. The post-1989 Sudanese government has been strikingly efficient in many matters, and tax collection is one of them. The permit fees collected on charcoal are a case in point. In addition to these labor, middleman, and permit costs totaling 3,000£S were miscellaneous expenses, including those for the sacks themselves, various bribes and payoffs, and loading fees, which together added an additional 500£S or so. The long-distance traders sold the charcoal delivered to Khartoum with a markup of 10 percent minimum. A sack of charcoal from Renk, then, was sold wholesale in Khartoum for about 3,850£S (or $7.50) in 1995. With the Renk woodlands moving to Khartoum at a fantastic clip, it was a lucrative trade especially for the long distance merchants (figure 4).

With this chain of events, the government of Sudan accomplished several ends—it collected a substantial income on the sale of permits to charcoal traders; it saved on limited hard currency that would be used to buy petroleum-based products—already the lion's share of foreign exchange expenditures—by ensuring an adequate, if less preferred, cooking fuel supply to urban areas; and it furthered its aggressive aims in the South, first by directing southern resources north—with most of the income accruing to northerners and northern institutions, and, second, by clearing southern forests, which the government handily claimed sheltered rebel fighters.

I will take these in turn. Using dated and thus conservative estimates of urban charcoal consumption, but assuming strict collection of taxes (i.e., that each kilogram consumed in towns was actually taxed—a feat even this government was not capable of accomplishing), I calculate that the government collected a minimum of $55 million a year during the 1990s in charcoalbased tariffs.[3] The reliance on charcoal for so much of its cooking fuel needs has enabled Sudan to hold down its potential expenditure on oil-based products, which have long consumed most of its export earnings. While in 1980 the government of Sudan spent 60 percent of its foreign currency earnings on petroleum products, this figure jumped to over 90 percent by 1981 for a number of reasons (cf. Kok 1992; Babiker 1981). Sudan remains an importer of oil and petroleum-based products despite its extensive oil reserves. These reserves have remained virtually untapped. The major oil licensee, Chevron, left Sudan in early 1984, a few months after the first salvos of the renewed civil war were fired (on them). The most recent figures indicate that in 1995 more than 72 percent of the total value of Sudan's export earnings was spent on petroleum products. Apart from the state's loss of foreign exchange earnings, but directly connected to the problem, petroleum-based cooking fuel supplies have been in short supply, out of reach for most urban household budgets or too inconvenient to procure for most of the past two decades in Sudan. At the same time, the use of charcoal throughout Sudan has increased tremendously. Thus, the steady supply of charcoal—even at prices of 4,000£S ($8.00) or more per sack—ensured that the urban population, already "infrastructurally challenged", had at least a reliable source of cooking fuel.

Finally, for the government of Sudan the cutting of southern woodlands was an act of war. In 1990, the Sudanese People's Liberation Army (SPLA)

---

[3] In an overview of energy use in Sudan, Abd el Bagi Abd el Ghani Babiker (1981) indicated that in 1979–80 charcoal consumption in Sudan was approximately 550,000 tons, almost entirely in the urban areas. By 1990, the urban population of Sudan had grown to 2.5 times what it was in 1979–80 (cf. Hassaballah and Eltigani 1995). To obtain rough figures for 1990 charcoal consumption, I conservatively assumed an equivalent growth in charcoal consumption and thus multiplied 550,000 by 2.5, which translated into 27.5 million fifty-kilogram sacks of charcoal. Collecting 1,000£S per sack, the government would earn £S27.5 billion. At 1995 exchange rates (500£S = U.S.$1.00), the charcoal trade would have netted the government $55 million in permit fees alone. This figure represents a formidable transfer of wealth to the government from the charcoal trade (cf. Ribot 1998). Literally "fueling" its war against the South.

controlled two-thirds of the South, including the area around Renk. In 1992, the government launched a major offensive, with the assistance of Iranian military advisers and Chinese weapons, and regained control of the northern part of the Upper Nile region (which includes Renk) and much of eastern Equatoria (figure 1). Seen in this light, the government's ambitions in the area around Renk went well beyond getting "the spoils". They were a means of creating and maintaining a steady presence there as well as commandeering southern resources for the North. The government and its agents have conveyed the message to people in Howa and villages like it that it is patriotic—even their duty—to clear forests around Renk and elsewhere in the South. The message has gotten through. In a number of interviews and casual conversations, people in Howa told me that these very woodlands were hiding guerrilla fighters. Given that in 1989 the civil war was declared a *jihad* and the village population, among many others in Sudan, appeared to be swayed by this claim, it is not too much to say that they constructed and understood the felling of the forests around Renk as duty to God and country.

Charcoal production was also a means of survival for the local population. Most men engaged in charcoal work gangs spent the better part of the dry season—in several week stints—around Renk, earning a substantial portion of their annual incomes. My research suggests that over a quarter of the men in Howa were engaged in charcoal production—a relatively higher number of younger men participated—most as cutters and loaders, but a small number—wealthier men with access to capital—were engaged as brokers and traders and were reaping fantastic profits from the vigorous trade. For most of the cutters and loaders, the sums were not great, but they did enable farmers—tenants and not—to survive outside, or on the margins of, the capitalist wage economy as quasi-independent producers.[4] While international and domestic labor migration has long been a common feature in Sudan of settlements along the Nile, in the Gezira, parts of the West and the South, it played virtually no role in the political economy of Howa prior to the mid-1980s. However, part of the expanded geography of Howa was witnessed in significant labor migration to either the Gulf States and Saudi Arabia or the urban centers of Sudan for the first time in its history. Even by the mid-1990s, only a relatively small number of residents migrated, and most of these were engaged in short-term return or cyclical migration. The charcoal trade was one of the economic supports that allowed villagers to

[4] Cf. Ribot 1998, which describes a complex patron-client relationship between cutters and merchants that operated in Senegal involving monetary advances to cutters, which ensured their sales to the lending merchant. While such crop mortgaging systems were common for farmers in Sudan, none of the charcoal producers or merchants in Howa mentioned a similar system for charcoal making. While such relationships may well have existed, they were probably relatively uncommon because the charcoal season followed on the heels of the harvest, when most farmers and farm laborers would have received at least some income to see them through the first weeks of charcoal work.

remain in the countryside or at least to straddle country and city as economic actors. As Peter Linebaugh (1976) argued with reference to forest thefts in nineteenth-century Europe, this sort of activity on the fringe guarantees the stability of capital accumulation—however halting—in the center. By employing a large sector of the increasingly landless and otherwise disenfranchised rural population, the southern charcoal ventures reduced pressure on the centers, their capitalists, and the government of Sudan to provide jobs, services, housing, and infrastructure. For young men in Howa and other villages within agricultural projects, itinerant charcoal work was also preferable to the government's other employment program, conscription in the "people's militia" to help wage the *jihad* further afield.

Conclusion

I have argued that contemporary Sudan is a country beset with crisis. Embroiled in a civil war, wracked by International Monetary Fund "structural adjustments", and almost uniformly ostracized in international arenas, the state and Sudanese population struggle to survive saddled with debt, inflation, and the effects of a corrosive war. These circumstances have fostered a peculiar, if not perverse, form of self-reliance that revolves around clearing old growth and other forests in southern Sudan. Woodcutting for charcoal production has become (1) a key means for the state to secure revenues, (2) one of the only reliable sources of dry season income for much of the disenfranchised and otherwise marginalized rural population of central Sudan, and (3) a hedge against the import of costly cooking fuel for those in the towns. The fundamentalist Sudanese state has authorized and even fostered this deforestation with patriotic appeals, assuring would-be woodcutters from the Islamic North that their work will help rout out southern guerrilla fighters hiding in the forests. Woodcutting has become an accessory to war.

War and the destruction of forests are not strangers. Agent Orange and Shakespeare's Great Birnum Wood are part of a long and sorry continuum. While scholars such as Paul Richards (1996) in Sierra Leone, Nancy Peluso (1992) in Indonesia, and Simon Schama (1995) in Northern Europe have shown how the forest itself can become an object of destruction in the course of war—whether to burn people out or destroy their livelihoods and homes—and others of a more Malthusian bent have suggested that the destruction of forests and forestry resources leads to war (an interpretation I reject), the government of Sudan has embarked on a different project. Having conquered the territory around Renk, the government asserted claims on it that helped to fund its continuing war against other parts of the South. The political ecology in this part of Sudan is one of a triangulated set of relationships that has helped to ensure (1) that the benefits of what Joe Whitney (1987) called "fuelwood mining" would continue to accrue to the North (and, I am suggesting, to its war chest in particular); (2) that contract farming areas of the North—themselves potential powder kegs of disenfranchisement—would remain relatively stable; and (3) that the urban areas of Sudan, most notably around Khartoum, would not be stretched further beyond their structurally adjusted limits.

The state's role in this configuration was substantial and deliberate. This configuration served capital accumulation, warmongering, and willful environmental destruction. It furthered immiseration, the loss of control over resources by local populations, death, and violence. The displacement of degradation to the South was proving to be a stealthy weapon of the government's war effort. But displaced degradation was only one part of the equation. Perhaps even more troubling was that the twinned practices of war and displaced deforestation were among the few options available in Sudan to absorb the increasing numbers of those from villages like Howa denied access to productive resources in their home environments thanks to the political-economic and ecologic squeeze produced by 'development' in the shadow of structural adjustment.

It is not in people's long-term interests in either the North or the South to allow these conditions of war, immiseration, and environmental degradation to continue. These circumstances heighten the poignant implications of the origins of the colloquial Sudanese Arabic word for collecting wood. The word, *fazaha*, is rooted in the Arabic word *faza*, which connotes a communal gathering in the face of danger. In the face of the present danger, which is grave, it is time for Sudanese northerners and southerners along with their friends and supporters to try to gather together and labor in concert to put their substantial resources to work for them in peace. For too long, the Sudanese people's resources and famous resilience have worked against them in war.

References

Ali, A.A. 1985a. The devaluation debate: A documentary approach. In *The Sudan economy in disarray: Essays on the crisis*, ed. Ali Abdel Gadir Ali. Khartoum: Ali Abdel Gadir Ali (distributed by Ithaca Press, London).

———. ed. 1985b. *The Sudan economy in disarray: Essays on the crisis.* Khartoum: Ali Abdel Gadir Ali (distributed by Ithaca Press, London).

Babiker, A.A. 1981. Energy in the Sudan: An overview. *Sudan Environment: Newsletter of the Institute of Environmental Studies, University of Khartoum* 1(3):3–4.

Barbour, K.M. 1961. *The Republic of Sudan: A regional geography.* London: University of London Press.

Cline-Cole, R.A. 1993. Wartime forest energy policy and practice in British West Africa: Social and economic impact on the labouring classes, 1939–45. *Africa* 1:56–79.

Hassaballah, H.O., and E.E. Eltigani. 1995. Displacement and migration as a consequence of development policies in Sudan. In *War and drought in Sudan: Essays on population displacement*, ed. Eltigani E. Eltigani. Gainesville: University Press of Florida.

Katz, C. 1991a. Sow what you know: The struggle for social reproduction in rural Sudan. *Annals of the Association of American Geographers* 81(3):488–514.

———. 1991b. An agricultural project comes to town: Consequences of an encounter. *Social Text*, 28:31–38.

——. 1994. The textures of global change: Eroding ecologies of childhood, New York and Sudan. *Childhood: A Global Journal of Child Research* 2(4):103–10.

Khairi, A.H. 1985. Call it conservation. *Sudanow*, July, 40–41.

Koch, W., and F. Bischof. 1982. Weed problems in irrigation schemes in the Sudan. In *Problems of agricultural development in the Sudan*, ed. Gunter Heinritz. Gottingen: Edition Herodot.

Kok, P.N. 1992. Adding fuel to the conflict: Oil, war and peace in the Sudan. In *Beyond conflict in the Horn*, ed. Martin Doornbos, Lionel Cliffe, Abdel Ghaffar M. Ahmed, and John Markakis. Trenton, NJ: Red Sea Press.

Linebaugh, P. 1976. Karl Marx, the theft of wood, and working class composition: A contribution to the current debate. *Crime and Social Justice* 6 (fall-winter): 5–16.

MacDonald, M., and partners. 1964. *Roseires Soil Survey, report 9: Dinder–Blue Nile Gezira Sennar to confluence, semi-detailed soil survey and land classification*. London: Hunting Technical Services.

Madut, A. 1983. A threat to our lives. *Sudanow*, September, 8–11.

Mahran, H.A. 1995. The displaced, food production, and food aid. In *War and drought in Sudan: Essays on population displacement*, ed. Eltigani E. Eltigani. Gainesville: University Press of Florida.

Peluso, N. 1992. *Rich forests, poor people: Resource control and resistance in Java*. Berkeley: University of California Press.

Planting Trees. 1982. *Sudanow*, August, 26–28.

Ribot, J.C. 1998. Theorizing access: Forest profits along Senegal's charcoal commodity chain. *Development and Change* 29(2):307–41.

Richards, P. 1996. *Fighting for the rain forest: War, youth, resources in Sierra Leone*. Oxford: James Currey.

Salih, M.A.M. 1989. Ecological stress, political coercion and the limits of state intervention; Sudan. In *Ecology and politics: Environmental stress and security in Africa*, ed. Anders Hjort af Ornäs and M.A. Mohamed Salih. Uppsala: Scandinavian Institute of African Studies.

Schama, S. 1995. *Landscape and memory*. New York: Knopf.

Schlesinger, W.H., and N. Gramenopoulos. 1996. Archival photographs show no climate-induced changes in woody vegetation in the Sudan, 1943–1994. *Global Change Biology* 2:137–41.

Sudan Update. 1995a. Vol. 6 (11) Dateline 12 July 1995.

——. 1995 b. Vol 6 (14) Dateline 12 September 1995.

al Tayib, G. 1970. *The southeastern Funj area: A geographical survey*. Funj Project Papers, no. 1. Khartoum: Sudan Research Unit, Faculty of Arts, Khartoum University.

Wani, K. 1983. Timber line. *Sudanow*, March, 38–39.

Whitney, J.B.R. 1987. Impact of fuelwood use on environmental degradation in the Sudan. In *Lands at risk in the Third World: Local level perspectives*, ed. Peter D. Little and Michael M. Horowitz. Boulder: Westview Press.

World Development Report. 1994. Published for the World Bank by Oxford University Press.

Producing Nature and Poverty in Africa: Continuity and Change

Richard A. Schroeder

Introduction

In the mid-1970s, Maasai residents of southern Kenya were abruptly relocated from land that was subsequently enclosed within Amboseli National Park, one of the continent's most famed wildlife reserves. In response, the displaced groups began a systematic effort to kill many of Amboseli's most prized tourist attractions, including dozens of leopards, elephants, and rhinos. This program of extermination was undertaken not for sport or profit but as part of a desperate protest campaign designed to counter the growing threat tour operations posed to Maasai land rights. Its effect was to expose the coercive conservation policies of the Kenyan state and force a temporary negotiated settlement more favorable to Maasai interests (Bonner 1993; Peluso 1993). In the mid-1980s, an equally oppressive set of state policies was adopted in Mali. Under the regime of President Moussa Traore, Mali's forest service implemented a series of draconian restrictions on rural residents' use of forest products. With fines for minor infractions against the new rules exceeding the rural per capita income in many instances and foresters entitled to 25 percent of fine revenues, implementation of the new policies was aggressively and ruthlessly pursued. Thus it was that forestry agents were "reportedly burned alive and chased from the countryside" (Ribot 1996) by rural residents when the Traore regime was overthrown in 1991. The upshot, once again, was a set of new policies that seemed to offer more hope to rural residents (Ribot 1999; Benjaminsen, this volume).

These episodes of political ecological conflict bracket a period of intense donor and state-led activity focused on the environment in Africa. While born of different circumstances centered on protecting endangered species and preserving dryland forest resources, respectively, the Kenyan and Malian case studies outlined above nonetheless share a number of common elements that shed light on the general upheaval of the period. Each involved conflicts between displaced or otherwise disenfranchised rural resource users and aggressive and acquisitive state agents acting on behalf of conservation interests. In both cases, policies explicitly geared toward securing important and potentially lucrative sources of state revenue provoked violent responses on the part of targeted rural groups. Finally, in each instance some measure of accommodation was eventually reached between the state and the groups in question (although the agreement involving Maasai land rights was short-lived). Thus, the case studies illustrate quite dramatically both the extent of state intervention in environmental management in Africa over the past two or three decades and the lengths to which Africans have gone to protect their interests against all manner of "outside" aggression. The conflicts in Mali and Kenya also suggest that new openings can sometimes grow out of the most unlikely of circumstances, leading to closer collaboration between historically antagonistic parties.

The contributors to this volume were brought together for a conference in 1996 to discuss the political ecological ferment of the two decades bounded by the events in Mali and Kenya. Specifically, conferees were charged with the task of analyzing the effects of recent policy changes on what the conference prospectus called "the topography of wealth and poverty on the continent." The essays gathered in this collection offer two broad readings of these changes. Some contributors see events such as the slaughter of the Amboseli rhinos and the reprisals against Malian foresters as the natural and somewhat inevitable consequences of a long and sordid history of heavy-handed environmental management on the continent. In these chapters, the emphasis is on the *continuity* of environmental intervention tracing back to the earliest stages of the colonial period, and the summary impression is that environmental policies have for decades done little to alleviate, and have at times contributed directly to the exacerbation of, poverty across the continent. Other contributors, by contrast, seem to be arguing implicitly that the events of the past two decades represent a significant departure from the past. They read the recent history of African natural resource management as an unprecedented escalation of environmental politics on the continent. And they have sought to analyze both the environmental interventions themselves and the specific responses different groups have made when they have felt their lives and livelihoods to be threatened. In these chapters, the net effect of policy changes has been more varied with local groups and outside interests prevailing by turns in the struggle to control natural-resource-based livelihood systems and the profits they generate.

These general approaches to the analysis of environmental change on the continent appear to be contradictory and incompatible in many respects, but they are in fact quite complementary. Indeed, both the long-range view of historical continuity and the more fine-grained political economic analysis of recent policy changes are crucially important if we are to properly understand exactly how nature and poverty are being coproduced on the continent.

Historical continuity

James Fairhead and Melissa Leach have taken the lead in challenging what they refer to as environmental orthodoxies in their chapter (see also Fairhead and Leach 1996, 1998; and Leach and Mearns 1997). Specifically, they document how mistaken or otherwise inadequate theories of equilibrium, population, and deforestation have had an enduring influence on the actual practice of environmental management in West Africa. They demonstrate how foresters in Ghana and Côte d'Ivoire have for decades misdiagnosed "problems" of forest change, exaggerated their scope, and prescribed inappropriate solutions. While they substantiate these claims in exhaustive detail in their chapter, they are perhaps most forceful in their contention that the science used to frame environmental problems in West Africa has shown striking continuity from the colonial period to the present. Moreover, the scientific models used to explain presumed patterns of landscape change in different parts of the subregion have, in their estimation, had persistent and deleterious knock-on effects on the social science of the region (cf. Cline-Cole's similar analysis of environmental policy in northern Nigeria). Most critically, they argue, these models have misrepresented the important contributions rural West Africans have made to forest enrichment and expansion in many areas.

The notion that contemporary environmental initiatives are fundamentally similar to colonial policies is shared by Roderick Neumann, who notes in his chapter the tendency to continue to structure analyses of local environmental practices against the backdrop of a discourse surrounding "good and bad natives". He assesses a number of contemporary practices in light of "their socioeconomic and political consequences" and sees a pattern that is clearly reminiscent of the colonial period (cf. Broch-Due's chapter). Thus, for example, he questions some of the recent rhetoric favoring broader political participation and more equitable distribution of the economic benefits of conservation. After a review of buffer zone projects organized around these principles, Neumann is forced to conclude that

> many of the projects sound alarmingly similar to the fortress-style approach to protected areas that they supposedly replace. Forced relocations, curtailment of resource access, abuses of power by conservation authorities, and increased government surveillance are reported more often than are successful integrations of local people into conservation management. ... In actuality, many buffer zones constitute a geographical *expansion* of state authority.

The compelling arguments Fairhead and Leach and Neumann offer regarding the remarkable continuity of environmental policies and practice on the continent over the past several decades are echoed to a certain extent in a number of chapters in this book. The struggle over territory between the North and the South in Sudan, for example, is currently, and seemingly always has been, about control over energy resources (Katz's chapter); the recent efforts to rehabilitate the Kondoa Eroded Area in Central Tanzania are perfectly consistent with colonial era soil conservation efforts dating back to the 1930s (Östberg's chapter); and the resolution of food and economic security issues in The Gambia continues to revolve around the task of mobilizing women's labor, just as it has since the onset of the groundnut boom in the mid-nineteenth Century (Schroeder's chapter; see also Carney and Watts 1991). Even the chapter by Broch-Due, which has an explicitly historical focus on the excessive spatial rationalization of ethnic territories in colonial Kenya, finds an unconscious echo in the types of ethno- or autoctonous territoriality purveyed by environmentalists in many parts of the continent today (cf. the discussion of *gestion de terroir* policies in Benjaminsen's chapter).

Giles-Vernick's case study of the Sangha river basin in the Central African Republic is instructive in this light. On the one hand, Giles-Vernick endorses the notion that environmental policies on the continent have demonstrated a remarkable consistency over the past half century. She acknowledges that powerful interests in both the colonial and postcolonial periods have sought to shape access rights and control the movement of Mpiemu peoples. In this regard, there is a continuity of sorts: the power to designate acceptable residence and migration patterns has remained in the hands of "outsiders" (cf. Broch-Due's chapter). On the other hand, however, Giles-Vernick demonstrates that the colonial government and the dominant actor in the Sangha basin in the contemporary period, the World Wildlife Fund (WWF), have had opposing prescriptions for the Mpiemu: the former wanted them to keep moving into the valley in order to create a labor reserve to service the colonial economy; the latter wanted them to move out in the interests of creating a "natural" park. Thus, both the colonial government and the WWF have sought to control Mpiemu movements but to very different ends. This brief example argues, in effect, that while we need to recognize broad historical patterns such as those highlighted by Neumann and Fairhead and Leach these should not obscure important historical and geographical changes taking place over shorter time scales. Only by means of careful conjunctural analysis can we hope to adequately determine the likely political outcome of a given set of environmental conflicts or where political openings such as those that emerged in the Amboseli and Malian cases might be found.

Contemporary policy changes

Several of the contributors to this volume have taken up the challenge of articulating the changes that have given contemporary environmental policies and practices a distinctive character. One issue that is dealt with in a number of chapters is the degree to which (and the particular ways in which) African environments have been commodified in recent years. The generation of state revenues by Malian foresters through the assessment of exorbitant fines is one case in point. Benjaminsen documents the fact that Malians frequently stopped gathering firewood altogether in the late 1980s in fear of the punitive sanctions that were being somewhat randomly imposed by the Traore regime under the Malian forest code. Instead, Malian citizens were forced to purchase wood at inflated market prices, a somewhat drastic measure that came at a substantial cost in terms of general economic security. A second example illustrating the environmental and economic costs of creeping commodification relates to the case of state-sponsored environmental degradation in Sudan described by Katz. She demonstrates how the deliberate destruction of southern Sudanese forests in connection with the fuel needs of the North has in fact met a series of northern political economic objectives. The assault on forests surrounding Renk provided "a key means for the state to secure revenues [through charcoal tariffs], one of the only reliable sources of dry season income for much of the disenfranchised and otherwise marginalized rural population of central Sudan, and a hedge against the import of costly cooking fuel for those in the towns." A third case in point comes from Cline-Cole's discussion of the "woodfuel gap" theory and its relevance to environmental policies in northern Nigeria. Cline-Cole argues convincingly that this theory bears little relationship to the actual dynamics of Nigerian woodfuel markets but survives nonetheless simply because it serves the economic interests of the Nigerian government. Specifically, the ability to claim conditions of scarcity in northern Nigeria has left policymakers in a position to secure rents from the successive infusions of development capital generated by donors to underwrite afforestation efforts.

While distinctive and often quite lucrative, the more idiosyncratic accumulation strategies spawned in the wake of environmental policy changes in the region pale in comparison with the intense activity surrounding the proliferation of different forms of nature and wildlife tourism. Honey's description captures some of the flavor of these developments:

> Nearly every developing country is now promoting some brand of ecotourism. ... Major international conservation organizations have initiated ecotourism-linked departments, programs, studies and field projects, and many are conducting nature tours, adventure tours, or ecotours for their members. International lending and aid agencies under the banner of sustainable rural development, local income generation, biodiversity, institutional capacity building, and infrastructure development, pump millions of dollars into projects with ecotourism components. The major travel industry organizations have set up programs, developed definitions and guidelines, and held dozens of conferences on ecotourism. ... [T]here are scores of magazines, consultants, public relations firms, and university programs specializing in ecotourism ... all this in just two decades. (1999: 5)

Indeed, it is in the tourism sector that the connection between the production of nature and the production of poverty is most apparent. The chapters by Neumann, Johnsen and Larsen, for example, all emphasize the deepening commodification of land markets at the behest of the tourist industry in Tanzania and Zanzibar. Where tremendous financial windfalls await investors in heavily capitalized tourist ventures, pastoralists and other rural groups that once derived livelihoods from territories set aside as tourist destinations are left with few economic alternatives. They are forced to either subsist on meager handouts as their conditions of poverty intensify—to "eat dust", as Johnsen puts it in her chapter—or "demonstrate to outsiders (i.e., Western conservationists) a conservative, even curative, relationship with nature while risking the loss of their land rights should they fail. ... Their lifestyles must allow them to do what immigrants and, significantly, Westerners, cannot—produce and reproduce in an ecologically benign way" (Neumann's chapter). The fact that many local groups cannot participate freely in profitable livelihoods of their own design and choosing is telling. Their dilemma rests in the fact that the legitimacy of residual land claims is often vested in their status as "traditional," "indigenous," or "local" residents. They are in effect locked into identity categories that help ensure their perpetual impoverishment.

As the quote from Neumann's chapter suggests, notions of traditional, indigenous, and localized attachments to land and other resources are frequently invoked as environmental planners seek to engage Africans in some form of environmental stewardship. The development of "participatory" approaches to natural resource management based on these and other forms of group identity (e.g., ethnicity) is rarely straightforward, however. Giles-Vernick's study of the Mpiemu in the Central African Republic once again provides an interesting case in point. She uses the history of Mpiemu residence patterns in the Sangha river basin to challenge the dual notions that "indigenous" groups are inherently sensitive in their use of the resource base while "migrants" inevitably act as environmental villains. In the broadest terms, Giles-Vernick's analysis begs the questions of who, precisely, is a migrant and who is a resident? What unit of analysis (historical, spatial) applies to these designations? Are all migrants, by definition, ecological threats? In more concrete, policy-related terms, who should be entitled to "local" benefits from park revenues when the residence history of a given group is short-lived or when that group has been involved in circular or interannual migration patterns? In "comanagement" scenarios, whose "traditional" patterns of resource use should be acknowledged and/or sanctioned? Whose claims to usufruct rights are most legitimate? These issues are not entirely new, of course, as Broch-Due's chapter suggests, but they have taken on a special salience in the context of recent political economic developments, including the international indigenous rights movement and the belated recognition by scientists of the value of indigenous technical knowledge systems.

Finally, contemporary policies directed at environmental management in Africa are further distinguished by their particular, and peculiar, attention to gender. My own chapter in this volume shows quite clearly that gendered policy interventions in both agricultural and environmental programs have come to the fore in ways that would have been unthinkable thirty years ago. While fraught with contradictory tendencies (are women to be seen as per-

petual environmental victims or as purveyors of invaluable environmental
knowledge? [cf. Schroeder 1999c]), these attempts have occasionally broken
new ground in the struggle to bring about greater gender equity. As the
Gambian case illustrates, however, women's "contributions" to environmental projects have often simply been reduced to providing cheap labor. In
such circumstances, the progressive objectives of greater gender equity have
clearly been subverted. Sadly, this sort of co-optation has occurred with a
number of potentially far-reaching social policies and practices. These include efforts to meet the goals of greater local participation in environmental
decision making, broader distribution of the economic benefits of conservation, and more systematic recognition of the value of indigenous technical
knowledge, to name but a few. Such co-optation is itself one of the hallmarks of the contemporary era (Schroeder 1999a, 1999b, and 1999c).

Conclusion

The contributors to this volume have articulated two broad perspectives on
recent policy changes related to African environments. The chapters by
Fairhead and Leach and Neumann make a point of setting contemporary
environmental interventions in their proper historical context. Fairhead and
Leach underscore emphatically that the "natural conditions" portrayed in
many of the most influential policy and environmental planning documents
are in fact wholly produced, if not fanciful. They insist that any attempt to
represent landscape change must adopt a longer term historical perspective
in order to ensure that the creative inputs of rural residents are not overlooked. Neumann in turn keeps our attention focused on a blunt reality,
namely, that the panoply of policies and practices adopted under the guise
of "new" approaches to environmental management have more often than
not done little to alter the balance of power in determining who gets access
to resources and who gets left behind to "eat dust." Indeed, his evidence
shows that contemporary natural resource management strategies premised
on marketing nature as a commodity have, if anything, only exacerbated
preexisting conditions of poverty in many areas.

The majority of the remaining chapters in the book offer a slightly different perspective on contemporary events. In these chapters, we see the commodification and commercialization of nature, the imposition of essentialized (traditional, local, community, ethnic, gender) identities and the co-optation of progressive political agendas emerging as critical forces in the
shaping of nature and African livelihood strategies alike. Instead of continuity, these authors emphasize an ever changing political ecological landscape,
one that forces both local groups and their erstwhile adversaries in state and
nongovernmental conservation agencies to continually adapt and adjust
their political tactics. Thus, the rulers of northern Sudan caught up in a civil
war with the South have made a patriotic virtue of deforestation; male
Gambian lineage heads have used seemingly benign agroforestry techniques

to reclaim highly valued lowlands from their market-gardening wives; and botanists working in the Sangha river basin have devoted their attention to the development of new classification systems that apply not to local flora and fauna but to different groups of migrants and indigenous communities. The relentless innovation in evidence in these brief illustrations speaks directly to the seemingly inexhaustible creative potential of capital embodied in development aid, state programs, and private investments. It also attests to the resilience of the localized political and cultural groups faced with these and other "outside" challenges.

In sum, the contributors to this volume have left us with two critically important messages regarding the recent wave of environmental interventions in Africa. The first is that policy analysts and critics cannot afford to work in a historical vacuum. A historical perspective demonstrates that many contemporary approaches to food production and environmental stabilization problems have in fact been tried over and over in the past with little success. This suggests that either the assessment of the problems themselves is wrong (Fairhead and Leach) or the proffered solutions have failed to adequately come to grips with the core environmental justice issues that fundamentally shape local practices (Neumann). In either case, political engagement around environmental issues cannot afford to be historically shortsighted.

The second key message is that environmental politics on the continent is far from overdetermined. Unpredictable political alliances have been, and will continue to be, struck between groups and individuals all along the local-to-global spectrum. As several of the contributors to this volume demonstrate, these new configurations represent both obstacles and opportunities to those who seek to reduce levels of poverty on the continent. It is thus incumbent upon all analysts and critics to carefully study the effects of contemporary policy changes lest the best opportunities for promoting economic and social justice be lost.

References

Bonner, Raymond. 1993. *At the hand of man: Perils and hope for Africa's wildlife*. New York: Vintage.

Carney, Judith, and Michael Watts. 1991. Disciplining women? Rice, mechanization and the evolution of gender relations in Senegambia. *Signs* 16(4):651–81.

Fairhead, James, and Melissa Leach. 1996. *Misreading the African landscape: Society and ecology in a forest-savanna mosaic*. Cambridge: Cambridge University Press.

———. 1998. *Reframing deforestation: Global analyses and local realities with studies in West Africa*. London: Routledge.

Honey, Martha. 1999. *Ecotourism and sustainable development: Who owns paradise?* Washington DC: Island Press.

Leach, Melissa, and Robin Mearns, eds. 1996. *The lie of the land: Challenging received wisdom on the African environment*. London: James Currey.

Peluso, Nancy. 1993. Coercing conservation? The politics of state resource control. *Global Environmental Change* 3(2):199–217.

Ribot, Jesse. 1996. Participation without representation: Chiefs, councils, and forestry law in the West African Sahel. *Cultural Survival Quarterly* 20(3):40–44.

———. 1999. Decentralization, participation, and accountability in Sahelian forestry: Legal instruments of political-administrative control. *Africa* 69(1):23–65.

Schroeder, Richard. 1999a. Community, forestry, and conditionality in The Gambia. *Africa* 69(1):1–22.

———. 1999b. Geographies of environmental intervention in Africa. *Progress in Human Geography* 23(3):359–78

———. 1999c. *Shady practices: Agroforestry and gender politics in The Gambia*. Berkeley: University of California Press.

About the Authors

Tor A. Benjaminsen is a senior researcher at Noragric, the Agricultural University of Norway. He holds a Cand.Scient. in resource geography (Oslo) and a Ph.D. in international development studies (Roskilde). He has for a number of years been carrying out research in Mali on environmental change and people-environment linkages in both the pastoral areas in the northern part of the country and the cotton zone in the south.

Vigdis Broch-Due is co-ordinator of the research program Poverty and Prosperity in Africa: Local and Global Perspectives at the Nordic Africa Institute and is a Visiting Reader in the Department of Anthropology and Sociology, School of Oriental and African Studies, University of London. She has held academic posts at the universities of Oslo, Bergen, and Washington. Her books include *Carved Flesh, Cast Selves: Gendered Symbols and Social Practices*, with I. Rudie (1993); and *The Poor are Not Us: Poverty and Pastoralism in Eastern Africa*, with D.M. Anderson (1999). She has published articles and reports on embodiment, gender, kinship, cosmology, poverty, colonialism, and development. Aside from completing an ethnography from northern Kenya entitled "Bodyscape and Landscape: Pathways through Turkana Life-worlds", Broch-Due has volumes in progress on topics related to the problems of wealth, conflict, and the politics of representation.

Reginald Cline-Cole is a geographer in the multidisciplinary Centre of West African Studies, University of Birmingham, United Kingdom. His research and publishing interests have revolved around the nature-society interface, forestry, rural energy, and north-south academic links. He is coauthor of *Fuelwood in Kano* (1990) and co-editor of *Contesting Forestry in West Africa* (2000).

James Fairhead is a Lecturer in the Department of Anthropology and Sociology, School of Oriental and African Studies, University of London. He has researched and published extensively on issues of ecology, knowledge and power, agriculture, food security, and conflict in the Great Lakes region and West Africa. He is the coauthor of *Misreading the African Landscape: Society and Ecology in a Forest-Savanna Mosaic* (1996) and *Reframing Deforestation: Global Analyses and Local Realities—Studies in West Africa* (1998).

Tamara Giles-Vernick is an Assistant Professor in the Department of History at the University of Virginia. She has conducted research on equatorial African conceptions of environment and history and has published in *Ethnohistory*, *Environmental History*, the *Journal of African History* (forthcoming), and the *Journal of African Historical Studies* (forthcoming). She is finishing a book entitled *Cutting the Vines of the Past: Environmental Histories of Loss in the Central African Republic, 1894–1994*. Her future research project focuses on the historical production of knowledge about malaria.

Nina Johnsen is a cultural anthropologist, general students' tutor, and librarian at the Centre of African Studies, University of Copenhagen, where she also periodically teaches. She has carried out fieldwork for her Ph.D. thesis in the Ngorongoro Conservation Area, Tanzania, on issues of ecology; health; dietary, ritual, and medical practices; and modernization processes among the pastoral Maasai in the Ngorongoro Conservation Area.

Cindi Katz is a Professor of Geography in the interdisciplinary Environmental Psychology Program at the City University of New York Graduate School. She has been engaged in ethnographic political ecology research in Sudan for over twenty years. She has published numerous articles on the subject, with particular attention to children's environmental learning, knowledge, and interactions, and done comparative work that addresses questions of social reproduction in rural Sudan and the urban United States. She is coeditor (with Janice Monk) of *Full Circles: Geographies of Women over the Life Course* and is completing a new book, *Disintegrating Developments: Global Economic Restructuring and the Struggle over Social Reproduction.*

Kjersti Larsen is an Associate Professor at the Centre for International Environment and Development Studies, Agricultural University of Norway. She has conducted fieldwork in Zanzibar, East Africa, and northern Sudan. She has published *Unyago—fra jente til kvinne: Utforming av kvinnelig kjønnsidentitet i lys av overgangsritualer, religiøsitet og modernisering* (1990) and numerous articles in national and international journals on issues of gender, knowledge and power, religion , identity, and the phenomenon of spirit possession. Currently, she is researching issues of forced migration in northern Sudan.

Melissa Leach is a social anthropologist and Fellow of the Institute of Development Studies, University of Sussex, where she leads the Environment Group. Her work focuses on ecological knowledge and power, forest history and policy, gender issues, and poverty-environment relations, particularly in West Africa. Her publications include *Rainforest Relations: Gender and Resource Use among the Mende of Gola, Sierra Leone* (1994); *The Lie of the Land: Challenging Received Wisdom on the African Environment*, edited with R. Mearns (1996); *Misreading the African Landscape: Society and Ecology in a Forest Savanna Mosaic*, with James Fairhead (1996); and *Reframing Deforestation: Global Analyses and Local Realities: Studies in West Africa*, with James Fairhead (1998).

Roderick P. Neumann is Associate Professor of Geography, Department of International Relations, Florida International University, and author of *Imposing Wilderness: Struggles over Livelihood and Nature Preservation in Africa* (1998).

Richard A. Schroeder is Director of African Studies and Associate Professor of Geography at Rutgers University. His published research on The Gambia, Tanzania, and Nigeria includes *Shady Practices: Agroforestry and Gender Politics in The Gambia* (1999) and a series of articles in such journals as *Africa, Annals of the Association of American Geographers, Antipode, Canadian Journal of African Studies, Economic Geography*, and *Progress in Human Geography*.

Wilhelm Östberg holds a Ph.D. in social anthropology from Stockholm University. He is Curator of African Studies at the National Museum of Ethnography, Stockholm, and is affiliated as a Research Fellow with the Environment and Development Studies Unit, School of Geography, Stockholm University. He is one of two editors-in-chief of *Ethnos: The Journal of Anthropology*. He is author of the monograph *Land Is Coming Up: The Burunge of Central Tanzania and Their Environments* (1995), which focuses on a region bordering the Kondoa Eroded Area. Currenty, he is involved in a study of natural resource management and agricultural intensification in Marakwet District, Kenya.

www.ingramcontent.com/pod-product-compliance
Lightning Source LLC
Chambersburg PA
CBHW082135210326

41599CB00031B/5990